# Invisible Solar System

When we look at a starry night sky, we are looking out through vast invisible expanses of our own Solar System. The planets, appearing as bright specks, have been revealed as worlds by space missions. However, the invisible spaces between them are equally interesting. Unseen forces, such as the effect of gravity, spiraling magnetic fields, and subatomic particles, originate from the Sun. Celestial bodies too small to see form unexpected patterns, while atoms and nuclei are hidden even if in our own bodies. Weaving the history of discovery with clear explanations, *Invisible Solar System* pulls back the cloak of invisibility under which myriad aspects of the local region of space are connected.

**Features:**

- Gravity, originally seen as an invisible force, is now revealed as a curvature of spacetime, and, even in its simple form, enables amazing patterns to form
- The smallest particles have other structures that enable them to interact, powering the present Solar System while also giving clues to nuclear events past and present
- Long-range forces of electricity and magnetism connect the Sun and planets, dominating the hot plasma gas of space while protecting us from cosmic rays via multiple layers of magnetic shields

**Martin Connors** is a Professor of Astronomy, Mathematics, and Physics at Canada's dominant distance education institution, Athabasca University. He is also affiliated with the planetary science group at Western University in London, Canada. He has authored numerous courses and scientific articles. His wide-ranging research has extended from the history of astronomy, through asteroids and their impact craters, to auroras and their magnetic effects. He has been a visiting professor at UCLA and at Nagoya University in Japan. When not doing scientific work, he reads about history, practices foreign languages, and blends photography with travel when possible.

# Invisible Solar System

Martin Connors

**CRC Press**
Taylor & Francis Group
Boca Raton  London  New York

CRC Press is an imprint of the
Taylor & Francis Group, an **informa** business

Designed cover image: Martin and Diane Connors, courtesy of Habbal, S. R., M. Druckmüller, N. Alzate, A. Ding, J. Johnson, P. Starha, J. Hoderova, B. Boe, S. Constantinou and M. Arndt (2021) Identifying the Coronal Source Regions of Solar Wind Streams from Total Solar Eclipse Observations and in situ Measurements Extending over a Solar Cycle, Astrophysical Journal Letters 911, L4, https://doi.org/10.3847/2041-8213/abe775, CC BY 4.0.

First edition published 2024
by CRC Press
6000 Broken Sound Parkway NW, Suite 300, Boca Raton, FL 33487-2742

and by CRC Press
4 Park Square, Milton Park, Abingdon, Oxon, OX14 4RN

*CRC Press is an imprint of Taylor & Francis Group, LLC*

ISBN: 9780367768515 (hbk)
ISBN: 9781032587783 (pbk)
ISBN: 9781003451433 (ebk)

DOI: 10.1201/9781003451433

Typeset in Times
by Deanta Global Publishing Services, Chennai, India

# Contents

# Preface

Specialization is a force in the pursuit of science much as it is in our everyday lives. One would not want to board an aircraft designed by someone who has ideas to test out, ranging from submarines to spacecraft, and doodled something that might fly. Similarly, knowing that your plane was serviced by a highly trained team, knowing everything there is to know about the specialized aspects of jet engines is the norm, is reassuring. In the days when the pilot greeted passengers, my mother would carefully size him up (they were men in her main flying days), making sure that he looked both competent and in good health. However, most people neither need to, nor wish to, specialize so much, although they would like to have a satisfying picture of how the universe works. It is at those curious and intelligent people that this book is aimed.

Science at the professional level has become a very competitive field, with the need to get grants leading to tightly honed scientific articles aimed to be at the top of their field, but, in most cases, accessible only to other specialists. University training in science fields sometimes allows a general first year of studies, but with the need to specialize upon entering the second year. Becoming very good in a certain field can impose a full load of specialized studies on students, leaving little time to indulge broad interests. One result of specialization is that modern space science has divisions. In studies of nearby objects, astronomy is now quite different from space physics, the former generally deals with planets, while the latter deals with plasmas (thin gases) and fields (such as magnetic fields). In the planetary sphere, some aspects are more like geology than astronomy, and deal with the surfaces of worlds that we explore, in some cases by landing on them. This book attempts to take a broad view, but with the theme of "invisible", so, to a large extent, geological aspects are secondary. This still allows the uniting of some traditionally disparate fields, and seeking common ground.

The 2023 Academy Award ceremony was dominated by a film named "Everything Everywhere All at Once". There is some danger in trying to write a book about many aspects of anything, especially in a science subject area where many threads are parallel but usually separate. In fact, it turns out that, to understand the present, we must examine the past, so we do not have the luxury of examining just the present if that is the meaning of "all at once". It also felt important to write words and not resort to the shorthand of science in the form of equations. The only equation, $E = mc^2$, is so iconic that it must appear, but hopefully in a role where its importance becomes clear. Among the words are numbers, in keeping with Lord Kelvin's maxim that "when you cannot express it in numbers, your knowledge is of a meagre and unsatisfactory kind". The huge range of sizes needing to be discussed requires "scientific notation", in which, for example, the approximate mean distance from the Earth to the Sun (the Astronomical Unit or AU) is not 150,000,000,000.0 meters, but $1.5 \times 10^{11}$ m. The 11 is how far back the decimal point is from where it would be in the fully written real number. Usually, we will give about three "significant figures" since at this level of discussion, more do not bring much more meaning. For example, the Earth's actual distance from the Sun ranges from $1.47 \times 10^{11}$ to $1.52 \times 10^{11}$ m in its slightly elliptical orbit. Rounding to $1.5 \times 10^{11}$ m gives the general idea of scale, while the range of distances using one more decimal place gives some idea that the Earth's orbit is not round. More than that is not needed for our purposes. In precision calculations by specialists, 10 significant digits or more might be used.

The writing level has been aimed at the "intelligent layman" and hopefully will satisfy a broad public wishing to learn about what is out there in the nearby part of space that forms our Solar System. Generally, "Solar System" is capitalized in this book. Adding "the" to the names of planets is sometimes done if it reads better. Some fairly deep topics are touched on, but an effort has been made to be clear. That effort may fail at times, but an effort also has been made to be correct while not oversimplifying. Readers finding errors are encouraged to report them.

A wide study quickly brings an author to realize the gaps in their own knowledge, and it has been fun filling those gaps. Hopefully, the material organized here can also bring fun into readers' lives by seeing how many seemingly disparate things have much in common. Perhaps that is the real "system" in Solar System. Many years ago, libraries brought trailer loads of books to outlying districts of my hometown, starting me on the path of assiduously learning outside the confines of school and curriculum. Now, knowledge is disseminated over the internet with encyclopedic websites such as Wikipedia, archival sites of public domain books like archive .org, and online libraries. The aim of a book like this one is to mine these riches to integrate and systematize. The value of the various sources cannot be overstated, and readers are encouraged to critically consult such sites and support them. The subject matter here is not very controversial, so even community edited sites are generally reliable.

I thank the many people who have been patient with me over the several years it took to write this book. This includes those working for the publisher, some of whom I have outlasted. My own family has put up with delayed repairs and chores with forbearance. Useful, broad-ranging discussions with Brian Martin, along with a review of the draft, have been invaluable. Discussions with professional colleagues have been stimulating. Athabasca University provided me time off in the form of a sabbatical year, and extra time based on my tracking of overtime work. While staying home to write this book during the COVID-19 pandemic cannot be compared to fleeing the plague as Newton did, governments and especially health care workers are thanked for getting me through it.

**Martin Connors**
Edmonton, Canada

# 1 Discovery of the Invisible Solar System

This book deals with aspects of the Solar System that cannot be readily seen by human eyes, nor by cameras on spacecraft giving views humans could see. We now perceive other bodies in the Solar System to be "worlds" since remarkable space missions carry cameras to show us what it would be like to be on or near them. Equally remarkable, but less obvious, are the invisible aspects measured by instruments. the data from which cannot be directly seen. These detectors operate in space, on the Earth, or in the atmosphere. Some things are invisible because they are merely hidden: the interiors of planets (even of our Earth) may be less accessible to direct study than even the remotest body. We will not discuss these but rather concentrate on things that are *hidden in plain sight*: thin gases, "force" fields, and exotic particles. This complex invisible Solar System is as beautiful and fascinating as the exotic surfaces of the bodies that surround us.

## 1.1 THE VISIBLE SOLAR SYSTEM

How can we convince ourselves that the Solar System is largely invisible? Our own perception can guide us. Imagine being in a beautiful natural location. Mountains and trees are a feast for the eyes. Brilliantly lit clouds dazzle with a multicolor sunset, and then clear as stars begin to emerge in the darkness.

Soon, one sees the "first star". Chances are that it is really a planet: in the western sky, possibly dazzling Venus, if elsewhere, likely Jupiter. The other planets are usually fainter than the brightest stars, except sometimes Mars, which varies enormously in brightness but usually has a telltale red color. In Figure 1.1, all of the planets of the outer Solar System visible to the unaided eye are shown. Uranus was discovered by William Herschel, in England, in 1783, using a telescope. It is visible to the unaided eye, but being very faint and starlike, was never noted as a planet by the ancients. One can tell a planet from a star simply because planets do not twinkle much. A bright planet has a very steady light, while the brightest star, Sirius, can put on a light show all by itself with its twinkling. Although a blue star, it flashes with multiple colors as the atmosphere distorts light on its path to our eyes. Planets put out broad beams of light, while stars have narrow ones that follow a varying path through our atmosphere, resulting in twinkling. To identify planets and stars, "modern" people may find an astronomy program or app helpful. Either by direct observation or using software (as Figure 1.1), one can note that planets are not randomly distributed in the sky but stay near an invisible line called the *ecliptic* (shown in orange in Figure 1.1).

Despite planets being large in angular size, compared with stars, only two Solar System bodies can be seen as discs by the unaided eye: the Sun and Moon. The Sun

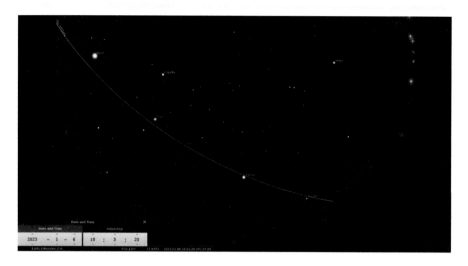

**FIGURE 1.1** View of the night sky after sunset on January 6, 2023, from a mid-northern latitude. The outer Solar System planets, Saturn, Jupiter, Uranus, and Mars, are visible. Jupiter is usually brighter than Mars and, on this evening, would have been the "first star" visible. The view shows that the Solar System is flat, concentrated near the apparent path of the Sun, the ecliptic (author, made with Stellarium, 2023).

is 400 times larger in diameter than the Moon, but 400 times further away: the Moon is smallish among "worlds" but very nearby. Its surface markings are familiar to us but were explained only in folklore until Galileo made the first use of a telescope (about 1610 CE) for astronomy and noted lunar mountains, craters, and dark plains (Figure 1.2). We now know the Moon as a world that has been walked upon, but the focus of this book is not the features of the worlds of the Solar System, rather what we cannot see. Suffice it to say that Galileo first deduced that the Moon was like Earth in the sense of having "geological" features, not being a perfect heavenly body as called for in the then-dominant theories derived from the ancient Greek philosopher Aristotle.

Similarly, Galileo observed features on the Sun, although ancient observers perhaps had some idea they were there. While viewing the Moon with the naked eye or through a telescope is harmless, one should never try to look at the Sun directly in any way. Galileo's "sunspots" (Figure 1.3) were part of what led Galileo away from notions of perfection in the heavens. In the Aristotelean model, as further developed by the Hellenistic astronomer Ptolemy, the sky turned around the Earth, bearing the Sun, to explain day and night; however, heavenly bodies were not supposed to show defects of the sort Galileo noted. The changing positions of the dark spots on the Sun showed that it also rotated. Since the Moon always turned, one face toward the Earth, it did not have to be regarded as rotating: however, the Sun clearly was. The luminous power of the Sun alone gives a clue about its central role in the Solar System. Even from our position, which we now know to be 150 million kilometers from it, the Sun delivers 1,400 watts, the equivalent of the output of an electric room

**FIGURE 1.2**   Galileo's 1610 Moon observations from the *Sidereus Nuncius* (public domain, Wikimedia-Galileo, 2023).

heater, to every square meter of the Earth's cross-section. All ancient peoples understood the importance of the Sun and tracking the seasons for agriculture or other outdoor activities, but the Earth's surface was the dominant frame of reference. Even for the Greeks, who knew not only that the Earth was round, but had a good idea of its size, the fact that, at all locations, its gravity pulled toward the center, led to regarding it in a natural way as the center of everything.

Returning to the night sky, planets are, on average, not impressive. It was mainly their motions that led them to be noted by the ancients. When we look at the Solar System with the unaided eye, there is not much to see: the Moon and Sun (about

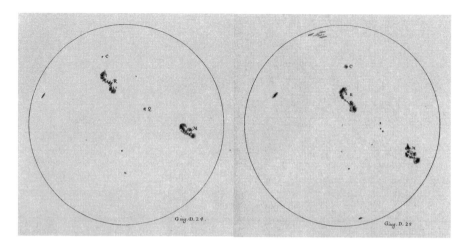

**FIGURE 1.3**   Galileo's observations of the Sun on June 24 (left) and June 25 of 1612. Rotation of the Sun is obvious (Galileo Project, 2013).

equal in angular size, as total eclipses prove) do not take up much of the sky; even the planets that are visible show no disks, and we can see right through to the background stars. We would be right in concluding that most of the Solar System, if indeed there is anything in it besides the Moon, the Sun, and the visible planets, must be invisible.

At some special times (as in Figure 1.1), several planets may be simultaneously visible, always near the ecliptic. The Moon remains near that line also. Furthermore, it moves a noticeable amount each night, progressing in basically an eastward direction in the sky. If initially viewed near the Sun at sunset, as in Figure 1.4, it will, nearly two weeks later, rise as a full Moon opposite to the Sun in the sky (which is the approximate configuration in Figure 1.1). Planets, too, move generally eastward with respect to the background stars, although very slowly compared with the Moon. Careful examination of Figure 1.4 will show Mars, Jupiter, and Saturn to be blurred slightly due to their apparent motion. Unlike the Moon, planets sometimes appear to reverse their west-to-east motion, to be backward or "retrograde". Although the Greek word that we have inherited for "planets" meant "wanderers", these wanderings were not random. The Solar System is basically a flat plane, and it is possible that the words "planet" and "plane" share a distant linguistic root. This is not necessarily due to early people having thought of the motion of planets as being in a plane. Indeed, the accurate description of planetary motion posed a challenge that defied human explanation for millennia. The simple description of planets going around the Sun that children are taught in school is by no means obvious, nor is proving that it is true within easy reach with only visual observation. A description of motion is called "kinematics", kine- being the same root as in "cinema", to denote "moving". That motion around the Sun, as is familiar to children, is called the heliocentric system. It sometimes is also known as the "Copernican" system, after Copernicus who first

**FIGURE 1.4** The sky at the same "star time" as in Figure 1.1 but over fourteen nights. The Moon, located in the northeast in Figure 1.1, was first visible in a dark sky 12 days earlier near the sunset, which in this January example took place in the southwest (near the SW marker). The Moon moves steadily eastward in its monthly cycle. The phase is not visible in this plot (author, modified composite of renderings from Stellarium, 2023).

described it properly in 1543. Certain Greek astronomers may have also suspected such motion, but their ideas did not take hold. Indeed, Copernicus' triumph over persistent but incorrect views that derived from Greek times is sometimes regarded as the beginning of the modern scientific age. At a yet deeper level, our modern understanding is based on the study of how forces move bodies in the Solar System. Isaac Newton can be said to have invented this aspect of the Invisible Solar System in the late seventeenth century, by understanding how forces move bodies and describing the force of gravity. A level of explanation involving forces is known as "dynamics". A Solar System moving based on the invisible force of gravity is a *dynamic* one!

If most of the Solar System is invisible, why is it even a "system"? The word itself originates from Greek, with the initial syllable being from "*syn*", meaning together, and the second meaning to stand. So, a system is something that stands together as one thing. "*Sol*" is the Latin name for the Sun, so the Solar System (always capitalized in this book) is something that is together and belongs to the Sun. We now know it to be heliocentric, where helio- is from the Greek root for the Sun, whereas a geocentric system is based on the geo- root for the Earth. That model dominated astronomy for about 1,500 years due to the efforts of Alexandria's Claudius Ptolemais (Ptolemy), who lived in the second-century CE. His fundamental book, renamed through Arabic transmission as "*Almagest*", was originally called "*Mathematike Syntaxis*", basically meaning a "mathematical ordering". It provided a means of calculation, but "syntax" need not indicate a physical "system". On the other hand, it was considered that there might be invisible crystalline spheres transmitting motion in the heavens, much like a machine, so we might infer that a universal system was indeed envisaged. Be that as it may, it is likely Galileo Galilei's *"Dialogo sopra i due massimi sistemi del mondo"* or "Dialog Concerning the Two Chief World Systems", published in 1632, that explicitly saw the Copernican and Ptolemaic models as "systems" or real things.

Earlier, Johannes Kepler, in 1619 in his *"Harmonices Mundi"* or "Harmonies of the World", had described the relation of orbits to one another with his Third Law. It relates the revolution periods of planets about the Sun to their mean distances from it. This was part of an overall approach, sometimes derided, to find geometric or even musical regularities in planetary motion. This certainly regards the whole Solar System as interrelated and as an entity. In 1686, Newton's master work, the *"Principia"*, had as its third book, *"De Mundi Systemate"*, or "The System of the World". The actual term "Solar System" appears to have been first used by the English philosopher John Locke, who moved in the London circles of Newton and may have been considered his friend. Near the time of the publication of *"Principia"*, Newton, with his rather prickly personality, wrote to Locke that he regretted having expressed the opinion that it would be "better if you were dead". In any case, this "friendship" allowed Locke to pick up enough about gravitation from Newton that he went beyond Newton's generalities about the mathematics of gravity to emphasize that the Sun dominated, and, in his *"Elements of Natural Philosophy"*, he was the first person known to use the term "Solar System", in 1704.

Thus, our modern view of a Solar System, largely held together by the Sun due to gravity as codified by Newton, with planets orbiting following dynamical laws devised by Newton, was firmly established about three hundred years ago. Gravity could act through what appeared to be a void, so that the Solar System was a large, empty place.

## 1.2 CLOCKWORK OF THE SUN, MOON, AND STARS

Resting on the achievements of previous generations, we can sometimes be smug in our knowledge. It is rare to question how we know what we know. For most people alive today, it is normal to have satellites orbiting the Earth, as that has been the case since they were born. Many young people do not know a time when telephones, and basically access at a moment's notice to all the world's knowledge, did not fit in one's hand. We all assume that matter is made of atoms, yet it is less than 100 years ago that one of the basic constituents of the nucleus was discovered, and only in 1911 was it even discovered that atoms have a nucleus. How many of us could devise a way to show that matter really is divisible into such tiny constituents? So it is also for the heavens and indeed the Solar System. How could we, based only on observations we can personally make, show with basic observations that the stars are very distant, or even find the scale of the planetary system? Yet, it seems that most introductory astronomy books are somewhat derisive in describing what the ancient peoples did achieve.

Due to lack of space, we will describe neither lore and religion associated with the skies, nor those accomplishments of non-Western civilizations that were not built in a direct way into our modern view. We cannot, however, omit to mention that Arabic and Muslim societies played an important role in transmitting ancient Western knowledge to the modern Western world at the time of its Renaissance, that word literally meaning "rebirth". Evidence of the Arabic role is that Ptolemy's book, mentioned above as codifying much of late Greek astronomy, is almost universally referred to by its Arabicized name *"Almagest"*. During Ptolemy's life from about 100 to 170 CE, his Egyptian city of Alexandria was under Roman rule. Egypt had for several hundred years been ruled by the Greek Ptolemy family, descended from a general of Alexander the Great, but likely not related to the astronomer. The last famous Ptolemy ruler was Cleopatra, who died in 30 BCE. Although Ptolemy lived in the Roman Empire, and indeed had a Roman first name, Greek cultural influence ("Hellenism") remained strong, so that Ptolemy may be regarded as the last great Greek astronomer of antiquity. Under the very clear skies in the Middle East desert regions, astronomy had developed in Babylonia but was best formalized by the Greeks, including in Egypt. At the time, the Arabic language had not yet developed. With the rise of Islam, starting in 610 AD in the Middle East, a vibrant culture interested in astronomy again arose, with Arabic as its main religious and cultural language. The same questions about the structure of the universe and how to predict the motions of the planets arose, with Greek manuscripts containing the best answers. As a result, many of these were translated into Arabic, and later re-transmitted back to the Western world in its intellectual rebirth as the Renaissance. Largely for this reason, we have a good insight into earlier, mostly Greek/Hellenistic developments in astronomy.

In many textbook accounts of the development of modern astronomy, much emphasis is placed on the ways in which the astronomy of the classical ancients was wrong and therefore had to be corrected. What, one might ask, was even the value of having classical knowledge passed on to us if it was so flawed? Reviewing the Solar System through the eyes of the ancients will give us insight into both basic phenomena and how models of the Solar System developed with time. Ultimately, the system arises through the action of gravity. We have a new view of that, which separates us from Newton as much as he was from the Greeks. However, "classical" Newtonian gravity is a very good framework for understanding dynamics in the regime applicable to our Solar System, the classical theory giving only very minor deviations from what is observed.

A basic tenet of ancient Greek astronomy was that the Earth was the stationary center of the universe. As every child learns in school, the Earth travels around the Sun. Simple calculations show that it does so by moving at 30 km per second, that is, over 100,000 km per hour (over 65,000 miles per hour). Common experience shows that rapid motion is "felt": it seems to defy our common sense from traveling in vehicles that the Earth could move so rapidly without us knowing it. Without knowing any differently, it is completely in line with common experience to assume that the Earth is stationary. We will return to this point below.

It is also taught in schools that the Earth is round and usually implied that this is a modern discovery. We seem to take pleasure in supposing earlier people to have had simplistic views. In day-to-day life, we seem to be at the center of all we see, and our normal state of motion is rest. From its surface, it is difficult to perceive that the Earth is not flat. Many changes came about as the result of Columbus' famed voyage of 1492, but the discovery that the world was round was not among them. This fact was well known to Greek astronomers associated with Pythagoras (who worked in *"Magna Graecia"*, southern Italy) in the sixth century BCE. We remember Pythagoras mainly for his theorem about the sides of triangles, but that merely emphasizes the deep knowledge of geometry that allowed accurate deductions about space near the Earth. One piece of evidence is that the Earth's shadow on the Moon was always round during a lunar eclipse (Figure 1.5), which was deduced to be caused by the Moon moving through the Earth's round shadow blocking the Sun's light. The size of the shadow, larger than the Moon, allowed an estimate of the size of the Earth to be made as early as Pythagoras' time. The size of the spherical Earth was measured more accurately by another Greek, Eratosthenes, resident in Alexandria in Hellenistic Egypt in about 200 BCE. By geometrical argument about shadows on the Earth's surface, and with a paid surveyor who walked and measured a several hundred-kilometer North-South path in Egypt, he determined the Earth's circumference with an accuracy of less than 1%. The modern value is conveniently close to 40,000 km, which is no coincidence since the Revolutionary French defined the measure of length in their new metric system in terms of the equator to pole distance being 10,000 km.

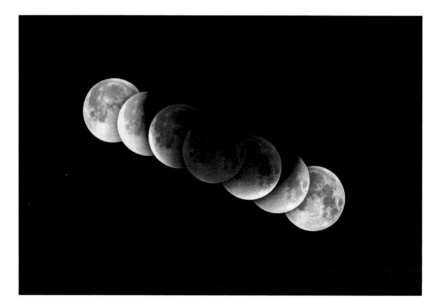

**FIGURE 1.5**   During a lunar eclipse, the Moon moves from west to east against the background of stars and is at a position where it must cross the ecliptic through the Earth's shadow. At various times during the eclipse, the curve of Earth's shadow allowed the Greeks to conclude that Earth must be spherical. In the deepest part of the eclipse, light from "all the sunsets in the world" results in a red color, also known as a "blood Moon" (courtesy of Dr. Sebastian Voltmer, www.voltmer.eu).

Being circular took on special meaning in ancient astronomy. The Moon and Sun (we again emphasize not to stare at the Sun, even when very low in the sky) could be seen to be circular in outline: the disk of the Sun (see Figure 1.6) was a god to the ancient Egyptians. The renegade Pharaoh Akhenaten, ca. 1340 BC, is sometimes credited with establishing the first monotheistic religion with his attempt to make the Sun god (Aten) the *only* god. His failure to do so is reflected in the changed name of his son, "king Tut", or Tut-ankh-*amun*, restoring the primacy of Amun as the chief god among a pantheon. Although the Moon has features ("the man in the Moon") visible to the unaided eye, the Greeks placed more emphasis on its round form alone, so both the Sun and the Moon seemed to have circular perfection.

Although the Chinese may have recorded sunspots as early as 1500 BC, the nature of their language gives ambiguities. Certainly, the Chinese were observing them at the most active time for early Greek astronomy; the Greeks, however, did not. Thus, the only objects in the sky that showed a definite form had a circular form. The remainder of the heavens appeared unchanging, apart from the motions of the few moving bodies, but made a daily cycle apparently around the Earth at the center. The very word "cycle" comes from the Greek "*kuklos*" for circle, so that attributing

FIGURE 1.6   The cow goddess Hathor bearing the sun disk, ca. 2500 BC. Plaster cast of original from Giza in Museum of Fine Arts, Boston (excerpt from Daderot, 2016).

an innate circularity to heavenly things seemed natural. Furthermore, the ancient Greeks, being mariners, easily understood that the sea horizon was due to the Earth being spherical, with, for example, the mountains of islands being visible first as they are approached, with the lower parts hidden by the curve of the Earth. The very fact that the Greeks knew that the Earth was spherical reinforced that it should be the center of a basically spherical universe. As noted above, they also knew that the Earth was very large relative to the scale of familiar objects. All objects fell vertically at all locations, which were on the surface of a sphere. Thus they fell toward a common center. It would seem natural to regard that as the center of all things.

The loss of ancient Greek knowledge played a major role in what historian Daniel Boorstin referred to as the "most influential miscalculation in history". The culmination of the Renaissance, in intellectual terms, indeed what philosopher Thomas Kuhn referred to as a "paradigm shift" (and indeed the first one), was the publication of Copernicus' view of the Solar System in 1543. In practical and more "down-to-Earth" terms, a date about fifty years prior looms large: 1492. Christopher Columbus, in this year, sailed to the New World and returned to what is now Spain claiming that he had reached Asia by sailing westward. That he was able to secure funding for his voyage rested on not one but two major miscalculations, one made about 1,300 years previously by Ptolemy. In addition to the *Almagest*, Ptolemy wrote the "*Geography*" (literally, this word simply means "writings about the Earth"). Unlike the *Almagest*, for which at least manuscript copies in Greek exist, the *Geography* survived only in Arabic copies which could be retranslated into Western languages. There is perhaps some irony in that Ferdinand of Aragon and Isabella of Castille, Columbus' sponsors, completed the long project of the reconquest of the Iberian peninsula from Arab rule, but used the fruits of Arab scholars in deciding on the project for which they are now most remembered. The original mistake of Ptolemy was that he ignored the meticulous work of Eratosthenes, his predecessor in Alexandria by about 350 years, in obtaining a very accurate size for the Earth. Although Ptolemy introduced the concept of latitude and longitude, he assigned to the degrees used to measure them values of only about 70% of the correct figure. As a result, the Earth in his authoritative *Geography* was believed to be smaller than its actual size, although spherical. Columbus' biographer Samuel Eliot Morrison stresses the great lengths to which Columbus went to promote his exploration project. He learned Latin to the extent that he could seek support in classical sources and misinterpreted biblical passages to suggest that oceans were much smaller than they were. This second mistake was soundly rejected by the Portuguese, who arguably had the greatest capability to assess his project. Isabella put him off by saying to come back only after the Reconquista war ended with the fall of Granada, which happened on January 2, 1492. This momentous date, on which western Europe first became fully ruled by Europeans again after the fall of Rome, is now usually forgotten beside Columbus' discovery of the Americas, later in that same fateful year.

The big surprise of the Columbian enterprise was that there was an unsuspected continent in the large space that most educated people (but not self-delusional Columbus) knew separated Asia from Europe. Although Columbus used distorted information to claim that the Earth was much smaller than it really is, making the voyage seem feasible, the concept of reaching the desired East by sailing west was easy to understand at the time, since it was already well known that the Earth was indeed a sphere. The occasional European (such as Marco Polo) had been to the Orient via the land route, but the astronomical techniques for determination of longitude did not allow its longitude to be determined, allowing the errors of Ptolemy to be compounded by Columbus. Ptolemy dominated Western thought in geography until 1492 and in astronomy until 1543 (or later, as Copernicus' innovations were only slowly accepted). We now return to the description of why his astronomical system was so durable.

Possibly because circles, already a perfect shape because of their symmetry and simplicity, were only clearly distinguished in the sky and not in natural things on the surface of the Earth, they became one of the aspects that set the sky apart as a perfect domain to the Greeks from periods well before that of Ptolemy. Knowing that many astronomical motions were cyclical, it seemed fitting to use circular motion as a basis for the ability to predict, or, although they did not think of it in those terms, to "model", motion in the sky. Only in the sky was motion perfect. Vehicles like wagons, even if based on wheels (circular), did not move smoothly on the surface of the Earth: moving between places on land was difficult, and indeed the natural state of wagons is not motion but rest. In the regions of ancient civilizations near the Mediterranean coast, the sea was a much better way to travel, especially with large cargoes, than was land. Again, ships are reliant on wind or muscle power to move. Only the heavens moved on their own and thus seemed to be a domain of perfection. How natural that circles be used to describe them!

Before the Greeks systematized astronomy, the earlier civilizations of the Fertile Crescent kept records from which patterns could be inferred. The keeping of records was one of the earliest activities of centralized states. Possibly, their first invention was taxation! This naturally led to a need to keep records. As ruling hierarchies developed, with centralized activities ranging from keeping the elites entertained to provision of water for the masses, everyone had to pay their share (or perhaps more than their share). Records were kept not only of who had paid, but of what was in the central warehouses. In many cases, we have very early records of this sort since they were written on clay that got cooked to stone hardness when cities were burned. Ironically, the death of a culture by conquest was often the immortalization of its records. As a result, from Babylon (referred to in the Old Testament by the possibly already archaic name "Chaldea"), we have detailed records of palace contents. It was not a large leap from recording storehouse contents to recording astronomical events and the priests of Babylon and its predecessor civilizations were adept at this and at finding patterns that allowed prediction.

Ancient peoples may have had a sense of connection to the sky, especially in dark places that characterized the entire world before artificial lighting, through noting that the skies had patterns. The stars appeared in fixed patterns to which names could be given, and the motions of the Moon and the planets were steady, even if they appeared complex in detail. Contrast this to the life of even the most powerful ruler, never mind a common person, which might be dramatically changed by unpredictable weather or rivers, or the enemy appearing at the gates. If indeed there was a connection between the sky and human lives, the ability to predict based on the sky had a vital importance. Thus was born ancient astrology. Indeed, the Babylonians were very good at making predictions based on tablets such as that shown in Figure 1.7. If an accountant could use records to tell the king how much grain was in the warehouses, then an astrologer who could forecast the next harvest and the possible arrival of enemies to take it was even more valuable. By studying patterns, the Akkadian forebears of the Babylonians knew, for example, that Venus, an important and bright body known as Ishtar, was present in both the evening and morning skies, and they knew its apparent orbital period very accurately.

FIGURE 1.7   The Venus Tablet of Ammisaduqa, ca. 1700 BCE record of the motions of Venus. The cuneiform text was made by pressing a stylus into clay (Fæ, 2010, CC BY-SA 3.0).

Evidence of Babylonian prowess in this area lives on in common modern speech in which the Hebrew expression "*mazel tov*" is used for "good luck". "*Mazel*" goes back to a Semitic root in astrology, so "*mazel tov*" really means "a good star", and this concept likely came into Hebrew due to the Babylonian captivity of the Jews. One can imagine that the power of the priests was augmented by the ability to predict eclipses, in which a terrifying blood Moon (Figure 1.5) could be seen over large regions, or, even more terrifying, a solar eclipse in which day turned to night. The latter occur only in narrow bands, yet the Babylonian astronomers could predict them. We still classify eclipses by the "*saros*" cycle, which is a Babylonian term. It appears that the story that the Greek astronomer Thales (from the city of Miletus which is in present-day Turkey) predicted an eclipse in 585 BCE, and during a battle, is dubious. His skill is usually attributed to having been learned from the Babylonians, but they do not seem to have developed this ability until several centuries later. Ranging slightly further from the descent of Western astronomy from Babylon, it appears that the story that Chinese astronomers were executed in 2159 BCE for failing to predict a solar eclipse is equally apocryphal. The cosmology of ancient China lives on, however, in that eclipses were attributed to sky dragons eating the Sun or Moon: the modern terms for eclipses in Mandarin directly translate to "sun-eat" and "moon-eat"! Suffice to say that eclipse prediction was of such importance that, whether correct or not, it was highly emphasized by ancient astrologer-astronomer observers.

The Babylonians made a connection, at least in some sense, between the planets and their gods. Ishtar or Venus was the goddess of both love and war, to us a perhaps curious combination. Marduk or Jupiter was an important god, the city of Babylon's patron. The seven visible moving sky bodies gave their names to days of their week. Coming through the Romans, this tradition persists down to our times, in which we begin the week with the Sun, followed by the Moon (Mon-). Tuesday is in English named for Zeus, perhaps indirectly through Latin in which Jupiter is derived from *Deus/Zeus Patera*, or Zeus the father. Other English day names come from Nordic influence, being Wotan's (Wedn-), Thor's (Thurs-), and Freja's (Fri-) days. In Romance languages, for which we use French as a prototype, Tuesday is for Mars (mardi), Wednesday for Mercury (mercredi), Thursday for Zeus (jeudi), and Friday for Venus (vendredi). We round out the week by returning to English, in which Saturday clearly refers to Saturn.

As astrology developed, the characteristics of certain gods and the placement of the related planets, with respect to stars, the Sun, or other planets, could lend a basis to prediction. Astrology did not need to be true; it simply needed to have enough people, including rulers, believe that it was true, to flourish. Within it, the concept of prediction was very important. It may come as no surprise that Ptolemy, whose *Geography* was a basis for that study for 1,300 years, and whose *Almagest* was the basis of astronomy for nearly 1,400 years, also was a noted astrologer. His astrological *Tetrabiblos* ("four books") has not survived but came to us via the now-familiar route of Arabic translation. It may be claimed that, in the earlier Classic Greek age, intellectual curiosity drove minds like that of Aristotle (d. 322 BC) to seek an understanding of the heavens. It was his student Alexander the Great who opened Greece up to other influences and ushered in the Hellenistic era. The motivation for observation-based work, culminating with that of Ptolemy, was not purely intellectual, but an attempt to use prediction of heavenly events in turn to predict their Earthly consequences. A model would have to incorporate some aspects of "theory" in the form of Aristotelian concepts, and observations, supplying what we would now regard as "adjustable parameters" and to be tested by its ability to predict future configurations and events that could be observed in the sky.

Based on common sense and basic observations, an astronomer in the Hellenistic environment could hardly develop an astronomical model without a stationary Earth, known to be spherical and large. Around this, heavenly structures would be created to match observations, but also to incorporate concepts from the Greek Classical Age. Circular motion would be fundamental but could be augmented in various ways. The most basic cycle of life is the daily change between night and day, with the Sun appearing to move across the sky from eastern to western regions (the Sun rises or sets *directly* in the east or west, respectively, only twice a year). Such noiseless motion could arise because it was distant, or because of some way in which the heavens differed from the Earth. So, motion of the Sun around the Earth seemed perfectly feasible. The Moon shows a similar daily motion but at a slower rate than the Sun. Each day, it seems to rise about an hour later (and it can often be seen in daylit skies). The result is that the Moon appears to change position relative to the Sun, coming back to near its starting point at the new Moon near the Sun about every month. The words "Moon" and "month" are related in English since the Moon's motion gave a convenient unit of time. The phases of the Moon are due to it being spherical and seen in various illuminated positions relative to the Sun. A little before Eratosthenes determined the size of the spherical Earth very accurately, Aristarchus of the Greek island of Samos used the phase of the Moon to deduce its distance from our planet, but not very accurately. The time from one phase of the Moon to the next similar phase is called the "synodic period". This has a value of about 29.5 days, comparable to the length of our civil months. For this reason, if you observe the Moon at some phase, about the same day of the following month it will have the same phase. Once again, simple observations can trick us, and the synodic "month" we use is *not* the time it takes for the Moon to travel once around the Earth with respect to the stars or, as we now know, in reality.

Before we explore how a month is not a month, let us return to the apparent daily motion of the Sun, which we normally call a "day". If a month is not a month, is a day a day? A good way to investigate this is with safe Sun observations, using a shadow. In a large band of the northern hemisphere, at some time of the day a vertical stick will cast a shadow from the Sun that points due north, and we call that time of day "solar noon". If we wait one day, the Sun will again be in a position to do so: this is the definition of the "mean solar day". Perhaps you wonder about the word "mean": it means "average", but why is it needed at all? It turns out there are slight complications in how fast the Sun appears to take to go around the Earth that we now know are due to the Earth's orbit around the Sun. Sometimes, solar time is "fast", compared with the time we can now accurately indicate on our clocks, and sometimes systematically "slow". A further effect we could note if we marked the end of the stick shadow on successive days is that the Sun's shadow at noon moves north and south. In northern hemisphere summer, the Sun is in the northern part of the sky and, in winter, in the southern part. The total excursion is about 45°, half of a right angle. These changes in celestial latitude (known as "declination") of the Sun result in seasons and are due to the tilt of the Earth's axis. The ancients knew of these variations well. However, these variations still do not really answer the question of the length of a day: is the "mean solar day", or the time the Sun *appears* to take to go around the Earth, actually the rotation period of the Earth?

To answer this basic question about the day, we can imagine the use of the same vertical stick to observe a star. Rather than rely on a shadow for safety, we could directly observe a certain star pass the stick while we look in some particular direction, let us say due south. The rotation period of the Earth (or equivalently of the starry vault of heaven around the Earth) is a "real" day. If we note the time at which a star is due south, we can observe one mean solar day later (this would be 24 hours on a clock). Facing south, we would find that the star was *not* lined up with the stick, but slightly and noticeably to its right. The mean solar day is not a "day" long if the day is the rotation period of the Earth! Lining up with stars gives a reference system that is not moving (or at least is moving *very* slowly) and is the best in which to measure. The *true* rotation period of the Earth is that which a star takes to repeat its passage past the stick and is referred to as the "sidereal" (meaning "star") day. It is about four minutes shorter than the mean solar day. If we wanted to see the star in the same position after a "day", we would have to come back four minutes earlier. This means that if we are using the Sun as our timekeeper, the stars seem to move westward after a day. In other words, the Sun, with respect to the stars, moves eastward. We also note that there are $24 \times 60 = 1{,}440$ minutes in a day. A motion of four minutes compared to 1,440 is the same as one part in 360. Since there are 360 degrees in a full circle, this means that the apparent motion of the Sun is in one day is close to one degree of angle. It goes back to the ancient Babylonians, well before the Greeks, to have noticed that a convenient number, near the number of days in a year, for a full circle of angle, was 360. Like the Greeks, they thought that the Sun went around the Earth in a year.

We concluded above that the Moon moves eastward with respect to the Sun, taking 29.5 days to return to the same position with respect to it (at the same phase). Very roughly, this means that the angle the Moon moves in the sky in a day is $360/30 = 12$ degrees eastward, with respect to the Sun. The Sun itself, in about 30 days, also moves eastward, by about 30 degrees. For convenience, let us try to imagine measuring the Moon from one new Moon to the next (although observationally this is hard to do since we cannot see the Moon near the Sun at that time). When the Moon comes back to the original position of the Sun, the Sun is no longer there! It has to move 30 degrees further, which, at the rough precision we are working at, takes about $30/12 = 2.5$ days. On the other hand, if we had measured with respect to a star, there would be no extra time: by the time the Moon came back to appear to be in the same position, it would really have gone once around the Earth. As a result, we expect the real, or "sidereal" month to be about 2.5 days shorter than the synodic month. In fact, the sidereal period of the Moon is about 27.3 days, so our math is a bit rough, but shows what is going on.

From this description, which covers only the Sun and the Moon, we see that there is some complexity to sky motions. Let us add in the apparent motion of the stars before summarizing, since that motion is relatively simple. The simplest would be if we simply observed four minutes earlier every night. In that case, the stars would seem to be stationary. However, as the Sun appeared to move eastward, eventually it would come to the position of the star we were observing, and they would no longer be visible. For this reason, the stars that we see in most regions of the sky change with the seasons as the Sun appears to move in front of them. There are certain stars near the pole that can be seen all year, but even they change relative position at night due to the motion of the Sun.

The ancients of all cultures could describe these sky motions well. Indeed, in many cases, their lives depended on being able to do so. Agricultural societies needed to know when to plant, and since the Sun controls seasonal rains or flooding, good calendars were needed and were usually the responsibility of the priests. The lunar cycle also received attention, as attested by the lunar calendars still used in some way by many cultures and religions. The daily cycle was all-important in times before artificial lighting (and, of course, it is warmer during the day for outdoor activities), the point being that the apparent (mean) daily cycle of the Sun, and not the real rotation period of the Earth, determines civil time.

## 1.3  THE CLOCKWORK SOLAR SYSTEM

We determined that the Solar System was mainly invisible by considering the other bodies visible in the night sky which are now known to be distant, the planets. Their apparent motions are more complex than those of the Sun, Moon, or stars. Of course, we now know that that complexity comes from observing them from the moving Earth as they go around the Sun. However, in the best-developed ancient system, that of the Egypt-based Greek astronomer Ptolemy, dating to about 200 CE, they could be regarded as moving around the Earth, much as did the Moon, Sun, and stars, but in complex patterns described by co-added circular motions. It may be that, in hindsight, we know that planetary motions are not centered on Earth, and in fact are not based on circular, but rather elliptical motion, but the Greek achievement with Ptolemy's model should not be underestimated. Use of circles was natural since it was known that the Earth, Sun, and Moon were circular. Circular motion repeats, a characteristic also of celestial motion. Aristotle considered that perfect circular motion was the only fitting kind to characterize the perfection of the heavens. This preference for circles may have become exaggerated when the Church played a large role in reviving astronomical knowledge one thousand years after Ptolemy's time. His system in fact worked quite well for calculating calendars and planetary positions as needed by the developing civilizations of Islam and the West.

The bases of Ptolemy's system were that motion was circular and around the Earth. Since the actual motion is neither circular nor around the Earth, the system was doomed to become complex. There is often simplicity to be had in a correct solution, and an incorrect one must add complexity to arrive at an answer. Basically, if a simple circular motion did not match observations well enough, Ptolemy's approach was to add other circular motions to it. We noted above that sometimes the real position of the Sun is "fast", or ahead of what its position would be if it were a perfect timekeeper, and sometimes "slow". The real reason for this is now known to be the ellipticity of Earth's orbit. However, if one simply wants a first approximation to this behavior, while having "perfect" circular motion, one can imagine the Sun moving at uniform (thus "perfect") speed on the circumference of a circle, but with that circle not being centered on the Earth, but rather being "eccentric". In this way, when nearer to the Earth, the Sun would *appear* to move faster (in fact, would move faster in angle), and, when more distant, slower. With just the one parameter, the eccentric offset, one could account much better for the apparent motion of the Sun than if once considered it simply to go in a circle around Earth. In a way, this system is less arbitrary than varying the speed on the circle itself in an *ad hoc* manner, and it lends itself to easy geometric calculation. If greater precision was needed, one again added motion on a circle, but this time on a small circle riding on the circumference of the large one. This smaller circle was called a "deferent". Again, the circular motion would be uniform, although possibly not at the same rate as the main motion.

Although an eccentric circular orbit could work well for the apparent motion of the Sun around the Earth, a deferent was usually used for the more eccentric and complicated lunar motion. The planets, compared with these two bodies which always appear to move overall toward the east, have a complex motion. This is illustrated in Figure 1.8, in which Mars appears to slow down and reverse direction. The normal eastward motion is called "prograde" and, by contrast, the backward motion is called "retrograde". One can indeed envisage that if Mars' normal orbit came from it moving on a circular path, a smaller circle moving in the opposite direction and carrying Mars could explain a behavior like this. The mathematical details of

**FIGURE 1.8** Retrograde motion of Mars in late 2022. Initially moving eastward, it reversed direction and moved westward and upward. It is above the winter constellation Orion. (Author, modified composite of renderings from Stellarium, 2023.)

all the circular motions needed could be worked out, and relatively accurate results obtained.

Ptolemy's system of linked circular motion may resemble complex clockwork, in which circular gears interact, partly because the ancient Greeks did have such mechanisms. At the technological cutting edge at the time, they must have been rare and expensive. We have only one example in our museums, preserved by being in a shipwreck. Among precious statues on the seafloor, a small brass mechanism was found that had gears and dials and seemed to be for astronomical calculation. A small model of the known universe was made with circular gears! Clearly, if a mechanical device could be made that reproduced the heavens with the circular motions possible with gears meshing with other circular gears, circular motion through complex gearing described in mathematics could be well described. Although the mechanical system of the heavens was not there to view, perhaps it was the same or at least similar. Clearly, gears worked by physical contact, so could there be perfectly clear materials used to build a mechanical heavens, engaged in the manner described mathematically in Ptolemy's circular calculations?

A remarkable analogy may be made between Ptolemy's system and the result of completely disparate studies done about 1,500 years later. Studies of heat in solid objects by French scientist Joseph Fourier, published in 1822, showed that the motion of heat (then thought of as a liquid called "caloric") through solid bodies could be described by, and the governing equations solved by, cyclical series. A series, in mathematics, is a sum of mathematical items. Fourier's series were made up of sine and cosine functions identical to those used to describe circular motion, co-added much like the terms in Ptolemy's system. Initially, Fourier's system was not accepted since certain features of known series at the time were elusive to prove, but it now forms the basis of much of our technological world. Fourier's system assumes a *fundamental* frequency and then adds various amounts of *harmonics*. As in music, the harmonics are at multiples of the fundamental. Any shape of a repeating curve can be synthesized in this manner. Although Ptolemy's system did not have harmonics in the sense that Fourier's does, it is remarkable that the Greek concept so strongly resembled one that came to be important for our own technological society.

A further common feature of Ptolemy's and Fourier's systems is their possibility of adding precision by doing more calculations. In the case of a Fourier series, terms at higher and higher frequencies can be added to get better agreement with a given function or with data (although a "ringing" effect, called Hill's phenomenon, can be a problem). With Ptolemy's system, equivalently, more "wheels within wheels" can be added to the calculations. However, as better and better astronomical equipment came into use, very large numbers of extra terms had to be added to "save the phenomena", making observation and theory match. Sometimes, the need to make a theory more and more complex is an indication that a better theory may be possible. Indeed, in 1543 CE, the year of his death, Polish astronomer Nicholas Copernicus published a theory that, by "demoting" the Earth to simply being one among many planets going around the Sun, better matched what we now know as reality.

Copernicus claimed that the Earth turned on its axis and was merely one of several bodies orbiting the Sun: a *heliocentric* theory. Objections from common experience that, on a moving body (such as a cart moving quickly), air flows by, or there is jostling, seemed to argue against a moving Earth. However, there was a well-justified, less intuitive reason to suspect that the Earth did not move. This is a simply demonstrated effect known as *parallax*. If a local motion takes place, objects at an intermediate distance will appear to move relative to those at a greater distance. If one moves one's head while looking at a finger at arm's length away, the position of the finger relative to a distant background will seem to change. If the Earth was moving, surely a similar effect would be seen, but it was not. Once the heliocentric theory was generally accepted, its actual physical proof had to wait for nearly *three hundred years* after its formulation to be proven using parallax. German astronomer Friedrich Bessel first observed the parallax motion of nearby star 61 Cygni relative to the distant background stars in 1838. The reason that parallax had to await such a long development of instrumentation and observing techniques is the immense scale of cosmic distances. Even the nearest star is so distant that parallax effects are tiny. In fact, we think of the distance scale to stars in *parsecs* (one parsec is 3.26 light years, an older distance unit). A star one parsec away has a parallax motion of one second of arc or *arcsecond*, i.e., 1/3,600 degrees. The distance from the Sun to the Earth ($1.5 \times 10^{11}$ m = 150 million km) can be referred to as the *astronomical unit* or *AU*. One parsec is 206,265 AU. Although we will revisit the scale of our big, empty (or at least mostly invisible) Solar System below, the huge scales of both the Solar System and interstellar space almost defy comprehension.

Copernicus' book was called *De Revolutionibus*, in the Latin of scholastic publications of the time. This means "Of Revolutions", referring to the annual *revolution* of orbital motion of bodies about the Sun (turning daily on their axes being *rotation*). Considered by the science historian Kuhn to be one of the few *paradigm shifts* that have characterized science, leading to a fundamental change of point of view, Copernicus' title also came into popular use to characterize a political uprising and change. Despite its revolutionary nature, Copernicus' theory retained circular motion. In fact, it required almost as many terms as Ptolemy's scheme, making it more of a philosophical than a practical achievement in astronomy.

The German astronomer Johannes Kepler developed a further key to understanding the motion of Solar System bodies after working for many years with pre-telescopic data. In 1609, he published the startling insight, a "law" that orbits could be best explained if they were not circular but elliptic in nature, with the Sun at one of the two centers (called *foci*) of the ellipse. As a second "law of areas" he found that planets move along these paths in such a way as to sweep out equal areas in equal time. This means that they slow down when further from the Sun and speed up when nearer to it. Although the first two laws referred to an individual orbit, his third law allowed comparison of different orbits. The period of revolution (often using symbol *P*) is proportional to the three-halves power of the orbital average distance from the Sun (known as the *semi-major axis* and usually represented by *a*). For example, Jupiter is about five times as far from the Sun as the Earth is (i.e., at about 5 AU), and to take the three-halves power, we take the third power as 125, then the square root of this. The answer, 11.2 years, is the approximate orbital period of Jupiter. More accurate figures are $a = 5.2$ and $P = 11.86$, in AU and years, respectively.

In moving from Ptolemy through Copernicus and Kepler, a kinematic model, perhaps originally based on gears, had developed that explained the motion of Solar System bodies well. A dynamic model, involving forces, could now be sought. Kepler was interested in mysterious forces that might activate the cosmos. Magnetism had recently come to the attention of the Western world and was known to characterize the Earth. Descartes in France advocated the action of cosmic vortices inspired by watching swirling liquids. However, the dark force that moves the Solar System lay in plain sight, requiring only the insight that could unite the mundane of the Earth with the celestial realm.

Based on work done over the preceding twenty years, Isaac Newton's book *Philosophiae Naturalis Principia Mathematica*, usually referred to simply as *Principia*, was published in 1687. It laid out not only the laws of mechanical motion but also the law of gravitation that allowed their extension to explain the now-known laws of planetary motion.

## 1.4 NEWTON'S DARK FORCE

Charming stories about falling apples may lead us to think otherwise, but Newton did not "discover" gravity. It is clear to a child, and well known to ancient thinkers, that unsupported objects fall. Going back to Aristotle, it seemed natural to think that objects had preferred states, and a simple look around shows that a preferred state for weighty objects is to sit motionless on the ground. Falling could thus be attributed to a body not being in its preferred state, but instead moving toward it. This idea of preferred state applied to other motions as well. A rock sliding on a flat surface comes to rest; a load of produce being taken to market has to be dragged or rolled. The concept of a lack of motion being a preferred state seemed only natural. In this sense, how different were motions in the heavens! The Moon seemed to move around the Earth every month, the Sun to go around the Earth every year, while both took part in overall daily motion, which also seemed to be about the Earth. Heavenly motion clearly did not obey the same rules as applied near the surface of the Earth, since it did not come to a stop. The further observation that heavenly motions could be explained with circles, the most perfect of forms, reinforced the concept of a divine and perfect realm of unimpeded motion, while the tawdry surface of the Earth was characterized by a lack of motion.

Even once Copernicus "moved" the center of motion to the Sun (a notion also proposed about 1,800 years earlier by the Greek astronomer Aristarchus), circular motion remained a dominant aspect, including many of the complications that Ptolemy's Earth-centered model used to preserve it. Hard cold facts based on exacting observation led Kepler to the startling assertion that ellipses were better than circles in describing celestial motion. The very name ellipse means "something that falls short" of a circle, as coined in about 200 BCE by Apollonius of Perga, the Greek scholar most responsible for their early study. Kepler's "war on Mars" in the early years of the 1600s derived from purposeful observations by Tycho Brahe in Denmark to determine whether fundamental differences in the layout of the Solar System in Copernicus' Sun-based model as compared to Ptolemy's Earth-centered model could be observed. Ironically, these observations failed due to the scale of the Solar System being about twenty times larger than was then accepted, meaning that Tycho's naked-eye observations were not up to the task. The method was to be "diurnal parallax", meaning that, as one is swept from west to east on a rotating Earth, one observes distant objects from vantage points separated by one Earth diameter. This method is able to give a highly accurate distance for the Earth's Moon, and Tycho had determined this very precisely to be about sixty times the Earth's radius. A similar method applied to eclipses had already allowed the second-century BCE Greek astronomer Hipparchus to get a similar value. For objects like Mars, well beyond the Moon, the small angular movements involved need a telescope to see. Kepler's "war", at the end of which he declared in 1609s *Astronomia Nova* (the "new" astronomy) that Mars had been taken prisoner, laboriously managed to extract from Tycho's data that the orbits of both Mars and Earth must be elliptical. Thus, one aspect of the difference between the heavens and the Earth, the perfection of circular motion, was removed from consideration.

Observations of parallax for Mars were inconclusive for Tycho, but similar observations led to his rejection of the "pure" heliocentric model of Copernicus. He could not detect any parallax motion for stars, and, if the Earth was moving, he reasoned that there should have been a parallax observable over half a year, much as the Moon had been observed to have a parallax when measured over half a day. Since other aspects of planetary motion were well explained by a heliocentric model, Tycho placed all planets, except the Earth, going around the Sun. Due to the lack of stellar parallax, the Earth was again placed, immobile, at the center in the sense that the Sun went around it! Tycho's basic mistake was in not knowing the huge scale of the Galaxy: even the nearest stars have a very small parallax of about one arc second. This is about sixty times smaller than Tycho's best observations which had a resolution of about one arc minute, which in turn is one-sixtieth of a degree of angle. When Bessel observed parallax in 1838, he used advanced visual observing techniques, but through a telescope. Kepler's adoption of elliptic orbits allowed the rejection of Tycho's model and brought our view of the planetary system back to the "purely" heliocentric Copernican form.

Part of Kepler's motivation for seeking to place the Sun at the center was religious: he felt that the Sun was to be equated to God, in the center and providing the power for movement of the whole system. He did feel that a motive force was needed, and that the God-like Sun would be the source of not only light and heat, but whatever this motive force might be. The orbit ellipses would have the Sun at one "focus", where the original Latin term meant "fireplace", as a central fire. Consideration of what the force might be was stimulated by the publication by Englishman William Gilbert of *"De Magnete"* in 1600, in which it was shown not only that magnetic forces could be systematically described, but that the Earth is, on the large scale, a magnet-exuding force. While his earlier works sought some relation to geometrical shapes or musical notes in determining the relations between the planets, Kepler's mysticism in the end blended with empirical observation, and, developing notions of forces, set the stage for Newton to consistently tie together laws of motion, force, and gravity, to make essentially our modern view of the Solar System.

Newton's insight into gravity, a "dark" force, arose from his development of dynamics, the science of forces. Unlike the ancient conception that the natural state of objects is to be at rest, Newton's is that steady motion (which can include being at rest) at constant velocity is the natural state of bodies. This steady motion is changed by forces. Forces resisting movement are very common. Friction with the surface below it will slow a moving object and cause it to come to rest. Due to friction being ubiquitous, rest came to be seen by the Greeks as a natural state which objects tended to assume. In Newton's view, an object in motion has a velocity, and friction is a force that diminishes that velocity. In our modern society, with good low-friction wheels, it is easy to think of objects moving rather freely on a smooth surface such as pavement. Since friction is low, little force needs to act, and the natural state of an object in motion is seen to be to stay in motion. In fact, an object will move in the direction of its initial motion. "Velocity" is the combination of speed and direction, and neither speed nor direction change unless a force acts. In the case of friction, the force is in the opposite direction to the motion, so the velocity decreases in size until it comes to zero. In the case of friction, once a body is at rest, there is no more frictional resistance arising from motion, so it stays at rest.

The change in velocity with time is called acceleration. In common usage, a decreasing speed is referred to as "deceleration", but, to a physicist, this is simply acceleration in the opposite direction to the velocity. A very important concept in dealing with orbits is that the acceleration (arising from a force) may not be in the same direction as the initial velocity or may even be opposite to it as frictional forces are. In the case of a sliding object with friction, the acceleration is in the opposite direction to the initial motion. In a circular orbit, the acceleration is *perpendicular* to the velocity. If we think, as Newton appears to have been the first to have illustrated, of a cannonball in circular orbit around Earth (either circular path in Figure 1.9), its velocity is always parallel to the Earth's surface, which makes it change in absolute space. That change in direction is an acceleration, due to a force toward the center (Figure 1.9).

That force is gravity, causing an acceleration that is directed toward the center of the Earth. To have a circular orbit, a specific speed is required. In a general elliptical orbit (like the outermost curve), the speed is high when near the Earth, and less when higher up. The elliptical motion is what Kepler deduced from observation, and the variation in speed is in accord with his "law of areas" mentioned above. Not shown in Newton's cannonball diagram is the very simple case of an object dropped straight down, which obviously heads directly toward the center of the Earth (acceleration and velocity are aligned in such motion from an initial state of rest). Newton made a link between such bodies falling near the surface of the Earth and those out in the universe, by considering the motion of the Moon. The acceleration of a body in circular orbit is proportional to the square of its speed and inversely proportional to the radius of its orbit (distance from the center of the Earth). With both the period of the Moon and its distance from the Earth being accurately known (the latter due to Tycho), its acceleration could be calculated. It turns out that this is the rather minuscule three mm per second each second (3 mm/s$^2$), the speed of the moon in its orbit being about one kilometer per second, but the radius being rather large at sixty

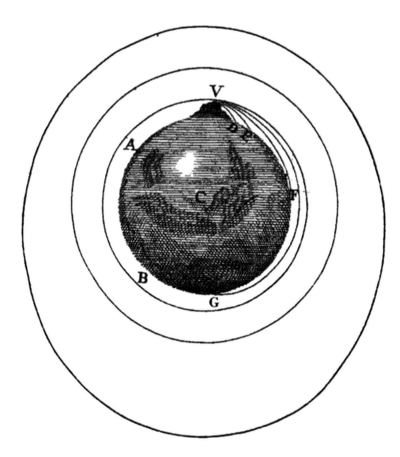

**FIGURE 1.9** Newton's Cannonball from Isaac Newton, *A Treatise of the System of the World*, 2nd ed. (Newton, 1731).

Earth radii (60 × 6371 km ~ 385 million meters). An apple (or cannonball) observed near the surface of the Earth would fall, accelerating at nearly ten meters per second each second, with about 3,500 times the acceleration of the Moon. This acceleration, which is nearly the same at all points on the surface of the Earth, is usually called *g* and more accurately has a representative value of 9.8 m/s$^2$. Newton's extension of a dark force into space would work only if the force of gravity fell off with distance from the center of the Earth to cause much less acceleration on objects at the distance of the Moon. Newton deduced that it had to fall off with the square of the distance ("inverse-square law"), which would mean that 60 times as far from the center as it is from the surface of the Earth, the force would be diminished by 60 squared, or 3,600, much as the rough numbers above indicated. Newton further realized that all bodies attract *all* other bodies with force proportional to the masses of each. In this sense the law of gravitation is "universal", applying to all bodies. Since it decreases with the square of distance, there comes a point at which effects become negligible.

For example, it would be difficult to measure the effect of the Moon on a falling apple near the surface of the Earth.

Newton's unseen force pervading the Solar System was thus to be proportional to the masses involved, inversely proportional to their separations, and always acting to draw them together. What came to be referred to as the "gravitational constant", allowing expression of the force in standard units, turned out to be rather small. It took over 300 kg of cleverly arranged lead to allow English scientist Henry Cavendish to measure gravitational effects between bodies near the surface of the Earth and effectively determine this constant, now usually referred to as $G$. This measurement was completed by 1798, about 110 years after *Principia* was published. Since the Sun is about 1,000 times more massive than any planet (Jupiter is the next most-massive body), its gravity dominates the Solar System, so it is not incorrect to say that the "system" arises from a collection of bodies following its gravitational mastery. In the vicinity of smaller bodies like planets, secondary bodies may be trapped, much as is our Moon. To compensate for the small (relatively speaking) masses involved, such natural satellites must be very near (again, relatively speaking) the "parent" planet.

We must make one small addendum to the above story of progression to a modern view. Generally, this story as told in traditional astronomy textbooks attempts to show "progress" from ancient times to the present. While we must realize that astronomical knowledge was not widely shared in the past (about half the people present in ancient Greece or Rome were slaves, for example), that which did exist in learned circles was surprisingly sophisticated. The Roman philosopher Seneca, in about the seventh decade of our era, wrote surprisingly modern assessments of meteors and comets, and in the same volume noted that "our descendants will marvel at our ignorance". About two hundred years before, but in the same intellectual tradition, Lucretius' *De Rerum Natura* (a poem) urged that the actions of deities not be sought to explain natural phenomena. Certainly, the mechanistic system of Ptolemy did not require them, although one of its main aims was astrological prediction. In many texts, some "odd" views of Kepler are stressed, while Copernicus and Newton are portrayed as neutral seekers of the truth, and Galileo as being in opposition to the Church. Nothing could be further from the truth. Kepler, while seeking "harmony of the spheres", was working on an intellectual problem that went back to the Pythagorean school two thousand years before him. They sought to extend physics-based (yes, experimental) findings on musical relations to the skies, setting the basis for Kepler's work. In both cases, this approach was unfruitful, but, in Kepler's case (as may also be claimed for the Pythagoreans to some extent), the harmony would have been evidence of the workings of God. Copernicus' view that the Sun should be at the center of the universe was mirrored in his view of a God around whom all things revolved. Galileo was a devout Catholic at a time when the Church was being challenged: he had an admirable devotion to the truth but to some degree was a pawn in a larger game, with many churchmen knowing full well that his observations did indicate that change was needed. Finally, Newton's statement "*hypothese non fingo*" (basically, I do not like to make guesses) about the actual nature of gravity did not indicate that he sought mechanistic explanations of all things. He was a deeply religious man and held to beliefs that, in a time of religious change, could have been dangerous. Rather than "marvelling at the ignorance" of the ancient forebears of the paradigm shifters of Copernican-era astronomers, we should marvel at how much they got right and their commonality of spirit with modern scientists.

Finally, let us return to how the Solar System appears as viewed by humans gazing out at night. We find planetary bodies separated by vast distances, dominated by the gravity of the Sun, whose effects can be calculated by the physical laws developed by Newton. The tiny parts of the Solar System that are visible to the naked eye swim in a vast unseen force field. If it took several thousand years to discover the unseen force of gravity, it subsequently took several hundred years to realize its implications, ending with the surprise that our very concepts of space and time were wrong from the beginning.

# 2 Astronomy of the Invisible

Astronomers lost little time in putting the invisible force of Newton to work in improving the accuracy of *ephemerides*, or tables of positions of the Moon and planets. This work was not driven only by intellectual interest. It is likely not coincidental that the scientific revolution took place mostly in Europe and around the time of European expansion to dominate the world. Most long-distance travel was done by sea: trade was a major activity, but its protection and expansion required the projection of power through strong navies. Mercantile or naval ships both needed to know their location to safely reach distant destinations and return. Practical needs drove the astronomy of the invisible.

The Earth rotates on its axis very precisely in one *sidereal day*. The Earth's axis points toward a position in the sky known as the *celestial pole*: as a result, that point appears fixed, with other stars appearing to rotate around it. Conveniently, in the Northern Hemisphere, the bright star Polaris is very near that point, making it easy to identify. The latitude of an observer may be determined quite easily in the Northern Hemisphere: it is the apparent *altitude* (in degrees) of the celestial pole above the horizon. With simple instruments, it is thus very easy to find latitude if the sky is clear at night. There is a bit of a complication in that the Earth's axis position changes slightly with time. Early in the era of sailing ship exploration, the pole was about three degrees away from Polaris, whereas now the difference is a near-negligible half degree. However, this could be compensated for by the use of a table giving the altitude of Polaris depending on the orientation of easily observable nearby stars. As mentioned in the last chapter, Columbus was much concerned with longitude, and in a critical view perhaps making sure that he got it wrong. About the same time, though, major north-south voyages near Africa were also being made, mainly by the Portuguese, so determining latitude became important and tables to aid in this were published.

Longitude is another matter. The east-west position of any object observed in the sky depends both on the time and on the longitude of the observer. In the early days of navigation, time was not able to be kept accurately enough to be used for navigation purposes. Determining longitude with the precision needed for safe navigation was not possible. The ultimate solution was the invention of the chronometer by John Harrison in the mid-eighteenth century. Until that time, however, it appeared that the clockwork of the heavens might provide means to determine absolute time, and large investments in positional astronomy were made by the competing governments of European naval powers. The ability to make good predictions based on known objects moving under Newton's gravitational forces was thus not only of intellectual interest but vital to national security and prosperity. This impetus allowed the

DOI: 10.1201/9781003451433-2

astronomy of the invisible to develop and be fruitful, as gravity was put into a burgeoning mathematical framework as much for practical as scientific purposes. Once the framework was developed, gravitational astronomy seemed to lose interest as an interesting field, a view that changed only recently. We will explore the new point of view but return to why the classical view suffices for many practical purposes.

## 2.1 THE MODERN RENAISSANCE OF GRAVITY

In February, 2016, an announcement was made that the era of gravitational wave astronomy had begun. Months before, twinned Laser Interferometer Gravitational-wave Observatory (LIGO) detectors monitoring dynamical distortions in "spacetime" had undergone changes in their four-kilometer-long laser reflector tubes that corresponded to periodic movement of about one-ten-thousandth of the breadth of a proton ($10^{-18}$ m). These tiny changes were inferred to have arisen from the merger of a pair of black holes about one billion light years away, making a new single black hole having a mass about four solar masses less than the sum of the masses of the original black holes. Subsequently, other black hole mergers have been detected, opening up a new way of surveying massive objects in the universe.

The need to view both space and time as one entity called spacetime will be discussed in more detail below, and is at the heart of a new way of viewing the universe. Known as General Relativity, this was developed mainly by Einstein early in the twentieth century. Gravitational waves are traveling distortions in this spacetime, moving at the speed of light. Their production requires moving objects to interact, rather like how electric currents must flow back and forth in an antenna to generate radio waves. A single symmetric collapse, no matter how violent, does not generate gravity waves. Although our technology, pushed to the limits of precision, now allows us to detect the minute effects of gravitational waves, the use of gravity to detect new objects in the Solar System dates back centuries. We will return to the modern view of the melding of space and time, but here note that, in our Solar System, speeds and masses are modest compared with those found in more exotic parts of the universe. As a result, Newton's gravity works exceptionally well and was a wellspring of discovery, including the use of indirect gravitational effects to detect new objects.

Newton's law of gravity states that all objects having mass produce forces on all others, proportional to the product of their masses and diminishing as the inverse square of the distance between them. An inverse square law bears an interesting relationship to the size of an object as shown in Figure 2.1. The apparent diameter of an object is inversely proportional to its distance: an object twice as far away looks half as big across as a similar object nearby. However, we largely perceive objects' sizes by their apparent area (in technical language, this is called *solid angle*). The apparent area is directly related to the square of the apparent diameter. When the apparent diameter angle is half as big, the apparent angular area is one-quarter as big, since like other areas it increases with the square. The force of gravity falls off at the same rate, so if the apparent area of similar objects appears to us to be smaller as they become more distant, the force of their gravity on us goes down in proportion.

**FIGURE 2.1** The apparent area of a small object in the sky is inversely proportional to its distance squared. An identical object at twice the distance has one-quarter the apparent area. This is the same ratio as applies for its gravitational force (author, from NASA images, US government public domain).

If we had two otherwise identical moons, as shown in the figure, one of them twice as far away as the other, the second would appear half as large in diameter, but that means that its apparent area would only be one-quarter as large. The gravitational force, compared to that of our present Moon, would also be one-quarter as large. This seems reasonable, as does the idea that a larger mass produces more gravity in the first place, and another mass reacts to that extra force in a way proportional to its own mass. Newton's law of gravity appears in all ways consistent with common sense and was amazingly successful when elaborated mathematically. As shown in the first chapter, Newton could support his new law with calculations which not only showed it to be mathematically correct but extended the familiar concept of gravity into the heavens.

## 2.2 FORCE AND MOTION

We must now plant some doubt, however. In addition to the law of gravity, Newton formulated his famous laws of motion. The first of these postulates that, in the absence of forces, a body continues in its original state of motion in a straight line. This in itself is not in accord with the commonsense observation of motion in every-day life that informed Aristotle. He regarded the natural state of bodies to be rest and to seek to go downward. Indeed, if we think of something as simple as a sack of potatoes, if it falls off a table, it plunges to the floor and comes to rest. If the bag breaks, the potatoes may scatter but then come to rest. How can Newton's view be reconciled with this simple counterexample? The answer is that not all forces are obvious. A sack sitting on a table is supported by material forces from the table: imagine, instead of a table, a trampoline. A sack on a trampoline would push it down in a way that the elastic straps making up the trampoline would be deformed, and, by deforming, would come to a position in which the sack can be held at rest. This deformation would be obvious. With a table, the deformation is not as obvious (although perhaps imagine a sack of potatoes on a flimsy card table, which might bend under the load). To Aristotle, the sack falling, if pushed off the table, simply has another natural tendency: to want to go downward. All heavy things seem to go down, so that must be their natural tendency to seek their natural state. Newton informs us otherwise. When no longer supported by a table, a sack of potatoes is attracted toward the center of the Earth by gravitational force. It moves downward (accelerated, as we shall soon see) until other forces come into play again, which are material forces from the floor rather similar to those from the table in the first place. Even if the sack ruptures, each potato will be held up by deformation of the small piece of floor under it, once it comes to rest. Again, we can imagine the potatoes on a trampoline: the elastic would not be much stretched, but it would indeed be stretched, more under each potato than where there is no potato.

The beauty, and the flaw, of Newton's view of falling, is that the gravitational pull on each object is proportional to its mass, but its acceleration is inversely proportional to that mass (as we will see shortly). So, the two effects *cancel out*: a falling potato of small mass and the whole sack of much larger mass fall at the same rate. In a famous, although perhaps apocryphal, story, it was Galileo at the Leaning Tower of Pisa who demonstrated this equal acceleration of differing masses, which contradicted Aristotle's common sense, well before Newton's time. Here, however, we plant the seed of doubt: why is it that the mass involved in acceleration, which is purely spatial, should be the same as the mass involved in gravitation, which gives rise, in Newton's view, to forces? Certainly, the mathematics works out in this view, but does it make sense? We will revisit this point later.

In a famous demonstration in 1971, David Scott, the commander of the Apollo 15 "Falcon" lunar module repeated Galileo's experiment. Figure 2.2 is a frame from a grainy video showing him dropping a feather and a hammer in the vacuum of the Moon to show them falling at the same rate. Gravity viewed as a force is proportional to mass, which also gives *inertia* or resistance to motion. Without the resistance of air to impede it, a feather moves much as a hammer does. However, the structure of objects arises from electromagnetic forces (physicists refer to this way of structuring as "chemistry"). Why should objects made of diverse materials, steel in the case of a hammer, organic materials in the case of a feather, be structured so differently by electromagnetism, yet be so uniform in response to gravity?

We alluded to acceleration. Newton's second law relates acceleration to force. In this, he went beyond Aristotle's view that the harder one pushes something, the faster it goes, in which force is related to speed. Instead, Newton makes force *build up* speed, the rate of the speed buildup being acceleration. Indeed, in a car, the pedal to "go faster" is the "accelerator", and one speaks of "zero to sixty in three seconds" as being how long it takes to get to sixty miles per hour (about 100 km per hour) in a powerful car. This indicates an acceleration since at the start the speed is zero and at the end sixty miles per hour. If the acceleration was uniform, an extra amount of speed of twenty miles per hour would have been added to the speed each second. The force to make the car accelerate came ultimately from the engine, although in

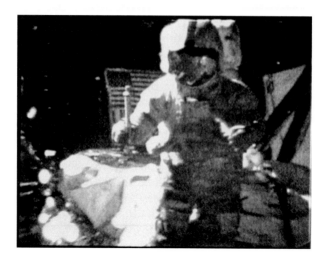

FIGURE 2.2    Hammer (left) and feather (right) were shown to drop at the same rate in a vacuum on the Moon (NASA).

a complicated way, involving road friction. We may not think of it much, but the mass of the car is as much involved in determining acceleration as having a powerful engine. Indeed, in terms of results, Newton's second law says that "acceleration is proportional to force applied, and inversely proportional to the mass it is applied to". A race car engine may be enough to generate the force to accelerate a truck (comparable "horsepower"), but, in a heavy truck, the acceleration would be much less than in a lightweight race car. So, big force, small car, lots of acceleration. Big force, big truck, small acceleration. Now, let us imagine that we drive both the car and the truck off a cliff! Under gravity, they have a force acting proportional to mass in Newton's view. So, small car, small gravity force, acceleration at the rate *everything* falls at. Big truck, big gravity force, acceleration *exactly the same*. The force equation is Newton's law of gravity, involving matter attracting matter. The acceleration law relates force to mass, but why *should* that mass be the same as the mass in the law of gravity? Again, food for thought about gravity!

Newton's third law of motion is the basis of using gravity for invisible astronomy. It states that forces are mutual: if you push on something, it pushes back on you. With gravity examples, such as sacks of potatoes, there can be some argument about how to exactly look at what is pushing on what. Imagine instead a ball bouncing off a wall. The ball, initially moving toward the wall, later moves away from the wall. Clearly, its velocity (speed in a direction) was changed by the wall, which caused an acceleration. Clearly, the wall exerted a force on the ball. Again, it is not totally obvious that the ball exerted a force on the wall. Here, the trampoline idea comes in handy. Imagine a vertical trampoline. Although this is not a common situation, it is easy to imagine that a ball hitting a vertical trampoline would deform its elastic straps until the ball bounced back. So, clearly, the ball did exert a force. Another example would be to bounce a ball off a wall made of weak metal like tin: dints would form from the force of the ball on such a wall. Even if the object acted upon moves, its force back on the object pushing it is the same as the force applied to it. In this application, we can find hidden objects by their force back on known objects. Since accelerations are inversely proportional to mass, a small planet accelerates a large one much less than the large planet accelerates the small planet. As an example, tiny Mercury is influenced, or *perturbed* by Jupiter, with a detectable response, whereas its effects on Jupiter can often be neglected. Indeed, sometimes, the whole inner Solar System's motions are neglected in calculations of the motions of the large outer planets: one simply folds their mass into that of the Sun!

## 2.3 THE ONLY EXACT SOLUTION

Newton's law of gravity, with its inverse square dependence on distance, allows an exact solution of the motion of one body around another, much larger one, and indeed of the slightly more complex problem of how two bodies move around their common *center of mass*. Part of our modern understanding of gravity is that it is an effect of the curvature of space. Before the advances of Einstein and others in the early twentieth century, the geometry of space was thought of as flat, without curvature. In flat space, the inverse square law was clear to Newton and demonstrated in *Principia* by geometric proofs related to the apparent angular area also falling off with the second power. Thus, even the elliptic orbits of Kepler arise in a natural way. The so-called *Kepler problem* is simply the orbital solution of the two-body problem, giving rise to eternal orbiting motion, unchanging in its details from one orbit to the next. In addition to planets, comets were known in Newton's time. Indeed, Edmund Halley, the namesake of Halley's comet, was one of Newton's contemporaries, who demonstrated that this comet had a highly elongated but periodic orbit. Some comets had such extreme orbits that they could have been on non-closed or parabolic orbits, and Newton could calculate these too, as will be discussed in a later chapter. We now know of some interstellar objects that, with respect to the Sun, have hyperbolic orbits, equally calculable. All such possible orbital paths are *conic sections*, known to the ancient Greeks, and described by *quadratic* equations (in which the maximum power of exponents is two).

Obviously, a Solar System with six planetary bodies (Mercury, Venus, Earth, Mars, Jupiter, Saturn) going around a seventh (the Sun) cannot be described by a two-body theory. Approximation methods must be used since the two-body problem is the *only* exactly soluble problem for general motion under gravity. With more bodies interacting with each other, as is this case, only approximations can be made, although mathematical developments allowed them to be made very precisely if one was willing to indulge in complex calculations. An approximation need only meet some needs and agree with the data to be useful, and one must balance the need for precision with the use. For example, if the need is to keep track of the seasons, and the data are the recorded times when the Sun's position gives rise to equinoxes and solstices, one just needs to think in terms of the motion of the Earth around the Sun. Even a model in which the Sun goes around the Earth may be fine for this purpose. At equinoxes, the Sun appears to rise exactly in the east and set exactly in the west. At solstices, it will be at either the northern (summer) or southern (winter) limit of its apparent motion. This might be determined with reference to some point on the ground, as is suspected to have been the function of massive stones in some ancient monuments. However, with sufficient computing power, very good approximations to planetary positions can be calculated. Some discussion about time and measurement are needed before further discussion of that and how deviations could be detected and used in the astronomy of the invisible.

## 2.4  A MATTER OF TIME

If the year is thought of as 365 days in length, a simple calendar in which one expects the same phenomenon (say an equinox) exactly 365 days later quickly goes out of synchronization. This defect was remedied by the Romans, who developed the "*Julian*" calendar. Roman calendar reform also produced the month of July, with Julian and July both being associated with Gaius Julius Caesar. The Julian calendar itself had issues and was replaced, but we still have July, and, in fact, August, named after Julius' great nephew Octavius, who became the first emperor of Rome and adopted the name Augustus. The reform tried to account for Earth's revolution period being closer to 365.24 days, which is referred to as the *tropical year*. Adding one day every four years takes account of 0.25 days for each year: this is the purpose of leap years. This is quite a good approximation, making the year effectively 365.25 days long (365 days each year, plus an extra day in leap years). However, the use of the Julian calendar far outlived the Romans, whose empire collapsed a few hundred years into the present era. By 1582, further calendar reform was needed, since even the approximation put in place by Julius Caesar had led to errors in the timing of seasons. Good information on seasons was needed for agricultural and religious purposes. A further calendar reform was instituted by the Church, adjusting dates and bringing in a new system in which only each fourth of century years (all divisible by four so thus normally leap years) would actually be a leap year. The year 2000 was the first century year since 1600 to be a leap year. Based on the name of the reforming pope, our present calendar is referred to as "*Gregorian*". The last country to adopt calendar reform was the Soviet Union, a now-defunct country that famously arose from the "October" Revolution in Russia in 1917. Not long after the Soviets took power, the calendar there was reformed, in February 1918. Due to that reform, the hundredth anniversary of their revolution was actually in November, not October, of 2017.

   The purely numerical system of keeping a calendar does not really need much of a theory behind it. Our orbit can be assumed to be circular, and, as long as we line up the counting to include that the Earth does not have an integral number of rotations (used to count days) per revolution around the Sun (or of the Sun apparently around us, that equally does not matter), all works out fine. We know to sow seed when it is a good time of year to do that, and Christmas can be snowy. Certain other religious feasts have somewhat complex computations, leading to the term "movable feasts" in the Christian church calendar. In this and several other religions, the changing dates reflect the use of the lunar calendar. The use of the Moon for calendars is for practical purposes now mainly reflected in the English word "month" being related to the word "Moon", and the fact that months are roughly the length of one Moon period. If you happen to notice some phase of the Moon on a particular day of a month, that phase will repeat on about the same day the next month. However, the pattern is quite quick to break down. The apparent (deemed *synodic*, a term related to the "meeting" of the Moon and the Sun each month at the new Moon phase) period of the Moon between identical phases is 29.5 days. Being shorter than most months leads to a given phase coming earlier in most months than in the one preceding, so the term "getting out of phase" applies eventually.

## 2.5  TO SOME DEGREE

As pointed out in the previous chapter, it is not a coincidence that we have a 365-day (or so) year, and 360 degrees in a circle. These two numbers are closely tied to our orbital motion around the Sun, which combines with the Earth's axis tilt (not dealt with in this book) to give us the seasons that are all-important for the latitude zones that many humans inhabit. However, they are not natural units for geometry. For small angles, astronomers use arcseconds, already introduced above. In analytical trigonometry, the unit of the *radian* is a natural one. In the *small-angle formula*, the angle of an object at the far end of a triangle much longer than its side is simply given, in the natural units of radians, by the physical size of the object divided by how far away it is (in the same units, of course). In the motion picture *Apollo 13*, astronaut Jim Lovell, played by Tom Hanks, blots out the Moon with his thumb as he imagines going there. An adult thumb is about 1 cm (0.01 m) across, and, at arm's length, is about one meter from the eyes, so the small-angle formula would tell us that the Moon must appear to be a bit less than 0.01 / 1 ~ 0.01 radians in angular diameter. The Moon has an elliptical orbit around the Earth, so its apparent angular size changes, but on average we can think of it as 0.5 degrees, or 30 arcminutes, in angular diameter. When dealing with larger angles, we find that the entire angle going around a circle, which for historical-orbital reasons we denote as 360°, is $2\pi$ radians. So, one radian is 360 / ($2\pi$) ~ 57.3 degrees. Conversely, one degree is $2\pi$ / 360 ~ 0.0175 radians. We find that the Moon's angular diameter, of a bit less than 0.01 radians, must be a bit less than 0.573 degrees, which is in accordance with what was mentioned above. To put things in context, there are 90 degrees from the horizon to the zenith (the "top" of the sky, directly above an observer), so this is about 180 Moon angular diameters. The Moon may look large to romantics, but it actually takes up a very small amount of sky! The situation is even worse if we consider angular area, also known as *solid angle*. Since areas (even angular ones) involve squaring lengths, the Moon takes up less than one ten-thousandth of the area of the whole visible sky (the actual calculation is a bit complicated by "spherical" trigonometry).

For the sake of completeness, we will mention two further aspects of angular measure. First, the degree-arcminute-arcsecond system for angles is like the hour-minute-second system for time with 60 minutes in the largest unit, and 60 seconds in a minute. When astronomers measure east-west angles in the sky, they count eastward, much as standard longitudes on Earth are counted. Unlike longitude, however, which is usually measured in degrees, they usually use time units and denote the coordinate *right ascension*. For convenience, residents in the Western Hemisphere (and in particular North America) often use "west longitude" instead of counting eastward. Astronomers have a corresponding "left ascension", counted backward in the sky, but this term is very infrequent. Much like longitude being marked at Greenwich, England, but in a less concrete manner, a "mark" in the sky is the zero point for counting right ascension. Due to the tilt of the Earth, there is a natural point for this, but it moves in a cycle called *precession*. Much like seasons, this topic is associated with the solid body motion of the Earth, so it will not be further discussed here. A line in the sky, the *celestial equator*, separates the northern celestial hemisphere from the southern and is the zero mark for counting. This angle, given in positive or negative degrees, is called *declination*, and lines up almost exactly with latitude. Small differences are due to the Earth not being exactly spherical. For all practical purposes, if you look directly upward (i.e., look at the zenith point), the declination you are looking at has the numerical value of your latitude. In the Northern Hemisphere, the celestial north pole, which has the relatively bright star (about the fiftieth-brightest star) Polaris near it, stands a number of degrees equal to the latitude above the northern horizon. These two facts make navigation and position determination using stars, relating latitude and declination, much easier than for longitude and right ascension.

Right ascension relates to the view from a point on the Earth's surface in a way that varies with time. At a given place, the right ascension overhead, and in the *meridian* dividing the sky in two, east-west, is the *sidereal time*. For navigators, the interrelationship of time, longitude, and right ascension makes determination of longitude more difficult than that of latitude. The importance of determining this, including for military purposes, led to hundreds of years of applied research which led to the development of accurate shipboard chronometers. For astronomers, however, the uniform rotation of our planet allows the determination of stellar positions from one location relatively easily, since good stationary clocks (usually pendulum clocks like the "grandfather" clock favored by antiquarians) have been available for a long time. In fact, ancient astronomers used dripping water clocks (called *clepsydra*, literally "*kleps-hydra*" or "water stealers", due to the steady loss of water drops) to enable mapping of the skies at night. As the skies appeared to move, times when objects passed a point in the sky, usually the meridian, could be recorded and in that way, positions were determined. This technique, called "meridian transit observation", was perfected by Tycho Brahe in about 1580. In a "user manual" entitled "*Astronomiae instauratae mechanica*", Tycho details a naked-eye observation procedure (Figure 2.3) that improved greatly when telescopes came into use.

While the great man directed, an observer watched stars pass through an opening in a wall and adjusted a nearly 2-m diameter quadrant to determine their declination. When they passed the meridian line of the sky, the time was noted from a clock and the observations were recorded. Once telescopes were invented (about 30 years later), and with improvements in clocks, positions in the sky more accurate than one arcsecond could be determined. On a minor technical note, since the Earth rotates through 360° in 24 (sidereal) hours, each hour sees 360 / 24 = 15 degrees pass the

FIGURE 2.3    Tycho Brahe (middle) supervising work with a large quadrant. An observer to his right sights through a slit high in the wall, a timekeeper watches a clock, and a recorder notes details of what was observed to pass through the meridian and when (Wikimedia-Brahe, 2023).

meridian. In turn, this is $15 \times 60 = 900$ minutes of arc in 60 minutes of time. At the equator, 15 minutes of arc go by in one minute of time, and 15 seconds of arc go by in one second of time. Even with a telescope, it is difficult to measure small angles, mainly due to the inability to mark time accurately. Tycho used a huge quadrant because the large size aided in determining angles in declination, and, of course, telescopes aided further with this. Much of the work of post-Tycho astronomers consisted of mapping stars in order to form a system of reference against which other stars could be measured relatively, and against which planets could be observed to move. It quickly became apparent that many stars had changed position since the time of Ptolemy's catalog, which was certainly a complicating factor. That topic is

**FIGURE 2.4** Early Greek *clepsydra*, in which the slow draining of water produced level changes in bowls to indicate the passage of time (Shutterstock).

beyond the scope of this book, apart from noting that 61 Cygni mentioned in Chapter 1 had a large "proper motion" of this sort, upon which the parallax motion observed by Bessel was superposed.

By the time of Ptolemy, clocks were mainly mechanical in nature. The first careful mapping of the skies was done using simple clepsydrae, as shown in Figure 2.4. In these early versions, the lowering of water level with time slows down when there is less water. Clepsydrae were augmented in later times by mechanical systems using gears, to make sophisticated and accurate timepieces. As noted in the first chapter, a mechanistic view of astronomy, like a gearwork, can hardly have been foreign to the Greeks since they did build geared mechanisms.

## 2.6 TIME AND TIDE WAIT FOR NO MAN

This section began by describing gravitational waves, the tiny effects of which in our tranquil region of the universe require our most sophisticated detectors and substantial statistical processing to even detect. In coastal regions, a much larger gravitational effect from a celestial source is obvious in the form of tides. In the open ocean, and in many locations on its shores, the amplitude of tides is about 1 m vertically. Near the shore, there can also be important horizontal tidal currents that can be quite rapid, and in certain places on Earth (like Canada's Bay of Fundy, see Figure 2.5) a local resonance of the natural sloshing period with the tidal forcing can produce giant tides. Tides reflect the difference in gravitational force across the Earth, largely due to the Moon, but also arising noticeably from the Sun. Once important mainly for safe navigation in harbors and near the shore, tides now take on an ominous role as warming oceans, with expanding water, make the sea level rise relative to the land. High tide times (known in some places as "king tides") can signal times of flooding. The effects are most pronounced when the Moon is nearest Earth, and lined up with the Sun. This highlights the fact that, for tidal calculations, a more sophisticated knowledge of the orbits is needed than for simple Earth calendar calculations. The Moon has a notably elliptical orbit so that its variation in distance from Earth is pronounced, which has a major effect on tides (as may be seen in variations in Figure 2.5).

Tides, even if only of the order of 1 m in height that typifies the open ocean, can be very important in determining hazard to ships in shallow areas. They are also

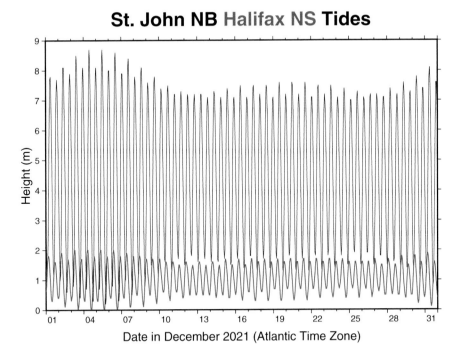

**St. John NB Halifax NS Tides**

Height (m) — Date in December 2021 (Atlantic Time Zone)

FIGURE 2.5   The contrast of tides at the open ocean port of Halifax, Nova Scotia, and St. John, New Brunswick, on the Bay of Fundy. A resonance effect makes tides much larger in the bay (author plot from Canadian government public data).

**FIGURE 2.6** Inchon Landings, Korean War, September 16, 1950 (Official US Navy Photograph, now in the collections of the US National Archives, 80-G-420027).

of extreme importance in harbor and military operations. The most daring operation involving tides was the landings at Inchon in 1950, which reversed the course of the Korean War. As shown in Figure 2.6, a very high tide allowed ships to sail into regions near the coast which became exposed when the tide went down, allowing rapid unloading of military supplies. The operation was felt by the enemy to be impossible due to the extreme tides, and thus not defended against. In addition, the highest tides come near the new Moon, allowing military operations to be undetected in darkness. A full moon also produces high tides, but not darkness, so is not favored. Tides are due to the *differential* effects of gravity, a topic that will be revisited.

## 2.7 DEVIATING FROM PERFECTION

Earth, like the Moon, and the other planets, has an *elliptical* orbit. Ellipses as orbits were Kepler's major discovery arising from his "war with Mars", and his first law states that planets move on such paths. His second law, the law of areas, also applies to the motion of one body in its orbit, whereas the third law compares motion in different orbits. Within an elliptical orbit, planets move faster when near the Sun than when further away. Their radius vector to the Sun sweeps out equal areas within the orbit in equal periods of time. When the planet is near the Sun, it must move quickly for it to do so, as compared with when it is further away (Figure 2.7).

Ironically, Earth is *nearest* to the Sun in the northern hemisphere's cold winter, so the distance to the Sun has little effect on temperature compared with other factors.

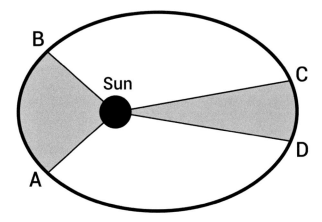

**FIGURE 2.7** Elliptical motion of planets according to Kepler. The Sun is at one focus of the ellipse, while the planets move fastest when near the Sun, so that the time going from B to A is the same as that from D to C, according to his "law of areas". The orange area in each wedge is the same (Shutterstock).

However, the speed in the orbit affects the Sun's apparent rate of motion in the sky. The solar day is longer than the true rotation period of the Earth since, each day, the Earth goes around the Sun a little bit in orbit, and so must turn a bit further to bring the Sun back to the same apparent position. When closer to the Sun, the Earth must turn even further, so solar time gets a bit behind. This effect can be compensated by a correction factor derived from the so-called *equation of time*. A good sundial has this built into the curves from which one reads the local time. It also is the reason that some regions apply "daylight savings time". In fact, no daylight is saved, daylight just gets shifted to times more convenient for human activity. But the reason one needs daylight savings time is that the Earth's orbit is not circular. If it were circular but the Earth was still tipped on its axis, day length would change, but the rising and setting times of the Sun would change equally, with no need for this adjustment.

This is a good place to mention that motions in the Solar System are generally toward the east, or counterclockwise. The hands on a watch or clock (those that still have hands, anyway) go "clockwise" because that is how the shadow on a sundial moves, with clock faces patterned after their flat dials. The shadow moves clockwise since the Sun appears to go counterclockwise over the course of a day. Similarly, the Sun sets in the west since the surface of the Earth is carrying us, as observers, toward the east. That almost everything revolves and rotates toward the east is a heritage of the collapse of the Solar System out of a dust cloud in which that happened to be the prevailing direction of motion.

Much as astronomical instruments by the eighteenth century had far surpassed sundials, the precision with which celestial motions could be measured had also improved. Already, Tycho's observations in the sixteenth century had inspired Kepler with the need for elliptical motion. With the invention of accurate clocks and with the telescope to enable viewing small angles, astronomers could study the interaction between planets, making going beyond the two-body problem possible. Approximate methods were developed by the great mathematicians of the post-Newtonian era, notably Pierre-Simon Laplace and Joseph-Louis Lagrange in France. Mathematical techniques allowed deriving solutions from series, that is to say, sums, of terms representing increasingly finessed approximations. Generally, these represented small changes, or *perturbations*, added to the exact solutions using ellipses. Although the gravity of the Sun did indeed appear to conform to the inverse square law very precisely, the pull of one planet on another is not spherical. Let us take the example of Mars, whose elliptic orbit caused Kepler great grief. Mars is perturbed by Jupiter, which, although the "king" of the planets, has a mass only about 1/1,000 of that of the Sun. Since Mars is roughly 1.5 AU from the Sun, and Jupiter is roughly another 3.5 AU further out, if both align on the same side of the Sun, the ratio of inverse squares of their distances is about 1/5. The pull is further proportional only to the masses of the two bodies, the mass of the Sun being about 1,000 times as much as that of Jupiter. So, the maximum pull from Jupiter on Mars, the planet capable of getting closest to it, cannot exceed about 1/5,000 that of the Sun. Nevertheless, eighteenth-century methods could readily measure the deviation of Mars' orbit from what it would have been in the absence of Jupiter, and many much smaller deviations as well.

The Solar System is rather flat, and the cumulative pulls of the planets tend to keep it that way. However, consider Mars once more, due to it being closer to the Sun than Jupiter and moving much faster in orbit than it. For about half of its orbit, Mars is catching up on Jupiter, and also being pulled toward it. Although it is receding from Jupiter's vicinity, it is still being pulled toward it. Such effects tend to make orbits more elongated or *eccentric*, and they also can make the ellipses swing around in space (*precess*). The alternating directions of pull, however, tend not to increase the overall energy of the affected planet, which is related to its semimajor axis (indicating how deep it is in the Sun's "gravity well"). Although the energy will increase as Mars approaches Jupiter, it decreases again while pulling away. So, in a nutshell, planetary perturbations tend to make orbits elliptical and precessing, without changing their semimajor axes very much.

Although the Kepler problem, with only two bodies making eternal identical revolutions, is boring in the long term, it is at least stable. It may be reassuring to think that we live in a stable Solar System. Once Newton realized that everything pulled everything else, the question about whether it could eventually all simply be pulled into the center worried him. Being a religious man, and not having the mathematical machinery (despite all that he devised) to answer the question, he assumed that if instability arose, divine intervention would be applied as needed. The French, especially after the Revolution, tried to answer the stability question without divine intervention. Laplace did succeed in proving, at least to a low order in his approximate series, that the Solar System could be stable. Over two more centuries of gathering evidence, we are now relatively sure that the present state of the Solar System resembles that which occurred relatively soon after its formation. In most places and at most times, stability prevails. This may even be a condition for us to be here to discuss the issue.

It is interesting to see the difference in style of mathematical astronomy in the 101-year period between the publication of *Principia* (1687) and that of Lagrange's *Mécanique analytique* (1788–89). Excerpts shown in Figure 2.8 show not only that the former was written in Latin and the latter in French, but that Newton used a very geometrical approach, with many figures serving as the basis for discussion in the text. On the other hand, Lagrange famously stated in his introductory text that "*On ne trouvera point de Figures dans cet Ouvrage*" ("One will find no figures in this work"), insisting that mathematical operations alone could lead to solutions. In Lagrange's work, calculus, in a notation very similar to that used today, is the primary tool. Newton (1643–1727) is now recognized as having co-discovered calculus, independently, but near the same time as his near-contemporary Gottfried Wilhelm Leibniz (lived 1646–1716 in what is now Germany). However, it does not play a prominent role in the *Principia*. In the crudest of terms, the most important mathematical techniques of post-Newtonian classical astronomy involve expansions of functions in terms of their derivatives, giving rise to series that must be summed. It is mainly through the study of series that the names Lagrange and Laplace are known to mathematics students, who may not be aware of their astronomical connection!

Based on the advanced mathematical techniques developed by the "mécaniciens", Laplace made a famous claim that, given sufficiently precise initial conditions in the

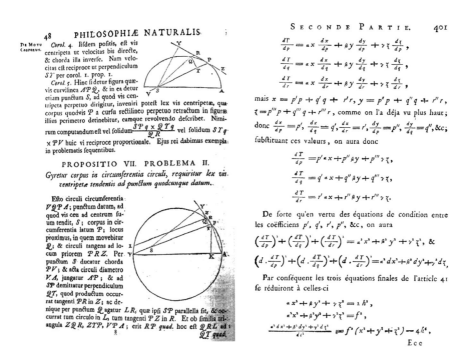

FIGURE 2.8   Newton's style as seen in the *Principia* (left) is geometry and proofs much as practiced by the ancient Greeks. By about 100 years later, Lagrange's style resembles modern algebra and, at least in his case, did not rely on geometrical constructions (public domain).

form of initial positions, velocities, and masses for all bodies, one could calculate the entire future development of the universe, forever. He considered that only actually performing the laborious calculations would be an issue. We now know that this view is optimistic in more ways than simply computing power being insufficient. It implies that mechanics is totally deterministic. At least in principle, this is not so. *Quantum mechanics* states that, at the microscopic scale, the universe is *not* deterministic. Einstein did not much like this fact, stating that "God does not play dice", but it is experimentally well established. On a large scale, quantum effects do tend to average out, making things closely approximated by classical mechanics. However, the *Heisenberg Uncertainty Principle* would allow calculation of how precisely one could specify both positions and momentum (mass × velocity) and, in principle, just how long Laplace's prediction could remain accurate. For that reason alone, and in ways that he could not then know, Laplace's claim is not accurate. A related effect is *chaos*. In physics, this is defined as a sensitive response to initial conditions. Newtonian dynamics in the gravity problem for many bodies is subject to chaos. Fortunately, for the large-scale structure of the Solar System, chaos does not prevail in general, at least not now. However, certain small bodies can display chaos. This affects numerical calculation of their orbits. No matter how precisely specified, small errors will build up rather than cancelling out, resulting in an inability to predict their future positions. Certain Earth-crossing asteroids, for example, can have their positions predicted only about one hundred years into the future.

Nevertheless, by the nineteenth century, mathematical astronomy had two intertwined and highly successful branches: that of the seen, exemplified by observational technique that determined positions with very high accuracy, and that of the unseen in the form of mathematical techniques based on the force of gravity. The triumph of astronomy in this epoch was the discovery of a new planet. The prediction of another planet, which could not be found, marked the end of the epoch.

## 2.8 UNSEEN PLANETS

By the late eighteenth century, mathematical astronomy's use of the dark force of gravity agreed very well with the results of optical observations. It seemed inconvenient that certain things, like the moons of Jupiter, were subject to effects from light travel time, but this was easily accounted for. The happy little Solar System of the Sun, six planets, and known moons seemed well described by Newton's theory enhanced by the mathematical techniques of those who followed him. Improved telescopes scanned the skies, studying for example smudges of light that were neither planets nor stars, but now known to be gas clouds or entire galaxies. The main interest remained "mechanical": comets had been discovered to sometimes be periodic by Newton's contemporary and benefactor Halley (pronounced with a soft "a") and thus new denizens of the Solar System. However, one such smudge did not look like those motionless "nebulae" to be avoided, nor enough like a comet to be one, to the trained eye of German-English musician-astronomer William Herschel. Follow-up observations showed that it moved, and, over a long enough time, it displayed a near-circular orbit more like that of a planet than the extreme ellipses of comets. Thus,

Herschel, in 1781, announced to his German-English sovereign, George III, that a new planet existed which would henceforth bear the name of the "Georgian Star". Herschel was following in the footsteps of Galileo, who, upon discovery of the moons of Jupiter, initially called them the Medician stars to please the ruling Medici family of Florence. As in that case, the name was not widely accepted, and Herschel's planet was renamed Uranus, a sky giant of the generation preceding Jupiter.

Uranus had not been predicted to exist since its effects on the planets inside its orbit were relatively small. Both Saturn and Jupiter are considerably more massive than Uranus, so its pull on them, although equal in force to theirs on it, causes less acceleration. In addition, the scale of the outer Solar System gets quite large, so that the inverse square law of gravity dissipates its effects. For mathematical astronomers, a few more terms could be added to the expressions, with the expectation that, in the future, predictions would improve. For others seeking a larger structure and ordering, however, Uranus was a stimulus. Certain regularities in the placement of the planetary orbits had been noted, notably under the form of "Bode's Law" about their spacing. Bode's Law worked fairly well, and Uranus lined up with it. Despite its important historical role in astronomy, Bode's Law is not a "law", to the point that we will not give its form. The human desire to find order and structure in the heavens once lent Bode's Law great appeal, although one should not impose structure where it is not in fact present. Johann Bode was Director of the Berlin Observatory and of the famous Astronomisches Rechen-Institut (now in Heidelberg), calculating the orbit of Uranus accurately and suggesting that name. It is ironic that the name of this great scientist lives on in a "law" that journals will not even entertain articles about. In the spirit of his times, it was noted that there was a gap at what we now know as the asteroid belt and where his "law" said that there should be a planet. Since it was clear that in this relatively nearby region between Mars and Jupiter, no large planet existed, telescopic sky scanning for a possible smaller planet came into vogue once Uranus' discovery showed that telescopes indeed had the potential to find new planets.

Twenty years of frustration in such searches came to an end on January 1, 1801, which, by strict accounting, was the first day of the nineteenth century. Scanning the clear skies of Sicily, Giuseppe Piazzi found a faint object, near the limit of visibility to the unaided eye, but very clear with a small telescope. These same circumstances applied to Uranus also, and, like that planet, it was not to be found in existing, laboriously made catalogs of fixed stars. A sure giveaway was that it moved, and like Uranus, it was in the ecliptic belt where planets might be expected. Unlike Uranus, when it was discovered, it was in a region of the evening sky into which the Sun was appearing to move due to Earth's motion and thus was not able to be observed for very long. More details of the observational circumstances are given in Chapter 5. Using the brute force techniques of celestial mechanics, its orbit could not be determined well enough from the few available observations to enable a useful future position to be predicted. "Useful" in this context means that, after the Sun passed through that region of the sky, which would take a couple of months, a predicted position would have to be well enough known to accurately point a telescope and "recover" the object.

The importance of finding a new planet was such that an appeal went out to the mathematical community to rapidly improve techniques and improve predictions. The theory of gravity was of course not much doubted, but the method of using data was not optimal. The project caused Carl Friedrich Gauss, of what is now Germany, to improve, not the basic methods of celestial mechanics, but rather the way of fitting data, indeed inventing methods of "optimization". Gauss was notorious for not publishing his techniques, but did accurately reduce the data using optimization, producing an accurate prediction which allowed the object to be found in the morning sky once the Sun "moved" past it. Having initially tried to add "Ferdinandea" to honour his king, Piazzi was able to name the object Ceres for the patron goddess of cereals and the island of Sicily. Eventually, many more similar objects in orbits between Mars and Jupiter were discovered and named *asteroids* due to the star-like appearance that their small size produces in normal telescopes. Ceres itself bears the number 1 in the catalog of *minor planets*, but, due to its large size, was reclassified in 2006 as a *dwarf planet* in the same reordering that demoted Pluto. Despite being discovered in 1930, Pluto ended up as a latecomer in the catalog of minor planets, since many were found before its reclassification, getting the number 134340. Ceres was in recent years the first target of the *Dawn* spacecraft, which orbited it, transforming it from a star-like asteroid into a world in its own right, as will be discussed in Chapter 5.

The lack of observable perturbations, along with its star-like appearance indicating its small size, indicated that Ceres was not massive. However, other matters in the domain of orbit calculations were defying the elaborately developed calculating algorithms. Uranus had an observable disk, which had led to its discovery in the first place. At its considerable distance, even the small observed disk indicated that it was many times larger than Earth, so clearly it was a true planet. The outer planets move very slowly, and most of their apparent motion in the sky is in the form of retrograde loops. Such loops in the case of Mars were complicated by its rapid motion, proximity to Earth, and elongated elliptical orbit. As mentioned above, Kepler's famous "battle with Mars" was a large factor in his decision that planetary orbits must be ellipses. In the case of outer planets, retrograde loops are easier to understand since to a good approximation the planet is stationary, and only Earth's back-and-forth motion in its orbit causes the apparent annual motion of the planet. The size of the loop is inversely proportional to the distance from Earth of the planet, so after about half a year, that can be determined accurately. A reasonable first assumption is that the planet's orbit is circular. With more observations, the orbit can be refined. Although the perturbations on Uranus from Saturn and Jupiter were small, these should have been able to be accurately calculated and factored into getting an even better orbit. However, the more one calculated and observed, the more its orbit increased its deviation from the predicted one. Due to the slow motion in the outer Solar System, the problem took many years to be realized. The most sensible interpretation, given the strength of belief in the characteristics of the dark force of gravity, was that another undiscovered massive planet lay further out than Uranus.

## 2.9  GRAVITY AS A PREDICTIVE TOOL

The tools of celestial mechanics allowed for the addition of small terms from known objects and summing of the resulting series to improve orbits through inclusion of effects from known planets. Newton's invisible force, as elaborated especially by Laplace and Lagrange, seemed to explain every observed motion with great precision. An unexpected interloper came to disturb the mechanical perfection of the late eighteenth century.

Uranus is not particularly faint. It is marginally visible to the unaided eye, although never noticed before Herschel found it in 1781 with a telescope. Examination of records after its discovery found many star charts with an extra star: many observations of Uranus well predating its recognition had been made. A problem immediately arose in that the positions were very different from those calculated using a simple orbit appropriate for 1781, with the discrepancy increasing greatly with time elapsed since the observation. The period of Uranus is 84 years, and the historical records dating back to about one revolution before seemed to indicate an orbit different enough that there was a problem either with the old observations or with gravity. The differences were nearly 300 seconds of arc, 6 minutes of arc, or one-tenth of a degree, as may be seen in the black curve of Figure 2.9. One degree of Earth turning takes four minutes (24 hours divided by 360°), so 0.1 degrees would take 0.4 minutes or 24 seconds. The meridian transit methods of Tycho, described above, were

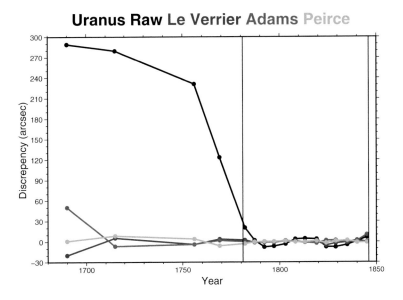

**FIGURE 2.9**  Position of Uranus in the case of no outer planet (black) and in the theories of Le Verrier (red) and Adams (green). Peirce (blue) values use a better estimate of the planet's mass. (author, data from Gould, Smithsonian Institution, 1850).

augmented by telescopes and good clocks by the late seventeenth century, and could not have been out by so much. Something might be wrong with gravity!

Some very fundamental problems with Newton's theory, like perhaps not falling off as an inverse square and thus not working properly as far out as Uranus, could not be completely excluded. However, the clockwork of the Solar System seemed to work with undeniable precision out as far as Saturn. For example, the interaction of Jupiter and Saturn gives a "wobble" with a period of 929 years. Newton did not take the right approach to explain this, and, being a religious man, suggested that gravity might get a bit out of kilter from time to time, requiring divine intervention to set things right. With the mathematical advances of over one hundred years, the French *mécanicien* Laplace moved the interaction of Jupiter and Saturn from being a puzzle to a triumphant example of Newton's gravity in action. Usually taken out of context, and likely apocryphally, Napoleon supposedly asked Laplace, who, in contrast to Newton, was not very religious, where divine intervention fit. Laplace supposedly replied that he did not have need for such a hypothesis. Taking his mathematical triumph in explaining celestial mechanics one step further, Laplace claimed that if there were a divine being, it could use the laws of mathematics to predict the future exactly, knowing the future equally as well as the past. So, around the turn of the nineteenth century, the stage was well set to suggest that if a planet showed deviations in its orbit, much as the two largest planets do, it could be due to the presence of a yet undiscovered further planet. The finding of Ceres in 1801 showed that Uranus was not alone: there could indeed be new planets. In creative astronomical minds, the perplexing task of how to deduce where a new planet might be, by using the invisible force of gravity and the powerful new mathematical tools, came to be a challenge. Much like the US and USSR had a space race in the 1960s, the two powers of the day, Britain and France, had a race to find a new planet. The main protagonists were the French successor of the famous *mécaniciens*, Urbain Le Verrier, and the recently-graduated Cambridge scholar John Couch Adams.

Le Verrier and Adams both made theories involving an outer planet. By luck or irony, Uranus was found about the time it would soon overtake Neptune, going faster because it was nearer the Sun. Without this fortuitous alignment, Uranus's orbit would have been well described using the perturbations from known planets, since the forces from Neptune would have been small (see relative force vectors in Figure 2.10). The older observations would not have been matched, though. Although I have declined to give details on Bode's rule, it loomed large in the minds of astronomers in the early nineteenth century because the recently discovered Ceres fortuitously fit it well. It turns out that the rule simplifies in the outer Solar System to simply say that planetary orbits double in size as one goes outward: roughly, Jupiter at 5 AU, Saturn at 10 AU, and Uranus at 20 AU. Thus, the first assumption of planet hunters was that the new planet would have a semimajor axis of 40 AU. As may be seen in Figure 2.10, the real value for Neptune (blue lines, thin one is modern) is close to 30 AU (and the orbit is nearly circular). Even nailing down the orbit to being in some region of space did not make finding a new planet easy. In the cases of Jupiter and Saturn, even if one needed to take into account mutual interaction with some mathematics that went beyond what Newton was capable of (and that is saying a lot), at least one

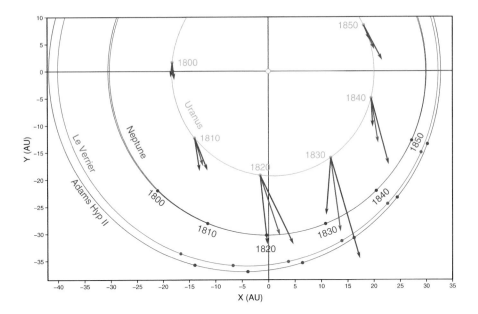

**FIGURE 2.10** Orbits of Uranus (orange) and Neptune (blue) leading up to the latter's discovery in 1847. The proposed orbits of Le Verrier (red) and Adams (green) were too large, but, by having perihelia near the discovery position, could represent the pull on Uranus quite well. The color-coded force vectors are also shown (author, calculated from historical orbit parameters, with modern orbits calculated from the *Horizons* system of NASA/JPL).

knew their basic positions in the first place. Here, the starting point was basically unknown, and the sky is a big place to search.

It is hard to emphasize not only how much creative thinking was needed to find a new planet by a combination of statistical and celestial mechanics means, but how difficult it was to do the calculations in an era before computers. Even the amazing accomplishment of Gauss in predicting where to look to again find Ceres once it came back into the night sky after having its path go near the Sun was based on at least some observations of the "planet" itself, and if Gauss had to invent the statistical procedures, at least they were small in scope. Here, Adams and Le Verrier, working independently, had only minute effects on the Uranus orbit to use, with the aim of predicting a search location within a few degrees to make searching a reasonable thing to do. Since no information was available about the postulated perturbing planet, orbit computation and improvement was a massive task. From an observational point of view, in the pre-photographic era, the best way to search was to watch the sky drift by the meridian, tabulating stars until something was found that was not on the charts. Furthermore, charts were not available for the whole sky, and often were made at individual observatories.

The English and French had quite recently been involved in a fighting war, with the Napoleonic Wars ended by the Treaty of Paris in 1814. At the time that Adams and Le Verrier began their computational projects in the mid-1840s, there was still intense rivalry between the rival imperial powers, including in the fields of commerce and science. There was little contact between the rival camps, but, nevertheless, personalities played a role. Adams was meek, whereas a glimpse into Le Verrier's personality may be had in that a recent book on him has as subtitle, "Magnificent and Detestable Astronomer". One insight into his personality is seen in his referring to the planet Uranus as "Herschel" long after it had been officially named. It seems that he was sure he would be the next planet discoverer, and this precedent would get the new planet named Le Verrier. The whole history of the discovery of Neptune could fill a book, and that book would read like a combined spy and mystery novel. Indeed, the whole story still does inspire historical research and new books. Suffice it to say that both Adams and Le Verrier made predictions based on orbits that were too large, with eccentricity of about 0.1 (see Figure 2.10). The mass of the perturbing planet was estimated by Adams as about twice what that of Neptune turned out to be, and, by Le Verrier, about three times. The mass overestimates compensated for the large orbit, making the hypothetical planet farther away than Neptune really was. In the figures, the color-coded arrows show the relative sizes of the real forces from Neptune and those of the two proposed orbits. Adams made several predictions, and his latest one is shown (Figure 2.10). The mathematical task was to make the observed position of Uranus best match observations, which on the figure means that the force arrows would line up as best as possible with the blue ones. Once Neptune was found, a few years of observation got a good orbit. Not only could new observations be made, but, once a basic idea was had, old observations could be used, much as had been the case for Uranus. In 1980, it was realized that such older observations included two by Galileo carried out in the winter of 1612–1613. The blue orbit shown in the figure, very close to the modern one (shown as a fine blue line), was determined by the mid-1850s. At this time, some pushback against the work of Adams and Le Verrier came out, claiming that their predictions leading to the correct place to look were mere chance. A moon, Triton, in what we now know to be a bizarre orbit around Neptune, was found within two weeks of the planet's discovery. The observations quickly allowed its mass to be determined as being much smaller than those used in the predictions. This added to the controversy.

In retrospect, it is likely that either of the predictions would have allowed the discovery of Neptune. In this sense, both Adams and Le Verrier must be regarded as having succeeded, even if later they had their critics. The main problem these theoretical astronomers had was in convincing some observers to use their results. This division into theoretical and observational persists in the world of astronomy today. Not getting any cooperation in France, Le Verrier took the chance of sending his prediction to Berlin, where the low-ranking observer Johann Gottfried Galle received his letter on September 23, 1846, and that very evening found Neptune, appearing about as bright as an eighth or ninth magnitude star, after about an hour of searching (Figure 2.11). Although an object this bright would readily appear in binoculars, and would have indeed been visible in thousands of spyglass class telescopes in the world at the time, Galle was advantaged by having recently made charts of the relevant region of the sky, which form the background of the figure. Neptune was only about 1° from Le Verrier's predicted position, and under 3° from that of Adams.

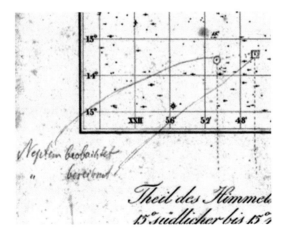

**FIGURE 2.11**   Map used to discover Neptune, likely annotated a bit after the discovery with "berechnet" (calculated, square) and "beobachtet" (observed, circle) positions on Sep. 23, 1846 (modified from Bremiker, 1845).

Of course, once Neptune was discovered, the case became celebrated, not only confirming the validity of Newtonian mechanics in a dramatic way but setting off a heated discussion about priority. John Couch Adams went on to a career mostly at Cambridge University, and subsequently, his most significant contribution to astronomy was in association with meteor orbits, an important topic in the late nineteenth century, mostly discussed here in Chapter 5. Le Verrier filled the shoes of his great predecessors Laplace and Lagrange, tackling notably the problem of changes in the orbit of Mercury that eventually led to the realization that Newton's theory was in some ways incomplete. This is ironic in a sense since Le Verrier's prediction of Neptune was seen as a verification that Newton's theory was correct if treated with adequate mathematical tools.

## 2.10  THE PREDICTED INNER PLANET THAT WAS NOT

Several asteroids had been found since 1801. The asteroids 2 Pallas, 3 Juno, and 4 Vesta were found in 1802, 1804, and 1807, respectively. Shortly before Neptune was discovered, 5 Astraea was found in 1845, and 1847 brought 7 Iris and 8 Flora. Starting in 1850, multiple asteroids per year were found, although it was also realized that these were quite small bodies, despite the term "planet" still often being used to refer to them. Their orbits lay between those of Mars and Jupiter, where the Bode "prediction" had placed another planet. Perhaps these small bodies were the remains of a disrupted planet? Even though many of them were found, they did not seem to be enough to make up a planet, and we now know that the asteroid zone is depleted of mass rather than having anywhere near the mass of a planet in all of its small bodies. This was already realized by the late nineteenth century in the face of huge increases in the rate of discovery of asteroids due to the use of photography. We will explore some aspects of asteroid orbits in the fifth chapter.

The next surprise in celestial mechanics came not from the outer Solar System but from the innermost planet, Mercury. It remains near the Sun in the sky to the point that it is quite difficult to observe. It is said that Copernicus, in misty northern Europe, never caught a glimpse of it. However, due to its short physical distance from the Sun, and its period being short (sidereal 88 days, synodic 116 days), it passes in front of the Sun's disc several times per century (Figure 2.12).

Transits transform the observation of Mercury from difficult and error-prone measurements through a thick and shimmering near-horizon atmosphere to easy ones through a filtered telescope pointing high in the sky. The standard methods of allowing the Earth's rotation to carry sky objects past a reference marker can be used to get extreme precision in such observations, much as had become the standard approach for star mapping at night. Mainly with this approach, it quickly became clear that Mercury was moving slightly too fast in the sense that its rather elliptical orbit was shifting its orientation in space. Planetary perturbation theory meant that such an "advance of the perihelion" was to be expected: the basic effect of the cumulative pulls from all other planets is to change the Sun's near-perfect central force, in which Kepler's immobile elliptical orbits are expected, to one which has small deviations from that perfect "two-body" motion. Indeed, with the techniques developed largely by the French *mécaniciens*, the complex sums needed to determine the perturbing effects could be accurately calculated based on the positions and masses of major Solar System bodies (and even in its inner parts, the effect of Jupiter dominates). For Mercury, this deviation is indeed an advance of the perihelion, by about 600 arcseconds, i.e., 10 arcminutes, per century. Since the Sun appears to be

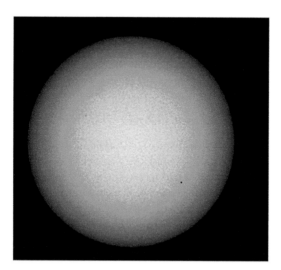

**FIGURE 2.12**  Transit of Mercury on November 11, 2019. Mercury at these times appears to be 10 arcseconds across (while the Sun is about 30 arcminutes, apparently 180 times larger). A discrepancy of 43 arcseconds/century would, however, be readily apparent during such events (photo by Dr. Brian Martin, https://martin.kcvs.ca/astro/course/gallery.html).

about 30 arcminutes across, in three centuries the position of Mercury would be off (mainly in the east-west direction) by about an observed solar diameter. This amount of advance was easily detectible with observational techniques of the mid-nineteenth century, and with very small error. The actual observed amount was however found to exceed that expected by about 7%, or 43 arcseconds per century, far greater than the observational error. Coming after the triumph of Neptune being discovered through gravitational effects, it was natural to assume that some other new planet, interior to its orbit, was perturbing Mercury. Never noticed from Earth, it could affect Mercury, the only planet to which it was fairly close, while having negligible effects on other planets. This hypothesis was taken so far that not only was this hypothetical planet named (Vulcan, for the fiery smith of the gods), but several claims were made of its observation. Le Verrier was heavily involved in these calculations, which were also taken forward by the Canadian-American astronomer Simon Newcomb. Unlike Neptune, similarly predicted using the unseen force built up from gravitational per-turbations, Vulcan never could be shown to exist, to the point that the elaborate and so far highly successful theory of gravitation itself came to be suspect.

The first foretaste of a new approach to gravity came from a clerk in the Swiss Patent Office. This clerk liked to think of unusual situations, some so extreme that they could only be done as mental or "thought" experiments. One we will return to involved trying to imagine if a person falling from a roof feels the force of gravity while they fall. In 1905, his imagination focused on what a beam of light would look like if one was traveling at the speed of light. Since, as we will see, light is electro-magnetic radiation, the formalization of these thoughts came out in a scientific paper entitled "Electrodynamics of Moving Bodies". The obscure author: Albert Einstein. This paper introduced the "Special" Theory of Relativity with the strange result that, even if one was traveling essentially at the speed of light, light would still be going past *at the speed of light*. If, for some reason, we wanted to measure coordinates of objects in a painting, a convenient reference could be its frame. Similarly, our refer-ence in making physical measurements is called a frame, and the above amazing statement is that the speed of light has the same value, "$c$" in all possible frames of measurement. The relation between observations in different frames, perhaps mov-ing relative to each other, is a theory of "relativity". The special theory resolved paradoxes that had arisen through the unification of the previously separate phe-nomena of electricity and magnetism in the late-nineteenth century, which will be discussed in the next chapter. The united theory of electromagnetism did not behave in a manner consistent with the theories of mechanics previously developed to such a high degree. The deviations became larger the higher the speeds involved. Thus, Einstein's linking of moving bodies and light came from dreaming of moving at nearly the speed of a beam of light. Mathematically, the deviations scale as the ratio of the speed of the body to the speed of light, or more often, as something involving the square of that ratio. We feel that Earth's motion of 30 km/s around the Sun is incredibly fast, since it is about 1,000 times faster than common highway speeds of vehicles. However, it pales in comparison to the speed of light of 300,000 km/s. The "$v/c$" ratio for Earth is 1/10,000. Since most effects deviate only in the square, those effects are one part in $10^8$, or one in 100 million. Even with precision astrometric techniques, such a small deviation would be difficult to observe. Although certain special relativistic effects can be observed with Mercury, the solution to the problem of its perihelion advance did not lie there. A change of approach to the structure of space and time was needed. Already, in special relativity, space and time are seen as having a strong connection, a view completely different from that of Newton in which time was absolute, passing everywhere in the same direction (toward the future) and at the same rate. Space in his view was intrinsically that of ordinary experience as laid out by the Greek Alexandrian Euclid, with measurements being independent of position and speed. The need to view a light beam as always moving at speed $c$, even if one also moved at a very high speed relative to the lamp that emitted it, meant that the very conceptions of space and time needed modification. The mathematical way of dealing with these conceptions quantitatively was to use scaling factors forming a *metric* in measuring distance and time. The metric of special relativity, like Newton's view of space, was Euclidean or flat. The correct way to measure distances between events in special relativity also folds in time, which may also be viewed as "flat". It thus uses the geometry of the Greeks, including that of Pythagoras (Figure 2.13) developed 2,500 years previously, in its formulation. However, where Pythagoras' axiom quoted as "the sum of the squares of the sides equals the square of the hypotenuse" applies in an evident way to small spatial distances, special relativity folds in time in calculating "intervals" of "spacetime". To have the units of distance for an interval, the metric rule multiplies time by $c$ before using it. A speed multiplied by a time is a distance: think 100 km/hour times 1 hour meaning 100 km of travel. It is common in everyday life to assume a speed and use a time to denote a distance. For example, one might say "the airport is quite far from the city, about an hour away". In relativity, we are more precise and always multiply by $c$ to convert a time to a distance.

Special relativity allows a more basic definition of time than does Newtonian physics. That seventeenth-century view is fine for an age in which good clocks were developed as a big improvement over Galileo using his pulse as a timekeeper. In this view, time passes at a steady pace and is measured by clocks, so one can arrive at a bit of a circular definition: time is that which is measured by clocks, and what clocks do is measure time. Special relativity gives primacy to the speed of light so that the best clocks and measuring sticks are based on $c$. If one has access to a nice flat part of spacetime, then a great clock can be made by placing a mirror one-half light-second (a distance) away and bouncing a light beam from it: the return time will be one second. In anticipation of what is ahead, spacetime on Earth is not flat, so this experiment, measured accurately enough, does not give a good clock. More practically, it is easy with inexpensive modern electronic components to measure one nanosecond ($10^{-9}$ s). In this time, light travels 30 cm, which is about the size of an adult foot (this happens to also be an antiquated measuring unit). So, a quite practical clock could be made with a mirror inside a shoebox (in fact only half the length of the shoebox is needed since we must count the time both out and back). The mirror will be 15 cm out, and a circuit will trigger the sending of a beam and count down until it comes back: that time will be one nanosecond. Let us imagine this experiment, possibly not an exciting one: we note a "0" on our clock, press a button, see a

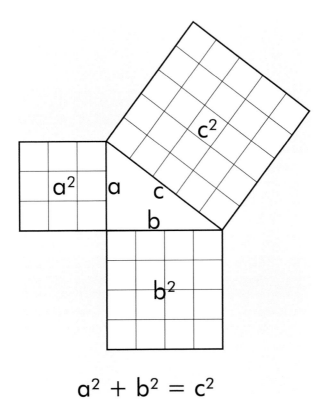

$$a^2 + b^2 = c^2$$

**FIGURE 2.13** The Pythagorean theorem, here shown in two dimensions, extends readily to three in everyday life, and to four in calculating separations of "events" in special relativity (Shutterstock).

flash and, next time we read the clock, we see a display of 1.0000 to indicate that a nanosecond had passed (Figure 2.14).

To have a more exciting time, we need to do a thought experiment since the practicality of getting our shoebox to move at near the speed of light is daunting. We do not

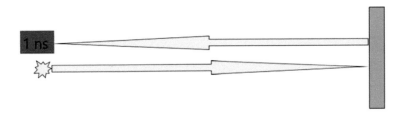

**FIGURE 2.14** Approximately life-size 1 ns "shoebox" light clock. A flash at left ("event") is set off as the timer starts, travels to the right, bounces off the mirror (blue), and reflects back to the timer, which clocks the interval as 1 ns (author).

want to work with mathematics and formulas, so imagine that the box is moved sideways on a "high-speed train" moving at three-quarters the speed of light. We make the shoebox light clock vertical, which is perpendicular to the motion of the train. We arrange it so that, as it passes us, the light triggers, which we see. When the light comes back to the original level, having bounced off the mirror in the shoebox, we photograph the clock on the shoebox (Figure 2.15). Indeed, that clock reads 1.0000 ns just as when we did the stationary experiment. We will call that time, as initially measured at rest, or as measured when the clock was moving but in the frame it was moving with, the "proper time". However, in our own frame, we feel something is wrong with the one-nanosecond timekeeping despite the timer indicating that there was 1 ns elapsed when moving with the train.

In the figure, for clarity, objects are shown only when they interact with the light beam. You can imagine that when the flash went off, the mirror was directly above the light bulb, and, as in the previous figure, the timer was right beside it and started counting. At the middle point, the timer was directly below the mirror. At the end, the mirror was above the timer. To get 1 ns light travel time, the mirror was 15 cm above the bulb and timer. If we held up a ruler as the apparatus went by, we would find that it was at that distance at all times. If the sideways speed was three-quarters that of light, then in the ½ ns for the light to go up and hit the mirror, the apparatus moved ¾ × 15 cm to the right, and in completing the path, the timer would move that

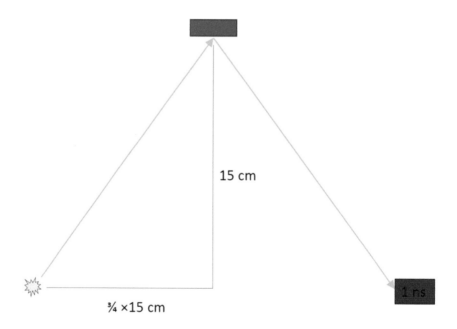

15 cm

¾ ×15 cm

1 ns

**FIGURE 2.15** Shoebox clock moving to the right at three-quarters the speed of light. The timer was originally beside the light flash but has moved by the time the light is received. Although the path in the moving frame ("on the train") is directly up and down, from the side of the track it forms the top of a triangle (author).

distance again. This makes up the short side in two 3-4-5 triangles, and the "5" side would be the path of the light. If the flash was aimed in a beam directly at the mirror and we had a way to see the beam, it would appear to travel on a path making up the two long sides of two triangles. If there was some dust in the air we would see this path light up as the light reflected from the dust. The total distance moved was not the 30 cm observed in a light clock at rest, but five-quarters of that, or 37.5 cm. Since this was moved at the speed of light, it took 1.25 ns for light, seen in *all* frames to move at speed $c$, to travel. All physical dimensions are known, so we must conclude that *time* changed: 1 ns for the clock was 1.25 ns for us as it went by at high speed. If we measure the time of events in a moving frame, the time in that frame slows down. This effect is known as *relativistic time dilation*, where dilation of course means stretching. For an observer traveling on the "high-speed train", the light clock is stationary and the experiment proceeded exactly as when we initially built and tested the apparatus. That observer agrees with the timer that exactly 1 ns had passed. The only way around this is for us to regard the *clock* as having run slower, as measured in our time units, in the moving frame. A similar experiment with the clock rotated 90°, so that now the timing pulse moves parallel to the "train" motion, would show that *lengths* in that direction, and in that direction only, appear to be shorter as measured in our frame. This effect, foretold before Einstein developed special relativity, is known as the "FitzGerald contraction". For uniform (non-accelerated) motion, formulas known as "Lorentz transformations" allow us to relate times and lengths in flat space.

Special relativity cannot explain the extra advance of the perihelion of Mercury, in part because planetary motion is accelerated while it deals only with flat space-time and uniform relative motions. It took Einstein ten more years to develop his "general" theory of relativity (abbreviated GR). Some of the same mathematical tools can be used in GR. The concept of metric adapts well, putting in scaling factors in calculating small displacements. In flat spacetime, it is usually expressed in Cartesian coordinates, but it can be placed into other coordinate systems as well. In the Sun-dominated Solar System, spherical coordinates centered on the Sun allow a good expression of quantities, such as distances, that involve the metric. Einstein's formulation of GR basically amounts to one equation, the Einstein equation, relating elements of tensors (which can be regarded as $4 \times 4$ matrices involving time and three spatial dimensions). There is considerably more complexity possible in GR than in the flat spaces of special relativity, especially since spacetime is curved, with the curvature generated by local mass and energy. GR can be reduced to have a form similar to that of Newtonian gravity in the weak gravitational fields of the Solar System. One of its most pronounced deviations in planetary motion is precession of the apsides, or rotation of orbits, exactly what late-nineteenth-century astronomers were unable to explain for Mercury (Figure 2.16).

Much as the Sun would be regarded as the source of gravitational force to hold the Solar System together in the Newtonian view, in GR it is a source of spacetime curvature causing nearby objects like planets to have paths through spacetime which are like orbits. Sometimes, a graphic is presented in which a funnel-like height is presented, and it is implied that this is analogous to spacetime curvature. This is misleading in that a funnel has a two-dimensional surface that may be seen to be curved in the three-dimensional space in which it exists. In real space, the entire space is curved and there is no further dimension into which it projects. The funnel approach also lends to thinking that space alone is curved, whereas the curvature of spacetime involves distortion of the time dimension as well. The exact solution of Einstein's equation for the case of a point mass was provided by Karl Schwarzschild in 1916, much to Einstein's surprise since he had published the equation to solve only in 1915. Schwarzschild was a physicist who volunteered as a German army officer, and died a year after his discovery, of disease contracted on the Russian front in World War I. A point mass makes a spherically symmetric distortion in spacetime, much as it makes a spherical "gravity well" in the Newtonian view. Angular coordinates are the same in both, in other words basically unaffected by how one treats gravity. The radial coordinate and time both have factors associated with them in the Schwarzschild metric which make the physics differ from that of Newtonian physics, although in the weak gravity prevailing in the Solar System, one may "parametrize" the GR solution to be like Newtonian gravity with small correction terms that cause precession like that of Figure 2.16.

In this approach, the radial dependence of gravity does not have exactly the inverse power two. In a Newtonian Keplerian ellipse, the power of two allows the equations for angular motion and for radial motion to both have the same period: in an elliptical orbit, the satellite body is at the exact same position after one orbit, both in angle and in radius. (This is also true of a direct radial dependence of the attracting force: an object bound by a linear spring, with the force back to the attachment point being directly proportional to how much the spring is stretched, can have perfect elliptical motion). As noted above, planetary perturbations can also result in deviation from the perfect power of two and generate about 600 arcseconds per century of rotation of Mercury's orbit (advance of perihelion). The small deviation due to general relativity adds 43 arcseconds to this. Einstein's 1915 paper had already shown this without the use of the Schwarzschild metric, but the result is more general and understandable if derived from it. Einstein used this agreement as a major underpinning for the validity of his new theory, as it continues to be to this day. The effect is more pronounced in stronger gravity (curvature) fields and increases with eccentricity of an orbit (as in Figure 2.16). Mercury, near the Sun and with the most eccentric orbit of any planet, shows it to a much larger degree than any other, although it has now been detected for other planets in our Solar System, and for many objects exterior to it. In some of the latter cases, the gravity field is so strong that the precession takes place over very short time periods. The exaggerated situation in Figure 2.16 would take place only in the strongest fields, like those near a black hole. Mercury's advance is tiny, needing very accurate observations over a long period of time to detect.

Hard on the heels of Einstein's demonstration that GR could explain the perihelion advance of Mercury, tests were made of GR's predictive value with an effect that had not yet been observed. GR predicted that the combination of distortion of radial distance and time in the Schwarzschild metric would result in apparent bending of the paths of light rays. The bending results in light paths shown in Figure 2.17, which, in the case of extreme mass concentration, can result in light "orbiting"

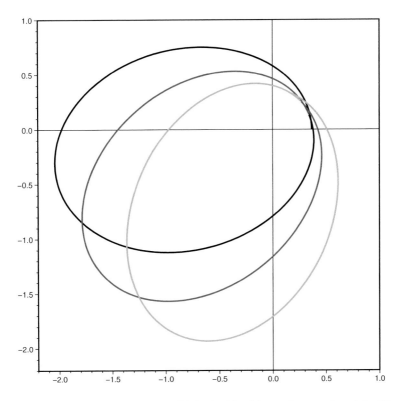

**FIGURE 2.16**   Rapid precession of an elliptical orbit with one focus at the origin. The first orbit is in black, the second in red, and the third in green. Precessing orbits are not fully closed curves (author).

or "entering" a black hole. In the Solar System, only the Sun has the concentrated mass able to make an appreciable (though very small) deviation of light. We are now familiar with this concept due to images of deep space galaxy clusters showing "gravitational lensing" effects as shown in Figure 2.18. Most of this effect is believed to have come from "dark matter", not yet detected in the Solar System and thus not discussed further in this book.

From Earth, we might imagine looking out toward stars that are near the ecliptic. Most of the time, and certainly when they are visible in the night sky, their light traverses paths that are essentially indistinguishable from straight lines. The metric in the space the light crosses is flat. Only when the Sun, in its annual path around the ecliptic, passes such stars, is their light deviated by its effect on spacetime curvature. If we could photograph stars in the night sky, and the same stars when the Sun is near their sight lines, we would find that their positions have changed due to spacetime curvature. In practical terms, of course, it is very rarely possible to photograph stars when they are near the Sun, yet these predicted deviations would be large only within about one degree of it. The only time that it is possible to take such photos is

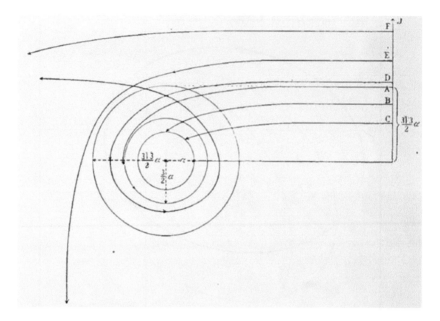

**FIGURE 2.17**  Bending of light rays (coming from right) by a massive body. Deflections in the Solar System, including those by the Sun observable during eclipses, are much smaller than those shown, which could be appropriate for black holes. The most relevant ray for our discussion is the top one, marked F. If viewed from Earth during an eclipse, the projection of this line of sight would be at an angle further from the Sun than if the Sun was not present (von Laue, 1921).

**FIGURE 2.18**  Gravitational lensing as imaged by the Hubble Space Telescope in 2016. A foreground galaxy cluster (white fuzzy galaxies) has imaged a distant galaxy into multiple arcs (yellow), mainly through the gravitation of its dark matter. A supernova (circled) has been imaged three times and was observed to change in time (STScI/NASA).

during a solar eclipse. In the late part of the second decade of the twentieth century, and the early part of the third decade, many eclipse expeditions were undertaken to try to detect this predicted effect of GR theory, which would form its second clear test as a gravity theory. The earliest, and very famous, positive result is that of Sir Arthur Stanley Eddington from an eclipse expedition pair to locations bracketing the Atlantic Ocean in 1919. The results were in good agreement with those predicted. The very small outward radial deviation of stars near the sightline to the Sun could be measured, with difficulty, on glass photographic plate pairs taken during the eclipse and of the same star field in the night sky. The tiny deviations are depicted graphically in Figure 2.19, as determined at a later eclipse expedition using a twin telescope of 4-inch (100 mm) aperture which had a focal length of five feet (1.52 m).

A third famous prediction of GR was first obtained somewhat unreliably outside the Solar System, then somewhat obscurely with laboratory techniques, yet now is an everyday part of our lives. We noted above that time seems to run slow in a moving frame as measured in one stationary with respect to it. The same is true for time when light emerges from the vicinity of a massive object. For example, spectral lines (see the next chapter: these are distinct wavelengths of light emitted or absorbed)

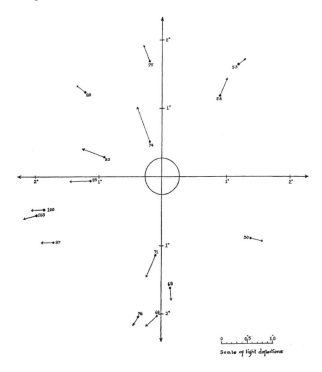

**FIGURE 2.19** Hugely amplified outward apparent deviations of star positions as observed at the Australian solar eclipse of 1922. The Sun is at the center and shown to scale with the axis markings, which are one degree apart. The scale of displacement arrows, starting at the positions of stars shown as dots, is at the lower right and corresponds to about a 3,000-times amplification compared to the overall scale (Campbell & Trumpler, 1928).

have longer wavelengths, i.e., have lower frequencies, when climbing out of a gravity well toward an observer than they would in their rest frame. Longer wavelengths are redder, so this effect is called "gravitational red shift", not to be confused with the red shift due to expansion of the universe. The effect was predicted to be too small to be observed with solar spectral lines. However, for an enigmatic object discovered in the nineteenth century, the prediction was expected to be able to be verified.

The brightest star in the sky, Sirius, is relatively nearby as stars go, about 2.6 parsecs away, while the nearest star is about half this distance. It has a companion star which is about 10 magnitudes, or 10,000 times, fainter than the primary star (Figure 2.20). Since they are at the same distance, this means that the companion star, Sirius B, is intrinsically (i.e., measuring its luminosity or power output, which as for the old way of measuring lightbulbs, would be in watt (W) units) 10,000 times less luminous. Sirius may have been the first star to have its spatial (proper) motion detected, by Newton's colleague Edmund Halley in 1718, although he may have mainly detected errors in Ptolemy's catalog, which he used as a basis for comparison. The spirit of the times had become one in which stars were no longer regarded as fixed to a sphere, so the idea that they could move relative to each other (the "proper" motion) was accepted. Friedrich Bessel's certain discovery of proper motion in 1838 was due to his innovations in precision astrometry (star measurement), which included atmospheric effects. He pushed techniques yet further so that he could, by 1844, find that there was a "wobble" in the proper motion of Sirius, with an amplitude of only about

**FIGURE 2.20** The brightest star, Sirius, and its white dwarf companion Sirius B (lower left) in 2003 as seen by the Hubble Space Telescope. Cross pattern is due to supports for the telescope's secondary mirror. At this time, the separation of the stars was about 5.5 arcsec (NASA, ESA, H. Bond (STScI) and M. Barstow (University of Leicester)).

10 arcseconds. Much as at this time the deviations of Uranus from its expected orbit led to the use of gravity as a detection tool for an unseen planet in our Solar System, Bessel concluded that the brightest star in the sky had an invisible companion. Within a few years, an orbital period of 50 years had been determined. We now know that the main star Sirius A has about twice the solar mass, and Sirius B has almost exactly the same mass as the Sun. The two stars orbit their center of mass with a semimajor axis of about 20 AU, although, due to a large eccentricity, the distance between them varies considerably. Recall that 20 AU is approximately the semimajor axis of Uranus in our own Solar System. Sirius B was not spotted until 1862 by the American optician and observer Alvan Clark. The large brightness difference makes observation of Sirius B difficult. In 1915, William Adams succeeded in getting its spectrum with the 60-inch (1.52 m) aperture telescope on Mt. Wilson, above Los Angeles, finding it essentially identical to that of the primary star. This meant, as will be discussed in the next chapter, that the two stars were of similar temperature.

Since the flux of light, or amount per unit area, given off by a star is primarily related to its temperature, Sirius B could be concluded to give off as much light *per unit of area* as Sirius A. Its 10,000-fold lower luminosity indicated that its surface area was smaller by that much in comparison, meaning its radius was 100 times less than that of the primary star. This makes it about the same size as our Earth, yet with a mass like that of the Sun, a density of matter never previously encountered

Such small stars came to be called "white dwarfs", although they cool due to having no major internal energy sources and they change color, even becoming red. One must be careful in that a small "normal" red star is called a "red dwarf", so the term "white dwarf" is used for such objects no matter what their actual color. Their extreme density, with a Sun-like mass packed into a planet-sized volume, was explained using quantum effects only in the 1930s. However, Eddington realized early on that the gravity near the surface of a white dwarf must be huge, with the potential for observing a gravitational red shift as predicted by GR. The spectral lines in white dwarfs are already quite spread out in wavelength due to among other things the high pressure in the stellar atmosphere, but it was claimed that the GR effect could be seen despite this. A further complication is that the orbital velocity for each star is several km/s, which would give a Doppler shift comparable to the GR effect expected. Eddington persuaded the Mt. Wilson observer Adams to again make a spectroscopic study, now using the larger 100-inch (2.54 m) telescope. These results, published in 1925, showed good agreement between Eddington's calculations and the observations. This appeared to confirm the gravitational redshift as well as the conclusion made with celestial mechanics and spectroscopy that Sirius B had a huge density. Unfortunately, both the calculations and observations proved to be in error and only spuriously in agreement. As white dwarf theory developed, puzzles arose which had to be resolved by careful Hubble Space Telescope measurements in 2003 associated with Figure 2.20, which was taken to aid in pointing. With such improved data, the gravitational redshift of Sirius B is equivalent to having it move away from us at 80 km/s, with less than 1% error. This brings the properties of Sirius B, which is one of the most massive known white dwarfs, back in line with the expectations of theory.

A second well-known attempt to show gravitational redshift used gamma rays emitted in a narrow line at near ground level on Earth along with an absorber nucleus constrained by an effect spreading momentum over the whole sample, known as the Mössbauer effect. By moving the absorber rhythmically, a very precise determination of the wavelength could be obtained, and it was found that, over tens of meters, the gamma rays had changed in wavelength by an amount that compared well with the predictions of GR. While the excess precession of the orbit of Mercury, the bending of starlight, gravitational redshift, and Mössbauer effect are verifications of GR that seem esoteric, one final example of a GR effect is anything but, initially being of huge military value and now having a huge economic value. This is the functioning of the GPS location and timing system, in which both gravitational redshift and special relativity play essential roles.

## 2.11 GENERAL RELATIVITY IN EVERYDAY LIFE

The Global Positioning System, more commonly referred to as GPS, was put into operation by the US military in the 1980s so that advanced technology would give a strategic advantage of instantaneous determination of position. Now, there are analogous systems run by other countries, in part since the civilian applications of global positioning are so important that dependence on the US has been felt not to be wise, and obviously in part since countries interested in military applications cannot depend on systems of a potential enemy. Initially, the military used a coding system called "selective availability" to degrade signals for all but those having a controlled receiver (almost all of which were owned by the US military) but the importance of global civilian use was found to be such that this was turned off in the early years of this century. If the military situation requires it, selective availability can be turned back on, allowing the US military to have far better geolocation than anyone else. The bottom line is that, with a complete and functional GPS system, any user can find their position anywhere in the world and continuously in time, to within about one meter, and time to a small fraction of a second, with a small and inexpensive receiver. Some of these are the size of a watch (Figure 2.21). With advanced techniques, usually involving two cooperating receivers, instantaneous position determinations to the order of about one cm can be achieved. GPS has become ubiquitous and is now a multi-billion-dollar industry with economic benefits likely in the hundreds of billions of dollars per year. And it all depends on general (and special) relativity!

The GPS satellites are in orbits of 26,560 km (about four Earth radii) semimajor axis at an inclination of 55°, which gives a 12-hour period (Figure 2.22). This high orbit and inclination mean that any given satellite may be detected at some time from any point in the world. The high, circular orbits also are not much affected by atmospheric drag and are very stable. A number of the nominally 32 satellites are in view from any point at any time. There are six orbital planes in which the satellites follow each other so that as one sets as seen from any point on the surface, another rises. Having at least four satellites in view at one time is enough to determine a position, but most receivers can handle many satellites and mix the results to improve accuracy.

**FIGURE 2.21**    A GPS smartwatch uses signals from satellites which must be corrected for both special and general relativistic effects to achieve timing precision measured in ns and corresponding spatial location to several meters (Shutterstock).

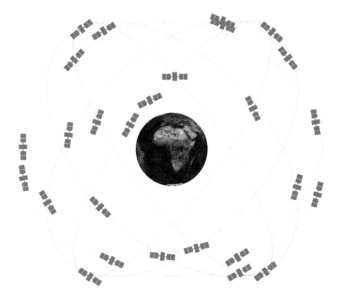

**FIGURE 2.22**    Orbits of GPS satellites with outsized figures of their locations in six orbital planes. Several satellites may be seen from any location on Earth at one time (gps.gov).

The general idea behind geolocation using radio or microwave (as in GPS) signals is not difficult to understand in three dimensions in flat space. An impulse from a satellite spreads out in three dimensions at the speed of light. The location around one satellite at which its signal of that short impulse can be detected is a sphere, of which three are shown in color in Figure 2.23. If, as is the case with GPS, an accurate atomic clock is on board, coded signals bearing the origin position and time can be

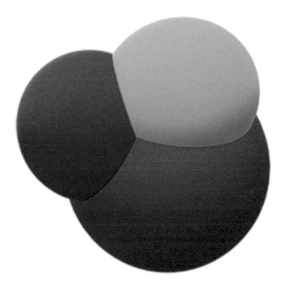

**FIGURE 2.23** Spheres associated with signal bundles expanding from their centers. The spheres get larger with time and the circular intersection lines of two spheres have equal distance to the center of one sphere while having a (usually different) but equal distance to the center of the other. One of two points at a determinable distance from the centers of all three spheres is at the intersection of all of them, near the center (author).

emitted. As is illustrated, points with known signals form spheres around the satellites. If a receiver detected a signal from one satellite (say, the one coded in blue), one could determine how far away it was and thus that one was on a sphere of a known radius from the position it broadcast. Receiving signals from two satellites (say, the blue and also the red), one could determine that one was on the circle where the two intersect, which is visible in the left of the figure. If, in addition, one compared the blue and the green, one would have to be on the circle where they intersect, and in turn, there are only two points where one could be on both the blue-red and blue-green circles. This ambiguity would quickly resolve itself with time, but in practice four fixes are needed in any case since time must be accurately determined at the receiver. Some added complications arise from the Earth's ionosphere (discussed in Chapter 7) whose conductivity delays, deviates, and spreads radio signals. Partly for this reason, a high (microwave) frequency is used by GPS as it is less affected by the ionosphere. In addition, the high frequency of GPS (GHz) allows for encoding more information per second into the satellite position broadcasts.

The flat space view above is complicated by two relativistic effects. We noted above the cases in which clocks in a gravity well (be it of a white dwarf star or of the Earth a few stories down from the detection point) appear to run slow due to general relativistic time dilation. The precision atomic clocks on GPS satellites are generally seen from deeper in the Earth's gravity well and thus appear to run fast. Since the orbits are circular, the correction for this at the Earth's surface is always almost the same, depending only on radial coordinate difference. In fact, the GPS clocks are intentionally run slow by about 39,000 ns per day to compensate for GR effects, so that from Earth they appear to run at the "correct" rate. If we recall that 1 ns is about one-third m of light travel time, we can see that, for this high-precision yet everyday application, omitting GR as "being a small effect" would result in an error of over 10 km per day! The second relativistic effect arises from special relativity, in which we saw that motion in a moving frame results in clocks appearing to run slowly. Unlike the GR effect, this one is symmetric: GPS would say that our clocks on the surface run slow, while we would claim that GPS clocks run slow. Correcting for special relativity is more complex than for GR, since the speed of relative motion must be determined to correct for the apparent time slowing. In any case, it is an effect of slowing and is of about the same size as the GR effect but needs calculation. Another effect to be considered is the Doppler shift, but this effect, changing the frequency of the microwave carrier signal, is not too important and can be thought of as the "train whistle effect" with an increase then decrease of frequency as a train passes through a station. Less obviously, but more importantly, the time-stamps in the GPS signals come in too early when a satellite's relative motion makes it come toward the receiver, and too late when it is moving away. Most of these corrections must be applied in the GPS receiver, which in modern times can be the size of a watch (as in Figure 2.21). Another common application is GPS trackers which can be placed in shipments of goods. These turn on occasionally during travel, take a GPS fix, and sometimes the temperature, and record the result. As mundane an application as tracking the origin and storage temperatures of goods depends on general relativity!

One final note about relativity is that the famous $E=mc^2$ equation does more than specify how much energy would be released by "destruction" of mass. It originally arose from special relativity, but the depth of the relationship between mass and energy is only revealed in its general relativistic sense. Energy generates curvature of spacetime much as mass does, but of course "scaled down" by the large factor $c^2$. Nowhere in the Solar System is there enough energy density to noticeably affect the curvature of space.

## 2.12 SUMMARY

In the previous chapter, we saw the development of understanding of planetary motion from Babylonian book-keeping to a Newtonian world order in which the Sun imposed a "system" of motion. One simple law of gravity displaced Aristotelian thought about the natural attributes of matter, and the conception of the Greeks that different rules applied in the heavens and on Earth was no longer needed. The same rules could apply everywhere. However, the Newtonian view was Aristotelian in one sense, perhaps even exceeding the assumptions of the Greeks: the flatness of space near the surface of Earth, and the uniform and universal flow of time, were included in Newton's theories. Improvements in instrumental and mathematical technique brought these assumptions into question: it was not so much that Newton was "wrong" but that his theory and some of the assumptions behind it were based on approximations. Through observations originally made in explaining the Solar System, it was found that the invisible "force" of gravity was best viewed as illusory, and time itself not fundamental. The level of precision now routinely used by our technological society makes the small deviations that eluded Newton of basic importance to us. As we, in turn, push back the scientific frontiers, the insight into the true structure of spacetime allows understanding of regions of the universe where mass-energy strongly interacts with it, and the larger structure in which small relativistic effects accumulate over large spans of time and space.

# 3  Let There Be Light!

In the Judeo-Christian account of the formation of the universe, God first created the heavens and the Earth. He then declared "Let there be light!". Being so high in the order of creation shows how important light was to early people. In an even earlier view, the heretic pharaoh Akhenaten (Figure 3.1) viewed the rays of the Sun coming to Earth with small hands to help the royal family and mankind. Clearly, early civilizations were aware of the significance of light, with its ability to make crops grow in nurturing warmth. Even at low latitudes, nights can be cold, and, in the absence of artificial light, hide dangers. Gods of darkness were *not* the ones that helped mankind. In northerly climes, the near-absence of light in winter led to celebrations on the shortest day, awaiting the new light of Spring.We can and will discuss darkness and shadow in our exploration of things invisible. On the other hand, how can light be part of the story of the invisible? Light is exactly what makes things visible! However, as we look outward into the starry sky, what we see of the Solar System is basically dark: light streams outward, but we can see it only if it is reflected to us from planets, or, more subtly, from dust. Light is thus invisible if not directed at our detectors, be they our eyes or electronic devices (in the early twenty-first century, photochemical detectors, such as photographic film, became obsolete). Light does not even interact with light: laser beams of huge intensity pass completely through

FIGURE 3.1    Pharoah Akhenaten as a Sphinx praising and making offerings to the sun god Aten, whose hands extend out on light beams to help and to accept offerings from the table at right (Hans Ollerman, Flickr, CC BY 2.0).

DOI: 10.1201/9781003451433-3

each other without mutual influence. However, the radiation field around the Sun, although not important in determining the basic structure of the solid parts of the Solar System, is fundamental in determining the properties of the thin near-invisible constituents, the small bodies we cannot see, and the planets' surfaces. Thus, unseen light streaming through it is an important part of the Invisible Solar System. To understand light properly, we must realize that it has in the classical view both electric and magnetic wave characteristics, and in the quantum view, a particle nature.

In the late nineteenth century, one of the great unifications of physics took place. Electricity and magnetism were formerly thought to be independent forces. The name of electricity derives from the Greek word "electron", meaning "amber", since it was known in ancient times that what we now call "static electricity" effects came from rubbing amber with cloth or fur. Describing what appeared to be a completely different phenomenon, Plato in 380 BCE wrote of the "stones of Magnesia" having strange attractive powers which could be transferred to iron, thus describing magnets. The Greeks had no reason to suspect that these two strange phenomena were linked apart from both being associated with stones. Since there is an electric effect from rubbing glass with silk, and the Chinese had both, it is surprising that they did not discover electrical effects at all, but they were sophisticated in the use of magnetism, especially in the form of practical compasses. Old Chinese maps usually have south at the top, and thus compasses pointed south. Aristotle's ideas on observed natural tendencies set back ideas about gravity in the West, but in China, a general perception of the flow of energy (chi) in the universe seemed to be consistent with gravity and magnetism and did not need further explanation. Only in 1831 CE, in the late stage of the Western Scientific Revolution, did the brilliant experimentalist Michael Faraday (Figure 3.2) show the most *practically* important linkage of electricity and magnetism: electric currents could be generated from changing magnetic fields. This observation can rightly be claimed to be the single most important discovery of the Industrial Age, as it is the base technology of electric generators. We are only now entering a post-Faraday age due to increasing proportions of our energy being generated directly by sunlight.

For nearly two centuries, electricity has been generated mainly by moving coils and magnets relative to one another, driven by waterpower, or steam from boilers powered by chemical (combustion) or more recently nuclear, heat. In more recent times, large windmills for power still have similar generators in them. Even as we transition to solar power, often stored in batteries, Faraday's Law is still used to change voltage levels in electronic devices by using "transformers" as necessary. Faraday's industrial contribution was finding electromagnetic induction, the basis of electric generators.

Prior to Faraday's 1831 demonstration that electricity could be generated from magnetism, it was known that passing an electric current through a wire or coil could create a magnetic field. The Dane Hans Christian Ørsted (Oersted) in 1820 noted that a compass needle near a wire carrying electric current would move to align itself perpendicular to the wire. The demonstration of the effect was extended to two wires carrying electric current by 1825, and the quantitative theory was worked out by the French physicist Ampère, and thus bears his name. Of course, that name

**FIGURE 3.2**   Michael Faraday holding a glass tube used to illustrate changes in light polarization due to the "Faraday effect" (photograph by Maull & Polyblank, Wellcome Collection, public domain).

is also familiar to us as the unit of electric current. As with all SI units based on names, ampere is not capitalized, and has no accent: it is often shortened to "amp"). Faraday, due to modest origins, did not have much mathematical training. He developed an intuitive model of magnetic fields as having "field lines". As we will see in later chapters, this intuitive model remains useful, although it can be misleading. He could make statements about quantities of field lines over an area, known as "flux", leading to the generation of electricity, but his knowledge of the linked phenomena of electricity and magnetism did not give rise to a general theory.

Faraday's field lines had a peculiar property that they wound up on each other: they had no beginning or end. It was also possible to think of electric field lines, but, distinct from magnetism, they could have a distinct source. We speak of "static electricity" usually in reference to sparks. Ironically, in a spark, electricity is in motion from one body to another, so it is not "static". A frequent way to observe a spark, under dry conditions, is to shuffle your feet on a carpet and then touch a doorknob.

In this way, electric charge built up on the body is transferred to the doorknob, with the voltage high enough to cause breakdown and glowing of a small quantity of air. Initially on our body, and, later, on the doorknob if it is well insulated from other objects, the "electricity" is briefly "static". However, when it flows, it is clearly not static. We now know that electrical effects are due to the flow of charged particles, but in the nineteenth century the nature of electricity was unknown, and it was usually considered to resemble a fluid. We associate the name of Benjamin Franklin with "atmospheric electricity" in the dangerous form of lightning. In Italy, Luigi Galvani noted the response of frogs' legs to electricity, resulting in the concept of "animal electricity". His colleague Alessandro Volta invented what came to be called a "voltaic pile" in 1800, resulting in the concept of "voltaic electricity". We now usually call voltaic piles "batteries", but Volta's work is immortalized in the unit of electric potential being the volt (small "V" but with V as its SI unit). For early experiments, a machine (Figure 3.3) that converted mechanical motion to electric current through "static" electrical effects was commonly used, so that type of electricity was referred to as "common electricity". In the early nineteenth century, it was not clear that all such types of electricity were the same. Until Faraday's discovery about the generation of electricity using magnetic field-based generators, voltaic piles became a common way to supply "the juice" (perhaps this expression reflects the concept of electricity as a fluid) for laboratory experiments.

**FIGURE 3.3** Generator of "common" electricity in the era of Coulomb and Faraday. Turning the handle resulted in charge being developed. This is now usually referred to as "static" electricity, but static electricity can flow if provided with a conducting path, so is not "static" (Shutterstock).

Charge was realized as a concept associated with electricity, but one should recall that fluids also can be stored. Placed on an insulated body, the charge stayed there and was "static", much as fluid sits still in a filled cup. By transferring electric charge to a small metallic object, the French physicist Charles-Augustin de Coulomb in 1780 showed that electric force diminished as the inverse square of distance. A good way to view this is as field lines radiating out from the small object uniformly into space. Spherical symmetry applies: we can think of lines penetrating spheres around the object, but the area of the spheres goes up with the square of the distance from the object. If the number of field lines is fixed, then their density per unit area on the spheres must go down with the inverse of that area, which is the square of the distance. Many things that go out from a center follow the inverse square law. Importantly, the very concept that the electric field lines originated on the object distinguishes them from magnetic field lines. We should state, and Coulomb also noted, that under certain circumstances, magnetic fields appear to originate in the pole of a magnet and show similar behavior in decreasing with the inverse square. However, the larger picture would trace them back to the *other* pole of the magnet, so that their overall behavior is different from that of electric field lines.

Due to his work with quantities of electricity, Coulomb's name is associated with the unit of electric charge, the coulomb. Linking back to the ancient Greeks, electric charge is associated with small particles, and the kind which most easily moves around to carry it is called an "electron", basically the Greek word for amber. For practical purposes, electricity can be regarded as a "fluid" since these charges are very small: $6.24 \times 10^{18}$ electrons are needed to make up one coulomb (C) of charge. One ampere (A) flowing for one second amounts to a transfer of one coulomb (C). A moderate heating appliance (such as a toaster) can use ten amperes, so roughly $10^{20}$ electrons per second flow in wires commonly found in the average kitchen. In the time that electricity was thought of as a fluid, there was debate about whether it was one fluid or two. Arguing for two was the fact that bodies could be made negatively or positively charged, depending on what they were rubbed with to charge them, or their own composition. On the other hand, could positive charge simply be an absence of negative charge? In our common technological use, it is, in fact, vast quantities of electrons that are the negatively charged workhorses, so we tend to often think in "one-fluid" terms, and that fluid is negative. We will see that, in space, other charge carriers (usually protons) can be important. In the nineteenth century, the process of electrolysis resulted in positively charged constituents of chemical solutions moving, so it was inferred that they moved as "ions", a word derived from Greek, meaning "travelers".

Returning to Faraday, we may consider that distinction often comes to those who got there first, not to those who made the greatest contribution. We sardonically note that a device storing "static" electricity is known as a capacitor, and if one of these devices can store 1 C of charge while showing a potential difference of one volt (V), it is referred to as having a "capacitance" of one farad (F). Thus, Faraday is commemorated with a somewhat secondary SI unit as compared to the amperes and volts of everyday use, and without his full name! Before Faraday invented electric generators, and when voltaic piles could not develop a high enough voltage (each element of the pile developed about 1 V so there was a practical limit), charge could be stored in the earliest capacitors, which were called Leyden jars (Figure 3.4), reflecting their place of invention in the Netherlands. A Leyden jar about the size of an instant coffee jar has a capacitance of roughly 1 nF or $10^{-9}$ farads. A typical modern hand-solderable capacitor of this size is shown in Figure 3.5.

Soldered to the surface of the electronic board, there are also many capacitors in this nF capacitance class. Capacitors are used in combination with resistors (which as their name suggests, oppose the flow of current in a circuit) for various purposes. Here, they protect from sudden voltage changes which might damage delicate chips on the circuit board. The small capacity of a Leyden jar meant that it could not store much electric charge, but, in all capacitors, the charge storage is proportional to the voltage. One could get a large charge stored by using high voltage, tens of thousands of volts. That combination of voltage and charge made Leyden jars potentially very dangerous. In contrast to the 1 nF of a typical Leyden jar, 1 farad is a huge capacitance. Attained in small devices only in recent times, so much charge can be stored in a so-called "supercapacitor" that they function almost like batteries. Such a modern capacitor is shown in Figure 3.5 but is limited to use at low voltage (up to the 7.5 V marked on it). The Leyden jar and simple capacitors like the red one work

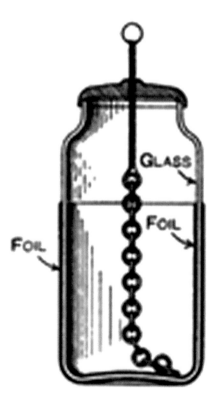

FIGURE 3.4   Leyden Jar, an early form of capacitor (Marchant, 1914).

FIGURE 3.5 Modern capacitors. The small red one at the bottom is 1 nF, about the capacitance of a Leyden jar. A 470 μF (μ means "micro" or $10^{-6}$) capacitor is used in power supply and audio circuits. The 1 F supercapacitor at the top has about 1 billion times the capacitance of the red one (author photo).

by effectively having charge (electrons in practice) on one foil and a matching lack of charge on the other, separated by an insulator. The lack of charge appears like an opposite charge, and, with charges, opposites attract, so the situation is stable, and charge can be stored. In the middle capacitor, a so-called electrolytic, a much higher capacitance is attained through having thin, rolled foil layers, and a material between them in which the molecules move in electric fields to enhance the storage. Supercapacitors take materials science to the extreme to achieve this desired charge storage effect. Due to the thin foils and the nature of the "electrolytic" material, any electrolytic capacitor cannot be charged to high voltage. Further, they are "polarized", as indicated by bands on their sides, and must have the proper polarity in a circuit.

By the early nineteenth century, Coulomb's view of electric field lines was well quantified, but electricity and magnetism were regarded as different things with possibly the same rate of decrease with distance, shared with that of gravity. The first hint of a deeper connection came when Hans Christian Ørsted (who happened to be friends with a Mr. Andersen with the same first names) noted the deflection effect noted above, which Faraday's work took further, but not to a complete theory.

In 1862, the great Scottish theorist, James Clerk Maxwell, took the full unifying step by first realizing that not only could an electric current in a wire create a magnetic field, but a *varying* electric field in space could do the same thing. This was known as a "displacement" current. Since electric fields could exist in empty space, the displacement current could too. We bypass some discussion of the time about whether empty space really was empty: there was much discussion about a possible "ether", which could never be detected. Since Faraday's Law related changes in magnetic field to electric currents in a coil, the displacement current idea, by allowing changing electric fields to create magnetic fields, led to the treatment of electricity and magnetism on a more equal footing. Most importantly, it allowed Maxwell to predict that electricity and magnetism could be intertwined in a form that would propagate through space as waves: *electromagnetic waves*. One could even combine two known distinct quantities determined from experiments for electricity and for magnetism separately, in such a way that they must be the speed of electromagnetic radiation. This, it turned out, was the speed of light!

The early demonstrations of Maxwell's prediction were done with antennas and coils using high voltage. Such experimentation had sparks flying in a way that inspired early horror movies. One early experimenter was Heinrich Hertz, whose name is now immortalized in the unit for frequency, the hertz (Hz) or cycle per second. Hertzian coils (Figure 3.6) could be made to spark when placed near other

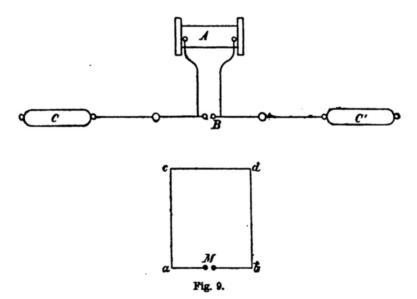

Fig. 9.

FIGURE 3.6 Hertz' transmitter and receiver. High voltages were induced by an "induction coil" operating on Faraday's principles (basically a high-frequency, high-voltage transformer) $A$ applying a voltage to a linear antenna $CC'$, the resonance of which made a spark at $B$ and broadcast radio waves. These were picked up in the moveable secondary coil at the bottom, as evidenced by sparks at $M$ (Hertz, 1893).

coils, transmitting energy from one point to another by what are now called radio waves. In German, radio is still called Rundfunk (circular spark). As in our radio tuners, frequencies for modern radio bands are typically in the hundreds of kilohertz (AM) to about a hundred megahertz (FM), with TV at yet higher frequencies. While modern apparatus operates at very strictly controlled frequencies, so that there is no overlap of adjacent channels, Hertz's experiments generated a huge spectrum of electromagnetic noise, much of it in the FM radio band. Ironically, Hertz did not envisage much by way of applications for his work. Simply by turning the device on and off, a "signal" could have been communicated to the secondary coil, and the leap to a telegraph would not have been a long one. An immediate domain of application was to ships at sea, which previously had effectively disappeared from the world for the duration of their voyage, having no direct communication with the shore. Of course, the practical applications, and refinement of technology needed for them, were not long in coming, and associated with such names as Marconi.

In Maxwell's theory, as demonstrated by Hertz, electromagnetic waves were of many wavelengths ranging from light to the radio waves that the coils copiously emitted. Light itself had been known to have wave characteristics since about 1800 when English polymath Thomas Young demonstrated diffraction. His "extensive sphere of research" included Egyptology, in the spirit of the times (as for the discovery of Neptune) done in competition with a Frenchman. He found that like waves entering a harbor, light will bend near an obstacle. It was known from Young's work that light had a very short wavelength, and we may regard the wavelength of typical light waves (green-blue) as 500 nm, i.e. $5 \times 10^{-7}$ m or 0.0005 mm. This is roughly to a millimeter as a millimeter is to a meter, i.e., small but not totally impossible to imagine. Fine human hair is not that much bigger than the wavelength of light and thus sometimes shows diffraction effects. As mentioned above, once the unification of electricity and magnetism took place, it was realized that the waves would travel at the speed of light. This important number is referred to as $c$ (from the Latin, *celeritas*, meaning speed, and related to the English word celerity). It is handily close to a round number in SI units, $3 \times 10^8$ m/s. In order to go past a fixed point at such high speed, many individual waves must rush past, each being one vibration or wavelength. This means a very high frequency of $6 \times 10^{14}$ Hz for light, roughly a million times higher than that of FM radio.

Radio waves themselves are ubiquitous and to carry large amounts of information, relatively high frequencies are used; 2.4 GHz is commonly used in wireless internet. Figure 3.7 shows what these waves would look like compared to a wood panel which is 1 light-nanosecond long. As pointed out in the previous chapter, a light-nanosecond is a handy unit, 30 cm long (thus very comparable to the quaint unit of the "foot"). Likely you will get a quizzical look if you go to a lumber store and ask for a piece of paneling 1 light-ns long, but that is shown in the figure along with a conventional tape. The nanosecond timer discussed before was 15 cm long due to the use of a mirror and operated with light, the wavelength of which is 250,000 times shorter than that of 2.4 GHz waves.

There are two ways to view a wave. The simplest is to sit still, let us say, on the antenna shown near the bottom of the figure as a gold metallic trace. Then it would

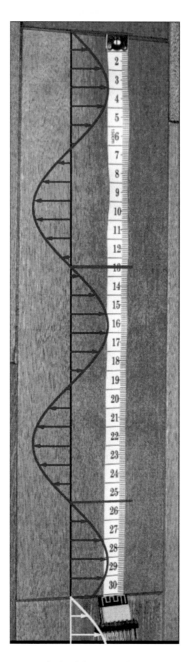

FIGURE 3.7    A 2.4 GHz wave typical of those used by Internet of Things devices (seen at bottom) as seen moving downward over a time period of one ns. (Author photo illustration)

look like the wave was rushing by at the speed of light, and an oscillating electric field (blue arrows with a blue outline of amplitude) would be felt by the electrons in the antenna, forcing them to move back and forth as a current, also at 2.4 GHz. The other way is to imagine the wave at one specific time when the electric field was zero but rising at the beginning of the tape. At that instant in time, points down the tape would each feel the electric field shown. The wave repeats and purple marks show the distance taken for the repetition, which is the wavelength. Since the wave is moving at the speed of light, it will have moved $c$ times one ns, one ns later. This is shown in yellow: the initial zero-but-rising field is now at the 1 light-ns (30 cm) mark. Although only the electric vectors are shown, an electromagnetic wave also has magnetic field vectors, at right angles to them. Both are perpendicular to the direction of motion (propagation) of the wave.

As with the "flow" of electricity, which appears continuous, one might ask if, despite the nice theory of Maxwell, the demonstration of diffraction by Young, and the use of radio waves by Hertz and his successors, whether light really "is" a wave. The ancient Greeks viewed light as emanating outward from the viewer's eyes, a bit like a superhero's heat ray vision. Going out to objects in the world, it would bounce back to give vision. Newton was an advocate of a particle theory of light despite having broken it into constituent colors using a prism, a result that can be explained quite nicely with wave theories. The very short wavelength of light is likely a good part of the reason that the very astute Newton did not pick up on this aspect despite writing a whole book ("Opticks") on light. Experiments around the turn of the twentieth century on "cavity radiation", that which comes out of a small hole in a heated, hollow solid object, showed that the distribution of light could only be properly explained if it came in small chunks, or "quanta" (a term adopted later). If one shines light *into* a small hole in a cavity, almost none of it comes out again since it bounces around inside and gets absorbed. For this reason, the cavity is called a "black body", and conversely, when heated, it is called a "black body radiator" even though that term seems contradictory. Many bodies that give off a small fraction of their stored energy are well described by "black body curves" in terms of the color of light they emit (the spectrum). Although the light is in chunks, they are indeed tiny. Each chunk, in the case of light, called a "photon", has an energy which is $h = 6.63 \times 10^{-34}$ times its frequency, in SI units of joules (J). Despite the very high frequency of about $6 \times 10^{14}$ Hz for blue-green light, this means that each photon has an energy of only $4 \times 10^{-19}$ J. A very rough number (which is good to remember) for the *solar constant*, or energy flux from the Sun, is 1,400 W per square meter, and since a watt is a joule per second, this means that 1,400 J flow through an area the size of a bath towel each second. If on average this is made up of green photons, there must be $1,400/(4 \times 10^{-19})$, or 3.5 $\times 10^{21}$ of them. Much as with electrons constituting a reasonable amount of electric current, the number of photons over a small area on a sunny day is a huge number, so that light seems continuous. An interesting comparison is that there are approximately $2.5 \times 10^{25}$ molecules in a cubic meter of air near the surface, roughly 10,000 times as many as the number of photons that would traverse it if it were a three-dimensional cube. On the other hand, if we regard the atmosphere as effectively about 10 km thick (it is thicker, but this roughly makes up for density decrease), then the number of photons each second is about equal to the number of molecules traveled past, each second. Air (in the absence of clouds) is nearly transparent, so most of the solar energy coming in as visible light penetrates to the surface.

It is somewhat inconsequential for the larger picture we wish to form, but the question of whether light is a wave, or a stream of particles, or both, has puzzled scientists since the days of Newton and Young and has greater implications. Radio waves are dominated by wave-like effects: perhaps you have detected some of them when a radio station fades in and out on an automobile radio. On the other hand, gamma rays, with a wavelength about a million times smaller, behave more like particles, and, in interactions with atomic nuclei, are usually dealt with as single photons. So the answer to the question as to whether various types of electromagnetic radiation are waves or particles in some ways comes down to the accountant's answer in a joke: "What do you want them to be?" The fact is that light is light, and if one thinks of it only as waves one will sometimes be wrong, and equally so if one thinks of it only like a particle. It shows "wave-particle" duality. Well, how about "particles"? Surely *particles* are particles? Not so, says modern quantum mechanics! Like photons, *all* "particles" also show a wave nature. The wavelengths of massive particles are in most cases very short, so that effects like diffraction are minimal. There are important exceptions: the diffraction of electrons has been an important tool for chemists and biochemists to determine crystal structure. Having discussed the nature of light, we now return to how it behaves in the Solar System. Of course, the dominant source of that light is the star at the center, the Sun.

Emission of electromagnetic energy from the Sun comes from a layer called the "photosphere", literally the "sphere of light". The Egyptians were quite accurate in depicting the Sun as circular, indeed in Figure 3.1, it is chiseled out to look like a sphere. The Sun may of course be viewed as a circular projection, being, along with the Moon, the only regular Solar System body with a shape visible to the unaided human eye. That circle marks where the Sun's atmosphere is no longer transparent, an effect that happens over thousands of kilometers but viewed from far away looks like a sharp edge. The circular shape is the photosphere viewed from the side. Above this level is an extended solar atmosphere that is the subject of later chapters. Because the quantity of energy in the gas at the edge of the photosphere is large compared to the amount that flows out each second, the Sun may be treated as a "black body", despite this term seeming contradictory for an object that gives off "white light". Newton's experiments with prisms and, in fact, using solar light, showed that white light may be broken down into constituent colors and reassembled with a lens and second prism into white light. In Figure 3.8, this breaking down into a spectrum is shown for the visible light in the solar spectrum. All of the colors of light travel at the speed of all electromagnetic waves when in a vacuum, the "speed of light". They slow down in other media: this explains how prisms work since the amount of slowing varies by color. In space, however, all advance at the speed of light, so that, in one second, $3 \times 10^8$ meters are traversed. The blue-green wavelength referred to above as having a frequency of $6 \times 10^{14}$ Hz advances in space while oscillating (in the wave picture) so that a number of oscillations lie along the path each second. By simple division, each oscillation, as seen in space, must be of length $3 \times 10^8/(6 \times 10^{14}) = 0.5 \times 10^{-6}$ m.

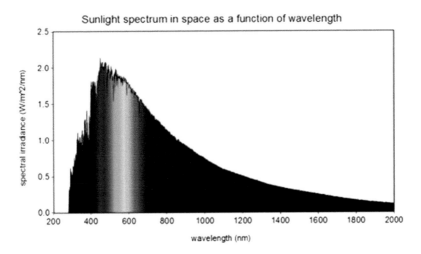

**FIGURE 3.8** Sunlight spectrum in space as a function of wavelength (Baird, 2013).

This is a wavelength of 0.5 microns ($\mu$), but, in the visible wavelengths, it is convenient to use nanometer ($10^{-9}$ m) units, so this becomes 500 nm, which is what is labeled in the figure. There, it can be seen that the spectrum peaks at about this wavelength, with longer wavelengths redder, and shorter ones bluer in color. It was mentioned above that the solar constant is about 1,400 W/m². This is a very important number for Earth: 1,400 W is comparable to what an electric room heater puts out. Every square meter of the Earth's cross-section receives that much energy input. At latitudes far from where the Sun is overhead, less is received per actual square meter of land, and, of course, there is none at night. We also must reduce this number a bit even at the equator due to the effects of the atmosphere, and locally by a large factor if cloud reflects the visible light into space.

The electromagnetic radiation of the Sun is produced most abundantly in those parts of the spectrum near the blue-green peak of emission. What we refer to as light, ranging from the ultraviolet to the infrared, has wavelengths ranging from about 400 to 700 nm, spanning the violet through red colors. It is likely not a coincidence that the human eye detects most efficiently those wavelengths at which the Sun emits most profusely. The solar constant of 1,400 W/m² can be deduced from the figure, which shows the incoming energy not only at its rate of arrival over one square meter, but also broken down by wavelength. For this reason, the irradiance (incoming radiation) is given in units of W/(m²-nm) (shown as W/m²/nm in the figure). To find the total power (energy per unit time) coming in, one has to add up all the contributions at different colors. This amounts to finding the area under the curve, or "integrating over wavelength". A very rough way to do this is to note that from the peak and to the right, the shape is (very) roughly triangular, and the area of a triangle is half its height times its base. Here, the height is roughly 2 W/(m²-nm) at 500 nm wavelength and the triangle extends to approximately 2000 nm, where we will consider it to be zero. This gives $2 \times \frac{1}{2} \times 1,500$, which is 1,500 W/m² (the nm unit associated with

1,500 cancels out the 1/nm part associated with 2, giving the final units). We may consider the dip in the curve to be filled by the part we ignored to the left, but with even rough integration, we see that all the electromagnetic radiation indeed adds up to a good estimate of the solar constant.

The longer wavelengths are found to the right in the plot and are not visible to the human eye. Certain animals, like rattlesnakes, can detect such heat radiation, which is called infrared or IR, an adaptive advantage for hunting warm-bodied animals at night. Some wavelengths of infrared can be felt by the human skin too. With our rough integration, we can draw another triangle, starting at 700 nm where the IR starts, where the value is about 1.5 W/(m²-nm). Trying to better approximate the area, we will claim that this triangle goes to zero at 1,800 nm, so that its area is 1.5 × ½ × 1,100 = 825 W/m². We see that roughly half of the incoming radiation from the Sun arrives as IR. Much of this never makes it to the surface, due to interaction with molecules in the atmosphere. At these IR wavelengths, the atmosphere is not clear, like at visible wavelengths. A further factor is that the heating of the surface by (mainly) visible light makes the Earth itself emit IR radiation. The complex interplay of incoming and outgoing power makes Earth adjust its temperature until the incoming and outgoing power are in equilibrium. Slight changes in molecular composition can affect that equilibrium. This is the physics behind climate change, mostly manifest in our times as "global warming" and could be the subject of an entire other book.

The shorter wavelengths from about 300 nm to 500 nm are the "ultraviolet" or UV to blue part of the spectrum, and our rough "triangle" approach would suggest that 2 × ½ × 200 = 200 W/m² arrives at these wavelengths (note that the plot may be incomplete below about 275 nm). In any case, we conclude that only a small percentage comes in in the UV-blue region, and, fortunately, almost all of it is absorbed by the atmosphere. Whereas IR interacts with molecules (and mainly only makes them vibrate), UV has enough energy to disrupt atoms and promote chemical reactions. That can happen in biological material, including our skin and corneas, making UV radiation biologically hazardous. The hazard is reduced by our protective atmosphere since the UV similarly interacts with atoms and certain molecules such as ozone ($O_3$). However, most people are still advised to wear protective clothing or skin creams (high SPF sunblock) to protect from the UV radiation on a bright day, especially if the Sun is high in the sky.

Earth has an overall reflectivity of incoming radiation of about 30%. Expressed as 0.3, this is known as the albedo. As Figure 3.9 shows, the albedo of the Earth varies considerably from place to place, mainly due to clouds, which of course move around. Land and sea reflect different colors of light. North America is visible as a brownish area at the top of the figure, whereas the dominant oceans, of which the Pacific fills most of the visible part of Earth, reflect blue light. This image was taken by a satellite at the $L_1$ Lagrange point between the Sun and the Earth and at the time of the new Moon so that the Moon was between the satellite and the Earth. Despite appearances, the Moon was not lined up well enough for there to be an eclipse, but the overall dark appearance of the Moon compared to the Earth is clear. If the Moon looks unfamiliar, it is because this image shows the "far side". Sometimes that side

**FIGURE 3.9**   Earth and Moon from the DISCOVR satellite on July 16, 2015 (NASA).

is called the "dark side", but this is an allusion to the Moon keeping one side always facing Earth, so the far side was unknown or "dark". In this image, the far side is fully illuminated by sunlight, and it can be seen that it does not have the familiar dark and light patterns that are present on the side facing Earth. Instead, the far side is an almost uniform highland region. On the nearside, the dark regions have an albedo of about 0.08, while the bright regions are highlands with an albedo of about 0.12. It is only against a dark night sky that such a dark object can appear brilliant, like a full Moon. The contrast between dark and light regions on the near side forms the pattern often known as the "man in the Moon". The far side has only one notable dark region, visible at the upper left of the Moon disk. It is called Mare Moscoviense (Sea of Moscow) since it was first imaged by the Soviet Luna 3 probe in 1959.

Solar radiation is the main energy transport mechanism in the Solar System, determining its thermal structure. Within that well-defined gradation of temperatures, our own planet is in the ideal *habitable zone* of the Solar System. Thus, we come to agree with the ancients on the importance of light, although we can examine it in more detail and with a more nuanced understanding.

## 3.1 LIGHT AS ELECTROMAGNETIC RADIATION

The speed of light plays an essential role in relativity theory, being the gauge against which velocities are measured, and the ultimate speed limit, as discussed in Chapter 2. Within classical physics, light is one kind of electromagnetic radiation, and all types of it travel at the speed of light when in a vacuum. Electromagnetic radiation has wave-like properties, in which the product of the frequency of variation and the wavelength of each variation (wave) is the speed of light. If you count waves going by, and each is of a certain length, enough will go by in a second to make the speed of light times one second. This is a large number by everyday seconds: 300,000 km. In Figure 3.9 the Moon is about 380,000 km "above" the Earth, or a bit over 1 second of light travel time (about 1.2 light-seconds). Only on astronomical scales (which in some cases includes bouncing signals from satellites) is light travel time noticeable. Mars at its closest has a light travel time of about three minutes, so that holding a "conversation" with astronauts who may go there will be difficult. Despite the constancy of its speed, electromagnetic radiation has widely varying properties. For short wavelengths, i.e., high frequencies, light has particle-like properties. For long wavelengths, like radio waves, these are secondary and wave characteristics dominate.

Light, of course, was known to early scientists. The Greeks regarded the fact that we see only when our eyes are open as indicating that light was emitted from the eyes and came back to them. Newton must be regarded as having had less success with his theory of light than with mechanics. Although he showed that white light could be broken into colors by a prism, and then recombined, he regarded light as being particles. Young interpreted diffraction, or alternating reinforcement and destruction of light intensity, as indicating waves adding or subtracting, and thus favored a wave interpretation. Within classical physics, Maxwell unified the previously separate domains of electricity and magnetism by showing that magnetic variations generated electrical ones, and vice versa, in such a way that waves could propagate. Even more convincing was the fact that the individual "strength" terms in the initially separate theories of electricity and magnetism combined to give the speed of light arising in Maxwell's theory. Basically, through a combination of the induction of electric fields through a changing magnetic field, and the change of that electric field being the "displacement current" that gives rise to a magnetic field, a self-regenerating situation arises. Although standing waves of light are possible (for example inside a laser), in free space the mechanism results in forward propagation at the speed of light. The simplest possible waves are polarized, which is to say that their electric field points in a certain direction, while the magnetic field is perpendicular to it. In low-frequency electromagnetic waves like radio, the electric field can cause electrons to move in a receiving antenna if the antenna is parallel to it (recall the small antenna at the bottom of Figure 3.7). It is thus natural to specify polarization by the direction of the electric field. In free space, both the electric and magnetic fields are perpendicular to the direction of propagation.

Both electric and magnetic fields carry energy, so, as the light propagates forward, it carries energy. In Newton's particle view, it was natural that the particles of light should carry energy. The modern wave-particle view calls the particles of light *photons*, although, under wave-particle duality, they also have wave characteristics. Each photon, which is also referred to as a *quantum* of light, carries energy proportional to the frequency. At levels of light our eyes can detect, the particle nature of light is not able to be seen, since so many photons are present that the light looks continuous. At very low levels, our best detectors can count individual photons. It is of interest to mention that some wave aspects of light take on strange characteristics at low levels: Young's famous double slit experiment still produces interference patterns at light levels so low that, in principle, only one "particle" would have been able to go through a slit. This somewhat amazing aspect is of little interest in our discussion of the Solar System, since sunlight provides huge numbers of photons.

Light radiating outward from a point or spherical source diminishes in *flux* with the inverse square law, much as does gravity. Flux is defined as the energy crossing a one square meter area, in the direction perpendicular to its surface, in one second. Falling off with the inverse square simply expresses conservation of energy, since, in Newtonian space, area increases with the square of the radius. So, for example, if we think of all of the energy from the Sun traversing a sphere of the radius from the Earth to the Sun (i.e., one AU), that same amount of energy would also go through a larger imaginary sphere at 2 AU. It would be spread out over four times as much area, so the flux would have decreased by a factor of four. But flux times area, the energy traversing that surface, remains the same. In physics, some common words have rather specific definitions: the energy traversing a surface in one second is the *power*. If the surface encompasses the whole object, then the power output is referred to as the body's *luminosity*, mentioned above in the context of Sirius A.

The flux at the Earth's distance is referred to as the *solar constant*. Determining to what degree it actually is a constant is an important matter since changes in solar input could explain changes in the Earth's temperature. For our purposes, we will regard the solar constant as 1,400 W/m². This means that one square meter in full sunlight intercepts *power* about equivalent to a room heater or a window air conditioner. Due to inefficiencies in converting power, and losses due to the atmosphere, a solar panel of about one square meter typically furnishes only about 100 W.

The Earth itself has the potential to absorb a very large amount of solar power. The atmosphere reflects a fairly large percentage of incoming light (mainly from clouds) and gives Earth a large *albedo* or average reflection coefficient of about 0.3. Earth, like all warm bodies, also emits radiation, mainly in the infrared part of the electromagnetic spectrum, at a much longer wavelength than the absorbed solar radiation. To calculate how much power traverses Earth's cross-section, we simply need to multiply that cross-section by the solar constant. The result is about $1.8 \times 10^{17}$ W. A large power station can produce about 1 GW, which is $10^9$ W. The power hitting the Earth, after accounting for the half or so reflected, is about equivalent to what 200 million large generating stations would produce.

Earth absorbs only a minuscule portion of the light traversing an imaginary sphere 1 AU in radius. The energy flowing out through this sphere in a second is the power of the Sun, its luminosity. The area of a sphere is $4\pi$ times its radius squared, and with the large radius of 1 AU, with 1,400 W flowing through each square meter, the luminosity of the Sun is $4 \times 10^{26}$ W. Old-style incandescent light bulbs give off power of about 100 W, most of which is actually heat. Greater efficiency in generating desired light is given by more modern lamps, such as LED bulbs, although the old designations hang on. This leads to incongruities like packaging stating the watt equivalent while also stressing the low power consumption. One finds for example, "60 W-equivalent using only 8 W". A more recent trend is to use actual SI units for light, such as lumens. We will not delve into those units further.

The interior of the Sun is not within our range of topics, being more in the "hidden" than the "invisible" Solar System, but all of its energy has to come from somewhere! It is now known that the majority of it by far comes from nuclear reactions. The direct evidence of this is discussed in a later chapter and is one of the most "invisible" aspects of the invisible Solar System. Surprisingly, nuclear reactions account, not for the heat of the Sun, but for its long life. As a large ball of gas, the Sun (or other stars) would be hot even without nuclear reactions, since gas balls must be hot to support their own weight through gas pressure. Instead, nuclear reactions allow stars to emit energy without changing their structure very much. The specific reactions will be discussed in a later chapter as the source of some particles filling the invisible Solar System, carrying off a small fraction of the Sun's power. Here, all we need to note is that nuclear reactions change mass to energy in accord with the one equation found in this book: $E = mc^2$. Since $c^2$ is a huge number, very little mass has to be converted to generate a lot of energy. Only a few tens of grams of matter are converted to energy in a nuclear weapon. The Sun's luminosity is so large, however, that about $4.4 \times 10^9$ kg of matter is converted to energy every second. With the density of rock being roughly 4,000 kg/m$^3$, this would correspond to approximately the 139 m high Great Pyramid being destroyed every second. This energy is primarily released as energy of motion of reaction products, and electromagnetic radiation in the gamma ray part of the spectrum. Not because of nuclear reactions, but consistent with them happening, the Sun's center is at a temperature of about 15 million K (kelvin degrees). The Sun is not transparent. Interacting with gas, light diffuses outward, remaining in equilibrium with the local surroundings as the temperature falls. Finally, near the surface, it gets to the boundary of the invisible Solar System. This boundary is the *photosphere* of the Sun. Rather than being trapped to bounce among the ionized gas inside the Sun, making a hot haze, light at the photospheric level stands a fair chance of escaping, almost all of it becoming "invisible" from our point of view as it streams untouched into interstellar space. Having originated as gamma rays characterized by a temperature of millions of degrees in the interior, light at the photosphere instead takes on its much lower temperature of about 5,800 K.

Blackbody radiation, which gives a good approximation to sunlight is characterized by one parameter only, its *temperature*. A curve, known as a *Planck function*, specifies how much of each color is present at a given temperature. Figure 3.8 resembles an ideal Planck function, although its small features, most appearing as local depressions in the amount of emission, are due to atoms and not characterized only by that ideal (more of these features are seen in Figure 3.10). A Planck curve has one peak, which is the color at which most of the light is emitted. The peak changes color in a way that is familiar to us: as the temperature increases, an object goes from being "red hot" to being "blue hot". Blue light is of a higher frequency and shorter wavelength than red light. In blackbody emission, the product of the wavelength of maximum emission and the temperature is constant, according to *Wien's Law*. At the temperature of the Sun, the maximum light is at about 500 nm wavelength. The Planck curve is wide enough that colors from ultraviolet to infrared are present, making white light, despite the peak being in the green region of the spectrum. The energy of an average photon of white light is about 2 eV or 3 $\times 10^{-19}$ J. For 1,400 J to flow through one square meter every second, about $5 \times 10^{21}$

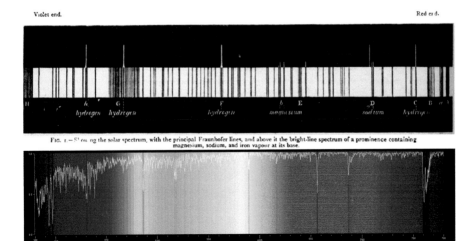

Violet end.                                                                                                                 Red end.

FIG. 1. — Showing the solar spectrum, with the principal Fraunhofer lines, and above it the bright-line spectrum of a prominence containing magnesium, sodium, and iron vapour at its base.

**FIGURE 3.10** Spectra of the Sun showing absorption and emission (top part of top panel, showing a few bright lines) by the solar atmosphere as "lines". The intensity plot (bottom plot, white line) shows that lines are due to reduced amounts of light at certain wavelengths (top from Lockyer, 1869; bottom generated by author from RGB Spectrum Generator, McNish, 2022).

photons must shoot through at the speed of light. Since the human body has an area of roughly one square meter, about this number of photons impinge on you each second if you are spread out to receive them, as you would be when sunbathing. Even though only a small percentage of these photons are in the harmful ultraviolet region of the spectrum, sunbathing even for a short time will expose enough cells that the body's defense mechanisms like tanning are activated. With so many harmful photons available, inevitably some cells are badly affected, resulting in damage outcomes like skin cancer if cellular repair mechanisms are overcome.

Some fraction of photons emitted by the photosphere interact with gas above the photosphere in the Sun's atmosphere. This gas has a diverse composition, although most of it is hydrogen. Light interaction with single atoms, or their electron-deprived counterparts, *ions*, is more complex than in black body radiation. The wavelength matters: atoms absorb and re-emit only at certain wavelengths. Often, an atom will absorb at a certain wavelength and then re-emit at the same wavelength or others, but in a different direction. In looking along a light beam toward the Sun, the absorbed wavelengths will seem to be diminished in intensity, that light having been redirected into other directions. Figure 3.10 shows this effect, with a white curve in the bottom panel showing the reduced intensities where the dark lines are. The upper panel is a spectrum of the Sun taken on August 7, 1869, as one of the first applications of spectroscopy to the Sun, published in the first issue of the now-famous *Nature* periodical. Notable here is the dominance of hydrogen emission in the spectrum of a prominence (Figure 3.10, top of top panel; an image of a prominence is shown in Figure 3.12). Unlike the dark line spectrum from the solar atmosphere normally seen absorbing the continuum light from below, a prominence is seen against dark sky and its lines are in emission. The dominant line in emission, likely not well registered in this early photograph at a time when only blue light was easily detected, is the red line marked C and also visible (but not lined up) in absorption in the bottom panel at 656 nm. This red color gives the layer just above the photosphere, visible during an eclipse, the name "chromosphere", which means "colored sphere".

To physically understand the origin of spectral lines, a partial energy level diagram for hydrogen is shown in Figure 3.11. As is the case for orbiting bodies in a gravity well, electrons bound to a nucleus have negative energy relative to a zero point, which is their energy if very far away. We start counting energy levels from the lowest one which is numbered 1, and since it is at −13.6 eV, not shown. At the lowest energy level, the electron cloud is symmetric and near the nucleus, resulting in the lowest possible energy. A fuller energy level diagram, including the lowest level, is shown in Figure 7.9. The diagram here shows three possible "paths" for the single electron to take to give an emission of light, if, for some reason, it had a higher energy than that of the $n = 2$ level. Electrons moving down in energy must conserve energy, the amount emitted as a photon of light being exactly that lost by the electron in moving "downward" into the electrical energy well. The possible paths are also limited by the need to conserve *angular momentum*, here simply considered to be given by the $l$ quantum number which has "Angular Momentum" values of 0, 1, and 2 in the figure. Higher angular momentum states start at higher energy, with the −13.6 eV level (Figure 7.9) having only zero angular momentum or $l=0$. The

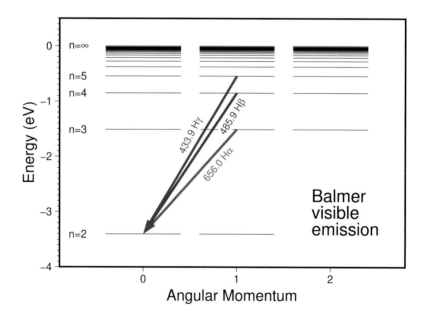

**FIGURE 3.11** Upper energy levels of hydrogen showing transitions of electrons from the third to second energy level (the lowest level is at −13.6 eV, off the bottom of the figure) (author).

$n = 2$ state shown here for $l = 1$ is the first state possible for that value of angular momentum. In a classical analogy, this makes sense since there is energy associated with rotational (i.e., angular) motion, even if the concept is not expressed that way in quantum mechanical formalism. The visible spectrum of hydrogen is dominated by lines of the "Balmer" series, which are given the abbreviation H with a Greek letter giving their order. Here, the first three Greek letters, $\alpha$, $\beta$, and $\gamma$, are marked to show the lines for $n = 3$, $n = 4$, and $n = 5$ transitions down to $n = 2$, respectively. If an electron instead absorbs energy from a "white" light beam like those from the Sun, it will follow the reverse path of the arrows shown, moving to a higher energy level. In this way, energy will be taken out of the beam. This fact alone does not give rise to the dark lines seen, because electrons do not spend very long at the higher levels, and instead fall back down to emit as shown in the figure. This re-emitted energy goes out in all directions, with only a small fraction being put back into the original beam. For this reason, we can see light emitted if looking from a direction *not* in the original beam. As mentioned above, this is the case with prominences seen above the limb of the Sun either during an eclipse or with a filter passing only their light. Conversely, we detect an *absence* of light at that specific wavelength if looking into the original beam, as we would be doing if we point our spectroscope at the Sun directly, instead of off its edge.

The instruments initially (as in 1879) used to detect spectra displayed light as line images of a narrow slit, so the small regions of the spectrum with missing light are known as *spectral lines*. During an eclipse, the chromosphere, as well as prominences, may be briefly glimpsed near the edge of the Moon. Apart from that fleeting glimpse under special circumstances, we may consider the chromosphere to be the lowest part of the invisible Solar System. Above the chromosphere is the *corona* (referring to its crown-like appearance during eclipses). The light of the corona comes mostly from scattering by electrons, giving it an eerie whitish glow. However, Norman Lockyer, the author of the Nature article referred to above, also noted a faint spectral line near the "E" line, which he could not identify but which had already been found ten years before. This green emission was initially considered to be due to a new element that received the name "coronium". It was not realized until about 70 years later that it simply originated from iron, but at an almost unbelievably high temperature. Another curious identification by Lockyer was marked as "?sodium". This did indeed turn out to be a new element, much discussed in the next chapter. Much of the structure of the corona arises from magnetic fields (see arrows in Figure 3.12), and it marks the beginning of the magnetized solar wind, which we will return to as one of the most important parts of the Invisible Solar System.

## 3.2 ENERGY DISTRIBUTION VIA LIGHT

As light streams outward from the Sun, it lights up the visible Solar System. The albedos of objects in it range from about 5% for dark basaltic bodies like Earth's Moon, to almost 80% for the bright clouds of Venus or the ice of certain outer planet moons. The albedo varies from place to place, showing details of changing clouds, or permanent geological regions. Carefully studying reflected light has brought about

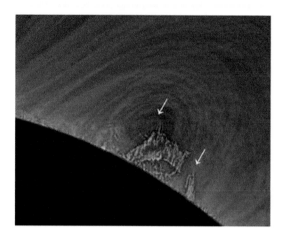

**FIGURE 3.12** Combined red and white (all wavelengths) image of the chromosphere and a prominence. Arrows indicate magnetically imposed structure (Habbal et al., 2021).

an understanding of many other planets as worlds, contributing in turn to our understanding of our own Earth. The domain is referred to as planetology, and now is being extended beyond our Solar System as other planetary systems with *exoplanets* are discovered at a rapid rate. We have learned that one of the most important characteristics of planets is temperature. The surface temperature of planets in a relatively old system like ours, 4.5 billion years after its birth, is mostly determined by the transport of energy as light. In young planets, heat left over from formation, including impacts that built them up, would also have been important. Even in the early Solar System, which was not as transparent as our current one, light was a major energy carrier and responsible for establishing zones of formation of planets made of different materials. The inner Solar System is dominated by rocky materials that could condense under hot conditions, and the outer regions by lighter gases that only condense into planetary blobs under intense cold conditions. Earth is in the happy middle, able to condense a rocky body, yet also able to retain atmosphere.

Nowhere else in the Solar System can light be in equilibrium with matter as it is in the opaque interior parts of the Sun. However, to a close approximation, planets emit radiation following blackbody rules. Since the temperatures are far lower than in the solar photosphere, Wien's Law informs us that the wavelength of their radiation must be long. In fact, it is in the far infrared. For example, Earth's emission peaks around 15 μm, near where carbon dioxide complicates matters by itself absorbing and re-emitting. This region is far off the long-wavelength end of Figure 3.8, which is at 2000 nm = 2 μm, and very far from the 500 nm wavelength at which most incoming energy is concentrated. Planets can come to energy balance by absorbing energy from solar light with a spectrum peaked in the visible as emitted by the Sun, then radiating it away at their blackbody temperature. This mechanism balances incoming energy, mostly dependent on distance from the Sun and albedo, with outgoing radiation, mostly dependent on the temperature of the emitting surface (or atmosphere) of the planet. The incoming energy retains a spectrum similar in shape to that of Figure 3.8 throughout the Solar System, but the absolute irradiance varies with distance from the Sun. The planet will adjust its temperature to emit and when in balance, it will have the temperature needed to do that. The balance is obtained due to the action of a second radiation law known as the *Stefan-Boltzmann Law*. Whereas Wien's Law deals with the peak wavelength of emission, the Stefan-Boltzmann Law deals with output flux as emission from a blackbody. Each square meter gives off luminous power proportional to the fourth (mathematical) power of its absolute temperature. Since energy is conserved, in the long run, a planet will attain a temperature allowing it to absorb light with a spectrum like that of the Sun, but diluted to lower flux as it travels outward through space, and then re-emit with a spectrum peaked at much longer wavelength. At the simple level of treatment here, we assume that the planetary temperature is constant and uniform. Ironically, the best example of how the mechanism works is also an example of non-constancy and non-uniformity. One can imagine spending a day in the desert. In the early morning, the desert is cold, including its rocks. However, with the sunlight pounding down, the rocks heat up. Even after sunset, rocks will be warm. As such, they radiate heat (infrared) into the clear night sky and cool down. In what follows, we imagine that heat flows around the planet efficiently, so that we can assign a global average temperature. On such an efficient planet, deserts would not warm up during the day and cool down at night; rather, an average temperature would be found everywhere and at all times of day or night. That temperature is in fact the one used to characterize thermal emission in the simplified blackbody approximation. The amount of such emission from each unit area is determined *only* by that temperature in the simplified world of blackbodies. A correction for the amount of incoming radiation uses the albedo: light which is reflected cannot play a role in heating the planet. The results of such a balancing calculation for the rocky bodies of the inner Solar System are shown in Table 3.1

Mercury is a rather dark planet, one of many respects in which it resembles the Moon. Being only roughly 0.33 AU from the Sun, it gets about nine times as much power per unit area from it as does the Earth. It is not surprising that its temperature is rather high. By contrast, Neptune, at about 30 AU, gets a mere trickle of light,

## TABLE 3.1

## Average Power Input Per Unit of Surface Area ($P_{ave}$), Albedo, Blackbody Temperatures of Planets, and Observed Temperatures

| Planet | Mercury | Venus | Earth | Mars |
|---|---|---|---|---|
| $P_{ave}$, $Wm^{-2}$ | 2290 | 662 | 342 | 145 |
| Albedo | 0.1 | 0.75 | 0.30 | 0.25 |
| $T_{blackbody}$ (K) | 437 | 232 | 255 | 209 |
| $T_{observed}$ (K) | 440 | 735 | 288 | 215 |

Data source: American Chemical Society.

about 1/1,000 the flux at Earth. This results in temperatures at which most common gases are frozen. The temperatures we find for Earth and Mars by this simple method are a bit low compared with those observed. That of Venus is drastically lower than its actual temperature. In addition to the details mentioned above, the effects of atmosphere, particularly the *greenhouse effect*, are important. This term has taken on a certain amount of notoriety in recent times, as a major player in it is *carbon dioxide* ($CO_2$), the concentration of which on Earth is being increased by human activity. Oxidizing, or burning, carbon-containing materials such as oil or coal, releases $CO_2$ into the atmosphere. This affects the transmission and emission of the atmosphere in the infrared portion of the spectrum where $CO_2$ has major *bands*. Bands are the molecular equivalent of lines in atoms, and major band influence distorts the application of the blackbody approximation. A higher temperature is needed to maintain a steady state with incoming radiation. The greenhouse effect on Earth is relatively small but critical. Without a greenhouse effect to raise the temperature above the blackbody value that would be here in its absence, Earth would be too cold for the development of life, the 255 K value shown being at −18°C, well below the freezing point of water. Since life is critically sensitive to temperature, it remains to be seen whether mankind's activities also render it uninhabitable by making it too hot. Venus is an example of a runaway greenhouse effect, with an anomalously high temperature in the Table 3.1. Its albedo is high enough to offset its closer distance from the Sun, making its blackbody temperature lower than Earth's, but its atmosphere is almost pure $CO_2$. The greenhouse effect is almost solely responsible for its anomalously high temperature. Its high albedo makes brilliant Venus one of the most beautiful sights in the sky as the morning or evening star, but perhaps it is more a beacon of warning than of hope.

The remaining planets generally follow a gradation in temperature, with surface gravity, dependent on density and size, also playing a role in being able to retain an atmosphere. The entire subject of planetary formation has infinite details. However, a good overview of the layout of the Solar System is provided by our simple consideration of the power balance, including how it got that way when the formation of the planets was determined by the original temperature profile, including that in the denser materials of the nascent Solar System.

## 3.3  IONIZATION, SHADOWS

Space is not empty. Interplanetary space near Earth has approximately 20 million particles per cubic meter under average conditions. While far from "empty", this also corresponds to a vacuum better than most that can be created in laboratories on Earth. Almost all of the particles in space have their origin in the Sun, being part of its rapidly expanding outer atmosphere known as the solar wind, the details of which are the subject of a Chapter 6. Numerically, 90% of the solar wind particles have the hydrogen-like composition dominant in the Sun. However, they are not hydrogen atoms, but rather are broken apart into protons and electrons. In this state, also common in the Sun, gases are referred to as being *ionized*, the individual atomic nuclei being *ions*. The second most common element in the solar wind is helium. Atoms heavier than hydrogen have more than one state of ionization: helium may lose one electron, then be referred to as singly ionized, or two electrons, to be termed doubly ionized. Inside the Sun, ionization is maintained through the constant interaction of particles with each other and with light. Ironically, in the solar wind, ionization is *maintained* for the opposite reason: particles have hardly any interaction at all. How particles in interplanetary space become ionized in the first place will be discussed in Chapter 6, but, once this does happen, it is hard for atoms to reassemble themselves. Although positive and negative particles strongly attract each other, they are so small and widely spaced that they rarely get close enough to *recombine* into atoms. Instead, the ionized gas, also known as *plasma*, has a very precise balance of positive and negative particles, a condition known as *quasineutrality*. In such a plasma, there is no overall tendency for the particles to move toward each other.

Being ionized is much of the reason that space is also transparent, at least at the high frequencies of light. Without any interaction between electrons and ions, there is no equivalent to absorption of light as there is in atoms. The topic of ionization will be addressed further in Chapter 7 since the interaction with planets of the solar outflow known as the solar wind is influenced by ionization in their upper atmospheres.

Because, in the vicinity of Earth, one square meter intercepts about 1,400 W of power, reasonably sized solar panels can capture enough power for a small spacecraft. New, small spacecraft called *cubesats* can be as small as cubes 10 cm on edge, with solar panels on perhaps several faces. Even such small solar panels provide enough power for a small satellite in a low-Earth orbit. Communications are demanding power since an electromagnetic radio wave must be generated with enough power to reach receiving stations on Earth. The present limit beyond which solar power becomes ineffective is in the asteroid belt, although the recently launched *Lucy* spacecraft will use solar power near Jupiter orbit (Figure 3.13). The *Dawn* spacecraft which recently went to asteroid 4 Vesta and dwarf planet 1 Ceres had a larger-than-usual electric power demand since its propulsion was also electric (*ion engine*, see Chapter 8). Its solar panel was about the size of a school bus. Beyond the asteroid belt, the inverse square falloff of available light power means that nuclear power systems, based on highly radioactive sources, often must be used. This topic is also discussed in Chapter 8. Conversely, spacecraft going toward the inner solar system must often be shielded from the intense luminous flux there. Even at Earth's orbit, luminous flux can pose a problem. Some telescopes, such as the James Webb Space Telescope (JWST), mainly sense in the infrared. Because bodies at 1 AU from the Sun are heated up to a few hundred K before coming into equilibrium with the light field, they would become emitters of the very radiation to be detected, if not cooled. In addition to cooling devices, the JWST has several reflective layers pointing in the direction of the Sun so as not to have the spacecraft itself emit in the infrared wavelengths to be observed.

Solar panels are based on the *photoelectric effect*, in which a surface emits electrons when exposed to light. Einstein's wonder year of 1905 led him to correctly describe the photoelectric effect as being due to individual quanta of light, the photons described above. Einstein received the 1921 Nobel Prize in Physics (awarded in 1922) in part for this work, with no mention of the theory of relativity described in

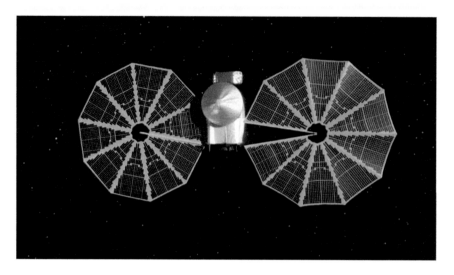

FIGURE 3.13   The *Lucy* spacecraft to Jupiter space has two solar panels each almost 7 meters in diameter (NASA).

the same year. To eject an electron, a photon must carry enough energy to overcome the retentive ability of the surface in which the electron sits (the so-called *work function*). Effective materials for solar cells release electrons readily, with excess energy that can then be put to work in the attached circuit. The electrons flow through the circuit and come back to the solar panel to replace voids left by their departure. In this way a solar panel remains neutral, making it a low-maintenance way to power a spacecraft. Of course, solar panels usually must be folded in close to the spacecraft when in the tight confines of the launch rocket. They are then deployed in space, usually a nerve-wracking portion of the initial commissioning of a satellite. The solar panels of the *Lucy* spacecraft, which will study the "Trojan" asteroids festooning Jupiter's orbit (see Chapter 5), are even larger in area than those of *Dawn*. As of late 2023, one had not fully opened (a detail seen in Figure 3.13) although this is unlikely to affect the overall mission.

The photoelectric effect acts on all parts of a spacecraft exposed to sunlight. Incoming photons, especially those of high energy in the ultraviolet region of the spectrum, cause surfaces to eject electrons at other places than in the carefully constructed structures of solar cells. Some of these electrons fly off into space around the spacecraft. However, each negatively charged electron that leaves, by carrying its charge away, leaves the spacecraft a little bit more positively charged. The positive charge attracts electrons. So, much as with light input onto planets heating in tandem with their blackbody emission cooling them, an equilibrium builds up. The spacecraft ends up with a positive electric charge and is surrounded by a cloud of negative electrons. In extreme cases, this effect can lead to a discharge, and damage to the electronics. The problematic time when a spacecraft enters a shadow can also be dangerous, with changes in the charge distribution causing electric currents in unwanted places.

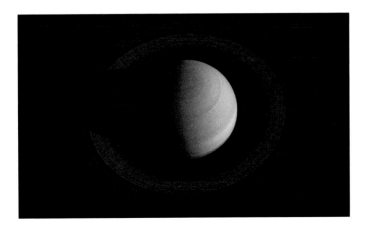

**FIGURE 3.14**   Image of Saturn from the north, taken in 2013 by the Cassini probe. Due to the rings, Saturn's shadow can be seen. In much the same way as Earth, Saturn has seasons, and the lit north pole indicates Northern Hemisphere summer (NASA).

Of course, entering a shadow is also not a good thing for spacecraft whose power derives from light. Low-Earth orbit (LEO) satellites have periods of revolution of about 90 minutes. In a general orbit, about half of this would be spent in shadow. Much as with a solar power system used by campers, during the sunlight period, energy is stored in a battery and then drawn out when in darkness. By using Saturn's rings as seen from above, Figure 3.14 dramatically shows the presence of a similar shadow near that planet.

Like planets orbiting the Sun, satellites orbiting Earth have periods that increase with the 3/2 power of their orbital radius, according to Kepler's third law. Although low-Earth orbit satellites have a period of 90 minutes, eventually there comes an orbital radius at which the period is one day. If orbiting in the same direction that Earth rotates, and sitting in the equatorial plane, such a satellite will appear to sit over one fixed point on Earth's surface. They are thus referred to as *geosynchronous* or *GEO* satellites. The orbital radius at which this occurs is about 6.6 times the radius of the Earth, or 40,000 km. Geosynchronous satellites are of great practical importance for communications and Earth monitoring, and several hundred of them are now in orbit. Being further from Earth than LEO satellites, GEO satellites do not spend as much time in Earth's shadow, although they do have eclipse periods at certain times of year.

## 3.4 ELECTROMAGNETIC WAVES IN THE SOLAR SYSTEM

*Occultations* are related to shadows, involving the point of reception of radiation not being in view of the source (or vice versa). Electromagnetic signals travel in straight lines at the speed of light in the Newtonian view of space (slight changes under relativity were discussed in Chapter 2). If there is an object between the source and the point of detection, the signal will not be received. A famous example is the Apollo moon flights, in which a critical engine firing had to take place on the far side of the Moon from Earth. Tension ran high until the Apollo spacecraft came out from beyond the limb of the Moon and could confirm success to Mission Control.

Earth's Moon has a large apparent size in the sky and travels rapidly as compared with other objects. It often passes in front of stars, thus occulting them. This allowed the orbit of the Moon to be well determined in the telescopic age. Conversely, in the pre-GPS era, occultation measurements could improve knowledge of positions on the surface of Earth. Occultations by other objects are rarer but can carry information not otherwise available. For example, a prolonged fading can indicate the presence of an atmosphere. Occultations of Jupiter's moons by the planet, or their entry into its shadow, could be timed very precisely and gave rise to the concept of the speed of light. When Jupiter is near opposition (for example at point L in Figure 3.15), and thus closest to Earth, the ephemerides of its moons may be accurately determined, giving a table of when phenomena such as occultations should occur based simply on some initial reference time and the period of the moon. It was found, notably by the Danish astronomer Ole Rømer that, at a later time, Earth being farther from Jupiter (point K in the figure), the observed phenomena lagged behind the tables. These results were published in French in 1676 and in English in 1677. Jupiter is still easily observable when Earth is about 1 AU farther from Jupiter and, in this case, the timing delay was about 8 minutes. It could be inferred that the phenomena were still happening at the same time but that light took a finite time to cover the extra AU of distance, allowing the speed of light to be roughly calculated. One limitation, of course, was that at that time the value of the AU was not known very accurately at all.

Bodies of star-like appearance in telescopes, such as asteroids, can be observed at multiple points on Earth, and the timing of disappearance and reappearance of the occulted star, as well as the limits of where the occultation is visible, can give information on the size of the asteroid. Earth is moving at 30 km/s in the prograde sense, and an asteroid, farther out, might be moving about 10 km/s in the same sense, so the relative motion would be of order of 20 km/s. Easily timed values in the one to several-second range would characterize how long the occulted star's light is blocked, allowing the size and shape of the asteroid to be determined. The *Lucy* spacecraft will fly by the Trojan asteroid Leucus in 2028, but shadow observations (Figure 3.16) of this asteroid at the distance of Jupiter already allowed its elongated shape to be determined ten years before the flyby. In some cases, two separate shadows are seen, and, in this way, it has been found that some asteroid-sized objects have moons.

If shadows can be used to determine the shapes of known objects which are too small to resolve with telescopes on or near Earth, can one hope to detect new small Solar System bodies in this way? The basic approach would have to be to look at star fields and check for stars momentarily disappearing. Very small bodies will not pass in front of very bright stars very often but, in recent years, both detection and data processing techniques have advanced considerably. Using archival data, the first claim for detection of an outer Solar System body being detected by its shadow was made in 2009. Such a body would have been relatively small, not only due to the

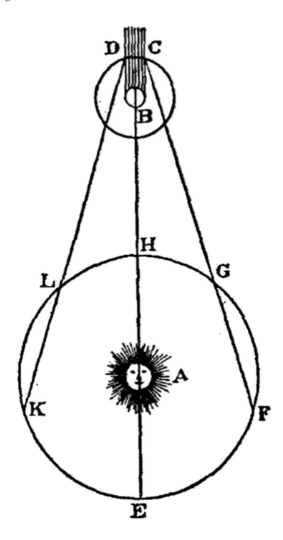

FIGURE 3.15   Ole Rømer's explanation of the timing variations of shadow entry of moons of Jupiter (top) as seen from Earth at various positions in its orbit around the Sun (A). In similar geometry at L and K and at G and F, the distance varies and so there is a time delay at the further points. The figure is not to scale (Wikimedia-Roemer, 2007).

minor effect that it had, but because statistically one expects more small bodies than large ones. Nevertheless, a dedicated survey with amateur-sized telescopes in 2019 detected a body likely about 1 km in diameter in the outer Solar System. Sixty hours of observing were done, with images taken about 15 times per second. Sufficiently high-quality measurements of the observable stars amounted to about 3 billion, with the detection shown in Figure 3.17 resulting. As a proof-of-concept, this is convincing, and once enough detections are made, statistical inferences about the outermost

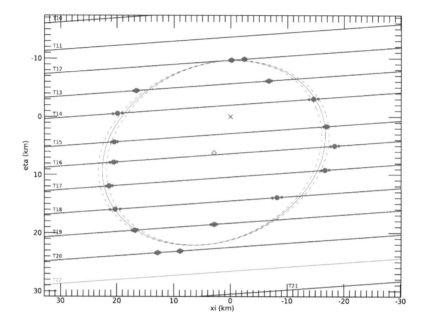

**FIGURE 3.16** The shape of the asteroid 11351 Leucus as determined by occultation observations from Texas on November 18, 2018. Best-fit ovals are shown in blue, but the actual outline may be more irregular as shown by the black curve, which passes through points determined by timing along lines parallel to the motion of the asteroid's shadow in starlight (Buie et al., 2021, CC BY 4.0).

Solar System, the Kuiper Belt, will be able to be made. In turn, this will constrain models of the formation of the Solar System.

## 3.5 SUMMARY

Light, or electromagnetic radiation of wavelengths about 300 to 700 nm, is emitted in large amounts from the outer atmosphere of the Sun and streams outward, the vast majority of it never intercepting any planetary body or even an atom. It determines the temperature of planetary bodies by heating them to the point that the incident light radiation is nearly in balance with outgoing infrared radiation. Heating due largely to light and IR radiation was also important in the original formation of Solar System bodies. Light carries information due to the interaction of blackbody-like radiation with the Sun's atmosphere, and with the atmospheres of the planets. Absence of light in the form of shadows is also important since spacecraft are charged by the UV portion of sunlight, and shadow entry causes a change in the charge distribution. While in shadow, solar-powered spacecraft are deprived of power. Shadows (occultations) can be used to determine the outlines of small and very distant objects. With enough observing time and processing power, occultations may be able to be used to find new objects and determine their properties.

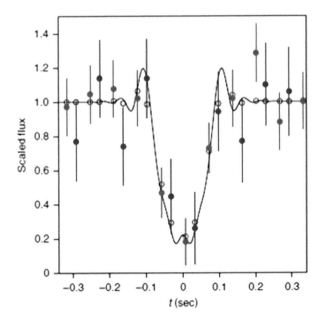

**FIGURE 3.17**   An observation of the shadow of a Kuiper Belt outer Solar System object about one km in diameter, made in 2016 (Arimatsu *et al.*, 2019).

# 4 Inner Space

The region beyond Earth's atmosphere is often referred to as "outer" space. In contrast, one dictionary defines "inner" space as being space near the surface of the Earth, or, metaphysically, one's inner being or perception of the world. Here, we will instead use the term to mean that which is unseen inside all matter due to its being too small. In the previous chapter, most of what was referred to as "light" arises from atoms, either singly or near other atoms. Single atoms, as we saw, have emissions at distinct wavelengths characteristic of their composition. A few atoms near each other may form a molecule, with more ways to wobble around, giving a more complex spectrum, often with features in the infrared, since more mass moves around and thus more slowly. Combining more atoms yet gives spectra distorted by their interactions, with the blending out of light into smooth blackbody curves being the limiting case. In this extreme, the composition does not matter anymore: we can make a blackbody in the traditional laboratory manner as a metal enclosure at a known temperature with a small hole, but even the dense part of the atmosphere of a star may, under most conditions, be treated as a blackbody radiator. All such atomic interactions mostly involve electromagnetic force, but on a small scale, which involves quantum effects. The beginning of the quantum revolution came when Max Planck published a method to explain the blackbody radiation curve if light came in small bundles, eventually called *quanta* as a general term, and, in the case of light specifically, the photons discussed in the last chapter. All of chemistry is governed by electromagnetic force as well, with quantum rules governing the structure and interactions of matter. One of those rules involves symmetry and the quantum mechanical property of *spin*. Although this rule plays an absolutely fundamental role in the structure of matter in the forms we usually see it, understanding it fully is beyond our treatment here. Further, it generally can be "enforced" with fairly simple rules and does not need to be much discussed. Even with spin included, electromagnetism is the best understood of all forces and is the subject of *quantum electrodynamics*, or *QED*, one of the most accurate and complete theories ever developed. The usual carriers of electric charge in everyday life are electrons, which QED regards as having no size. Quantum mechanics assigns free electrons a wavelength, and, within atoms, they may be thought of as forming probability clouds. Although the issue of size and location is a bit hard to pin down, due to this *wave-particle duality*, electrons have a definite charge, whose value could already be determined from chemical experiments by Faraday and other nineteenth-century workers on electrolysis under certain assumptions. Electrons also have a definite mass which, being very small compared with that of a proton, was hard to determine even with some assumptions. Here, we will mainly be concerned with electrons in their interaction with atomic nuclei as early views showed it. The modern view of this interaction features a new, invisible particle that pervades the Solar System, but this is deferred to a later chapter.

DOI: 10.1201/9781003451433-4

Gravity, to the extent that it can be regarded as a force, is not fully integrated with quantum mechanics. Everyday life has gravity as a force that seems important for a paradoxical reason. If regarded as a force, it is by far the weakest. On the scale of atoms, electromagnetic forces are *far* stronger. To the extent that we can regard a bound electron as being a small particle going around a much more massive one (a proton or more complex nucleus), we can compare the gravitational and electric forces since they both decrease with the inverse square of the distance. For the moment we consider only static particles, which do not feel the magnetic part of electromagnetism. We find the astonishing result that the electric force is $10^{40}$ times stronger than gravity! We also note that the gravitational force is always attractive because all bodies have positive mass. The much larger electric force in an atom arises because the particles have *opposite* charges. This huge force means that matter around us is exquisitely balanced in terms of the amount of positive and negative charge it contains. Enough "static" charge can be placed on only flimsy things like balloons to have an influence on the motion of a tangible object. The larger amounts of charge needed to move a more massive object would create such strong electric fields that electrons would flow in (or away) to quickly remove the charge. No such neutralization can happen with gravity since there are no "negative" gravity effects to cancel out its apparent attraction. Even "antimatter" always has a positive mass and thus generates spacetime curvature, which we call gravitational force, much as ordinary matter does. It has been firmly established experimentally that antimatter particles have the same mass as their ordinary matter counterparts. For one thing, the amount of energy liberated when matter and antimatter meet and annihilate into pure energy is exactly as expected. Of course, in our "ordinary matter" region of the universe, enough antimatter to directly observe its gravitational effects does not exist. There being no cancellation possible for gravity, the weakest force has far more effect on the *large* scale than one $10^{40}$ times stronger! In yet more irony, in plasma, which must like almost everything remains electrically neutral (the technical term is actually "quasi-neutral"), a secondary relativistic effect from electric charge is often more important than gravity. That effect is called magnetism.

By 1900, based on studies of gravity, electricity, and magnetism as combined into a full theory of electromagnetism only a few years before by Maxwell, and bulk concepts involved in thermodynamics, the story of physics seemed at an end. That year, however, turned out to be the one in which the combination of thermodynamics with electromagnetism as "blackbody theory" was overthrown. Hesitatingly, German physicist Max Planck realized that energy exchange had to be "quantized". It occurred in small packets, not in continuous amounts. Almost concurrent developments about the structure of matter revealed new and previously unsuspected types of energy. By 1945, this new nuclear energy had been released in both controlled and explosive forms, changing the world forever. It had taken millennia for the story of broadly felt gravity to be unraveled, and centuries to fully understand electromagnetism. In contrast, it took only decades to master (but perhaps decades more to understand) the nuclear force. Rather than binding things together on merely the atomic scale, already hidden from us due to its small size, the nuclear forces act on a scale at least ten thousand times smaller and are vastly more powerful. To complicate matters, the nuclear forces that are partly discussed here are more complex and come in two varieties, completing the "four" fundamental forces. These are the "strong" and "weak" nuclear forces. Much like bonding forces which can hold molecules together even though their atoms are directly bound by electric attraction between protons and electrons, the strong force binds nuclei together indirectly. The essential constituents of nucleons (Figure 4.1) are *quarks*, with interactions described by *quantum chromodynamics*, a more sophisticated analog of quantum electrodynamics. The strong force binds these constituents but somewhat overflows from each nucleon to affect others, much like the electric force overflows from otherwise neutral atoms to allow some types of chemical bonds. Whereas electromagnetic forces evaded our view by being very long-range and strong, so that opposed charges cancel and neutralize each other, nuclear forces are very short-range. Their elusiveness is related to the short distances over which they act with enormous strength. Much like with electromagnetism, these strong forces often form strong and stable structures. For electromagnetism, these structures are atoms, and for nuclear forces, nuclei (and strictly speaking, their nucleons). In our relatively cold and dark part of the universe, chemical change is continuous but usually slow, and nuclear change unusual. For this reason, an unsuspected and hidden world only came into our consciousness about

**FIGURE 4.1**  Although previously regarded as an "elementary" particle, the proton is now known to be made of basic particles (larger fuzzy spheres) known as "quarks" interacting via exchange of "gluons" (small paired spheres). Interquark forces increase with distance, so that protons or neutrons cannot be broken into free quarks (Argonne National Lab, US government).

the year 1900, the year physics was incorrectly expected to be fully explained. The long-held notion of stability of matter itself came into question since matter could be "active": radiating energy, particles, and electromagnetic waves, certain types could be "radio-active".

Gravity, always being attractive (apart from recently discovered gravitational waves), does not seem to be "radiation". In contrast, electromagnetism, as fully developed in the late nineteenth century, has radiation as an essential feature. It can be emitted from a source and radiate away from it. Our word "ray" can reflect this meaning, and of course, the radius of a circle goes out ("radially") from the center. Already the Latin verb *"radiare"* had a meaning like our "radiate" and was used by ancient authors to describe the light from planets, and even the supposed emission from eyes as seen in the glowing eyes of panthers in the dark! The first step in the discovery of nuclear forces came with "radioactivity". In the early stages, it was clear that something unknown was coming out from substances, thus radiating. More amazing was that normal-looking substances, previously judged to be inert on their own, were "active" in the sense that they gave off emissions of various kinds (Figure 4.2). Pioneering French physicists Marie and Pierre Curie coined the term "radioactivité" in 1898. Other substances could of course be "active": gunpowder is an example of the emission of energy and gas on a short (explosive) timescale, while yeast acted slowly in some of the most important reactions (fermentation, giving beer or bread) in human history. In both cases, the activity had to be induced by

**FIGURE 4.2** While a now common, if frightening, aspect of our nuclear age, the concept that inert materials like uranium ore rocks could emit invisible radiation without apparent change was amazing at the turn of the twentieth century (Shutterstock).

ignition or by the addition of water, and resulted in a change of the substances, a characteristic of chemical reactions. Refined uranium ores emitted energy seemingly on their own and seemingly from nowhere, without apparent change to themselves. The strength of nuclear forces means that tiny rates of reaction could emit prodigious quantities of energy with consumption of only tiny amounts of material, as was eventually elucidated. We now so routinely use the term "radioactivity" that it is hard to imagine the wonder caused by the realization that seemingly inert substances could be "active". While it appears that statements from famous physicists at the turn of the twentieth century that physics now was complete and need only strive for more precision are at best misquotes, certainly it would have been difficult to conceive then of the fundamental changes about to come. A lump of metal derived from uranium that stayed warm all on its own was just a hint of things to come.

## 4.1 IMPROBABLE PHYSICISTS

William (Friedrich Wilhelm) Herschel, mentioned in Chapter 2 as the discoverer of Uranus, was born in the small German state (an "Electorate", due to its role in the Holy Roman Empire) of Hanover. Its ruler as of the time of his emigration (1757) to England was Duke and Prince-Elector George, who also happened to be King George III of the United Kingdom. Herschel was a successful musician in the Southwestern English city of Bath, but his discovery of a new planet attracted royal patronage, perhaps partially because he initially named it "Georgium Sidus", or George's star. George III is often regarded as the "mad king who lost America" due to his mental illness and the fact that the United States of America successfully revolted against his rule, with very defamatory indirect remarks about him in their Declaration of Independence. In fairness, he was the first ruler of a major power to show serious interest in science and created what is now the British Library by posthumous donation of thousands of books he had collected. George was the longest-reigning British king, ruling for nearly 60 years, although in the last ten with his son as Regent. In a break with the characteristics of most monarchs, he had a faithful marriage and no known mistresses. He sponsored Herschel's further work, appointing him as "King's Astronomer" a year after his discovery of Uranus in 1781. In this special position (there already was an "Astronomer Royal"), Herschel received a salary of £200 per year (about £26,000 in current value) allowing him to continue and expand his astronomical work, including observation of the Sun. Required to be available to the King on demand, he moved to Slough, near Windsor Castle. With any telescope, it is necessary to reduce the light from the Sun to see its visible features. Being an expert (although self-taught) instrument maker, Herschel experimented with various filters. We stress to readers that it is unsafe to look at the Sun, even without a telescope, without a certified filter used properly. He found that some filters, while reducing the light from the Sun greatly, allowed a sensation of heat to be felt. While it would surprise no one that the Sun has a heating effect, this was the first clue that this effect was not only from its visible light. Herschel went much further by passing sunlight through a prism, much as Isaac Newton had done, but taking careful temperature measurements at the various colors, and beyond the range of the visible light. Figure 4.3 shows schematically that he found the invisible light beyond the red end of the spectrum to provide heat, making the temperature there higher than room temperature as measured by a thermometer not affected by the light at all. He initially attributed the heating to "invisible light". He backtracked on this basically correct concept, eventually settling on the idea of "radiant heat". Since the main idea about heat flow at the time was that it was a fluid called "caloric" that could flow between bodies, and since Newton's then-prevalent view of light was that it consisted of particles, the inference was that heat, like light, could travel through space as particles. These results were published in 1800. Effectively, while getting some essential details wrong, Herschel had expanded the range of known light into the invisible on the red side of the spectrum. Only in the latter part of the nineteenth century did the term "infrared", now often abbreviated to "IR", come into use.

Having read of Herschel's discovery of heat at the red end of the spectrum as measured by thermometers, a young German resolved to test his concept that the blue end of the spectrum might instead have a cooling effect. Johann Wilhelm Ritter, a twenty-five-year-old physicist and chemist, had been studying electrolysis in the eastern German city of Jena, later famed for the Zeiss optical company. Finding that Herschel, using thermometers, had found no effect beyond the blue end, he instead used a method that was the forerunner of photography to study that region of the

**FIGURE 4.3** William Herschel's discovery of infrared (IR) radiation was based on measuring its heating effect using a thermometer placed beyond the red end of a spectrum from sunlight. A control thermometer, whose temperature is just that of the room, lies nearby (Shutterstock).

spectrum. Before digital imaging took over, photography was a chemical process in which silver chloride (AgCl) was exposed to the light of an image and reacted to it. When exposed to light, this substance dissociates into silver and chlorine. The AgCl salt turns black when illuminated, that being the color of very small grains of silver formed by the reaction. This was known as far back as the time of alchemists, having been described by John Dee, alchemist, astrologer, and spy, serving Queen Elizabeth I in about 1600. Perhaps more relevant here, it is found naturally as a mineral called chlorargyrite in a mining district (the Erzgebirge or "Ore Mountains") not far from Jena. In the case of photography, further chemical treatment ends the reaction to light and makes a permanent image. Much as Herschel had placed thermometers along the spectrum and beyond to sample the heating due to light, Ritter exposed a strip of AgCl powder to a spectrum. The result was clear and striking, allowing Ritter to publish perhaps the shortest article ever announcing a major scientific discovery, in the German *"Annalen der Physik"* in 1801:

> Am 22sten Febr. habe ich auch auf der Seite des Violetts im Farbenspectrum, außerhalb desselben, Sonnenstrahlen angetroffen, und zwar durch Hornsilber aufgefunden. Sie reduciren noch stärker als das violette Licht selbst, und das Feld dieser Strahlen ist sehr groß. … Nächstens mehr davon. Ritter.

> (On the 22 February I also found sunlight on the violet side of the color spectrum, outside it, using Hornsilver (AgCl). It reduces even more than violet light itself, and the area of these rays is very large...more to come. Ritter.)

His other articles in 1801 were about electrochemistry, including that of silver salts, such as AgCl. Clearly, he made the leap to investigating light, having the materials on hand and thinking in an original way. Ritter also made contributions in other fields related to electricity, although he died at the age of 34 of tuberculosis, so that one can only speculate that he might have been a more familiar name in physics in happier circumstances. There were now two forms of invisible light, but Ritter's form, eventually called ultraviolet (UV), played a large part in the first step to subatomic physics taken at the end of the century. We now often think of UV as "black light", causing fluorescence like that in Figure 4.4, and that phenomenon played a role in what is being led up to. We need to stop in the middle of the nineteenth century to understand it further.

Sir George Stokes was an Irish-born English physicist whose name crops up often for students of mathematical physics but is not as well known to the general public. In the middle of the nineteenth century, he was appointed to the Lucasian Chair in mathematics at Cambridge University, a post held by Sir Isaac Newton and, more recently, by Stephen Hawking. His reputation was already great in the field of fluid mechanics, where the "Navier-Stokes" equations describe fluid dynamics in a way analogous to Newton's second law for particles. The fluid dynamic solutions are more difficult to solve than Newton's due to the close intercoupling of all parts of a fluid. His studies in optics are alluded to by the use of "Stokes parameters" to characterize polarization of light. Among numerous other contributions, "Stokes' Law" about fluid viscosity was later to play a vital role in characterizing the electron as a particle,

FIGURE 4.4 Fluorescence in daily life. Left: dried detergent on a bottle lid under UV light. Right: Labels fluorescing (author photos).

the first to have its properties determined. At Cambridge, Stokes, whose polarization work involved crystals, explained the effect of "fluorescence" as observed in the mineral fluorite. This mineral's name derives from its use as a "flux" in making steel, while the ending "-escence" comes from Latin, basically meaning "activity". The fluorescent effect in this mineral, as well as in liquids containing quinine (such as tonic water, a popular mixer for gin), was strong enough that it could be seen in sunlight. Stokes showed that the glow arising was due to the ultraviolet part of the solar spectrum. A similar effect can be seen in modern books with blanched paper (Figure 4.5) or whitened fabrics. Fluorescence adds brightness under sunlight and,

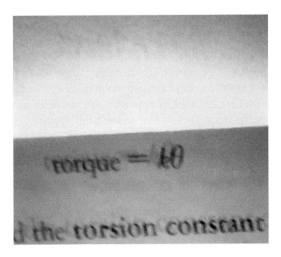

FIGURE 4.5 A sheet of blanched paper placed over top of a page of a book printed on unblanched paper fluoresces under UV light. There is enough UV in sunlight to make blanched paper look very white (author photo).

of course, can produce novelty effects under UV light illumination, now available from LED sources.

A related effect to fluorescence is called "phosphorescence", where "*phos-*", related to "photon" is a Greek word for light, while "*-phor-*" means carrier. The two effects are formally differentiated by allowed transitions at the molecular level, and, while it is not completely accurate to say so, historically phosphorescence was regarded as long-lived or at least active after another light source was removed, while fluorescence acted only while another light shone on the material as in Figure 4.4 and Figure 4.5. To produce visible light, a shorter wavelength must activate the glowing substance, which is called the "Stokes shift" as a general effect reflecting higher-energy photons going in and lower-energy ones coming out. As a tantalizing note, we will mention that, although phosphorescence is named for the violently reactive element phosphorus, which emits a glowing smoke, the effect in phosphorus is actually "chemiluminescence", with the activation caused by energy released in the chemical reaction of oxidation. Uranium ores and salts are sometimes phosphorescent, which does not have a deep connection to their radioactivity except in the sense that this property allowed radioactivity to be discovered. In the nineteenth century, both phosphorescent and fluorescent effects, which lie purely in the domain of electromagnetic interactions, were under active investigation. Fluorescence had further been found to be able to be stimulated by a mysterious new phenomenon called "cathode rays".

These two physical effects were supplemented by technological progress in the nineteenth century, including the ability to generate and apply high voltages, purify gases, and generate low pressures. Since scientific demonstrations were also popular, novel sources of light were used (bearing in mind that "Edison-type" incandescent lights were only invented in 1879). Humphry Davy, Faraday's initial sponsor, found that a continuous spark could be maintained to form an "arc lamp" about 1810. Such studies led to encasing the discharge and lowering the pressure. When it became possible to lower the pressure enough, in a "Crookes tube" as in Figure 4.6, what we now know as electrons could travel long distances without striking molecules. Not coincidentally, this took place at pressures matching those of our own atmosphere at or above about 100 km, where particles also move relatively freely. If a high voltage was applied across a Crookes tube, it appeared that particles moved in essentially straight lines from the negative terminal toward the positive terminal. Using terminology from electrolysis, the negative terminal was called the cathode and the positive terminal, the anode. In "conventional current flow", electricity was regarded as being from positive to negative, a convention still used today. The term cathode was related to the Greek term for "descent" because the electricity was viewed as descending into it. However, the particles in question appeared to emerge from the cathode, and thus were called *cathode rays*. Despite many years of experimentation, the first major result to come from studies with tubes making cathode rays, leading to the first Nobel Prize in Physics, had little to do with cathode rays. This 1895 discovery extended the spectrum of light much further than Herschel had in 1800, and past the short wavelength region first shown to exist by Ritter in 1801.

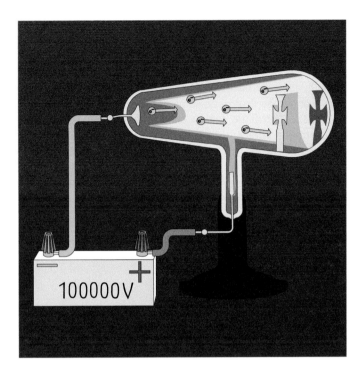

**FIGURE 4.6** Crookes tube for studying cathode rays, emitted into a near-vacuum from the left (Shutterstock).

A then 50-year-old professor of physics at Würtzburg, considered to be part of the southern German state of Bavaria (in the late 1800s a kingdom forming part of the German Empire), Wilhelm Röntgen, had had an unremarkable career. He was systematically studying cathode rays with a Crookes tube, as were many other researchers. Sometimes, fluorescent material impregnated into paper (as in Figure 4.5) allowed cathode rays to be detected. Having such a paper detector nearby, Röntgen noticed that it gave off light when the Crookes tube was operating. Thoroughly determining that the effect could not be due to cathode rays (which were known to be incapable of penetrating the thick glass walls of a normal Crookes tube), Röntgen considered his discovery of a new form of radiation so amazing that he called them X-rays, X for "unknown", which, in 1895, they certainly were. Röntgen, being a modest person, resisted the efforts of others to name them for him, and it is only in German-speaking countries that they do now bear his name (Röntgen-Strahlung). As he systematically researched the properties of this new form of radiation, he discussed it only with his wife. Once he found that the rays penetrated materials of low density, even opaque ones, she became his guinea pig, holding her hand still between a photographic plate and the device. The results did not impress her, since in the skeletal image she claimed "I see my death". Röntgen's results were communicated to the Physikalisch-medizinische Gesellschaft zu Würzburg in late 1895 and made such an impact that they were translated and published in the American journal *Science* in early 1896. The impact there was likely enhanced by the publication of an X-ray of what appears to be the hand of his wife. Figure 4.7 shows a comparable image of his colleague's hand. It is fitting that the first presentation was to a society interested in physics and medicine since the medical applications were immediately apparent. In worldwide news about the discovery, some newspapers declined to publish the first X-ray image since it seemed too unbelievable. Only by meeting with the Kaiser, and giving a demonstration with a live viewing of his colleague Albert von Kölliker's hand, could Röntgen's work be taken as more than science fiction and indeed its huge medical value be exploited. Röntgen's early work included checking whether X-rays were affected by magnets, which was known to be the case with cathode

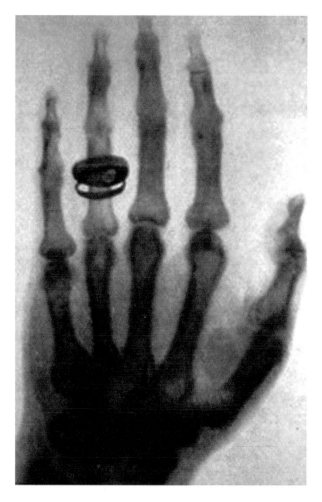

FIGURE 4.7    Early X-ray image: the hand and ring of Professor Albert von Kölliker, taken by Röntgen (Röntgen, 1896).

rays. Since they were not affected, and had other characteristics in common with light, he concluded that they were a form of light. Röntgen's wife may have feared death because of poor health, and he withdrew from life, not taking a patent to gain from the applications of his discovery, donating the Nobel Prize money to his university, and somehow managing to avoid giving the traditional Nobel Lecture. At his death in 1923, he had his personal and scientific correspondence destroyed. Acting as a person of great integrity, Röntgen did not benefit from his discovery, nor did he directly contribute much more to science, but X-rays inspired the next step into the world of the atom and nucleus.

Röntgen's thoroughness did extend to testing whether, in addition to other common properties with light, X-rays could be focused. He found that they could not be, which is not surprising given that they would pass right through most normal lens systems. As a result, X-ray images, including those made today in dental and medical offices, are generally "shadowgrams". In most small modern machines, a point-like source of X-rays is used, with a heavy and durable metal, such as tungsten, also used in incandescent light filaments (Figure 4.8) as a target for a focused beam of electrons. In the more than one hundred years of development of X-ray technology, ways to recreate images from multiple shadow measurements now allow "scans" to be made to isolate depth. It is now also possible to focus X-rays with mirrors that look like a set of rings well ahead of a detector (Figure 4.9). In this way, astronomical X-ray images from the Sun and other celestial objects can be made. Röntgen also noted that stopping X-rays required dense materials: our typically thinking of lead as stopping them is due to its high density, so that small thicknesses of it can do so. It is often important to protect people from X-ray exposure in settings where they are used: lead is often used as "shielding", not only since it is a dense and common metal, but because its low melting point allows it to be cast in desired shapes.

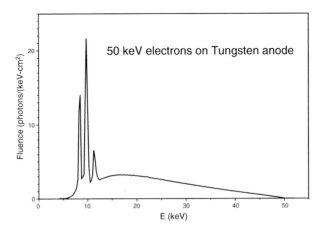

**FIGURE 4.8** X-ray emission typical of a mammography machine. The line spectrum of tungsten and *bremsstrahlung* background are visible (author).

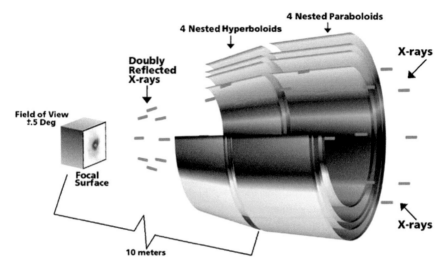

FIGURE 4.9   Cut-away schematic of the Chandra X-ray Observatory's X-ray telescope. Four nested, co-axial, confocal, grazing-incidence mirror paraboloid-hyperboloid pairs, the largest is 1.23 m in diameter, focus cosmic X-rays onto the telescope's focal surface about 10 m away. Chandra is used for faint cosmic sources (NASA/CXC/D. Berry).

With modern techniques, we can determine the spectrum of X-rays as shown in Figure 4.8. In a modern medical machine, as in the case of the Crookes tubes used in the initial discovery, electrons are accelerated with a potential of 50 kV or more toward an anode, which they strike. Such anodes are often made of the durable heavy metal tungsten. The energy deposited results in emission of X-ray photons, whose energy is often designated in keV. The distinct emission lines are due to inner, very energetic electron shells of atoms (such as tungsten) of high atomic number, that bind inner electrons very strongly. The continuous part of the curve reflects a different form of radiation due to acceleration (more properly, deceleration) of electrons. This emission is known as *bremsstrahlung*, which, in German, means "braking radiation". X-rays of 10 keV have a wavelength of about 0.1 nm, 5,000 times shorter than that of visible light.

## 4.2 THE SUN AS AN X-RAY SOURCE

The Sun's electromagnetic emission is dominant in the visual part of the spectrum, with the solar "constant" indicating optical power at Earth being over 1 kW/m². In contrast, X-ray emission would rarely be over 1 mW/m², one million times weaker. Nevertheless, X-rays can interact with Earth's atmosphere, and, if an astronaut is exposed long enough, they may build up a "dose" which is harmful to life. Earth's atmosphere is affected largely by having its ionization at altitudes from 70 to 100 km increased, affecting radio wave propagation. The solar emission of X-rays is mostly from regions in which the magnetic field of the Sun is both strong and tangled. Usually, these regions are above sunspots and are regarded as being in the lower corona of the Sun, discussed again in Chapter 6. Active regions can emit "hard" X-rays similar to those from an X-ray tube (blue in Figure 4.10), softer X-rays of lower energy of roughly 1 keV, and far-ultraviolet light. The latter is not as concentrated in active regions as the former, and typically has wavelengths of less than 20 nm. However, the tangling of magnetic fields can store a large amount of energy and, if the geometry is right, it can be transferred to particles and radiation very efficiently through a still-poorly-understood process called "magnetic reconnection". This is the main process driving the non-steady X-ray emission of the Sun, largely in "solar flares". In these, the emission increases by factors that can exceed 10,000 in a matter of minutes. Studying the configuration of magnetic fields in active regions can give some indication that a flare is about to happen, but prediction is not possible with great accuracy. Figure 4.11 shows flare activity as recorded on GOES spacecraft at geosynchronous orbit 6.6 Earth radii out from the center. Since X-rays are not affected by magnetic fields, the same flux of X-rays would be detected anywhere above Earth's atmosphere. The atmosphere does absorb them, mostly in a layer tens of kilometers thick between about 70 and 100 km, protecting the surface. As Röntgen pointed out, denser materials absorb X-rays, so a few cm of aluminum around a spacecraft would be good protection. A spacesuit, made of fabric, is not adequate, so that X-rays from large solar flares make "extra-vehicular activities" impossible.

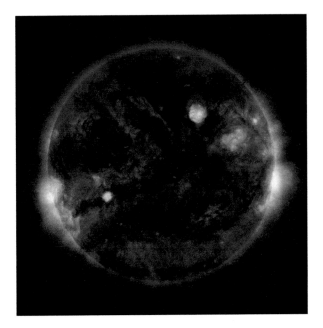

FIGURE 4.10 X-ray (blue) and extreme UV (green and orange) Sun. These high-energy emissions are concentrated in active regions (NASA/JAXA).

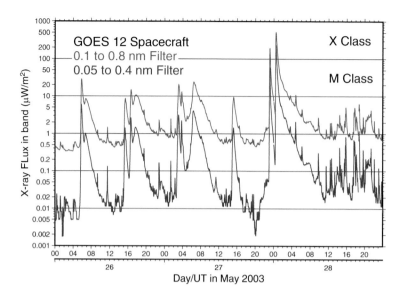

FIGURE 4.11    GOES X-ray flux during a very active period at solar maximum (author, data from NOAA SEC, 2003).

Changes in the upper atmosphere occur only where X-rays directly impinge on it from the Sun and are strongest where it is vertically overhead. Lower radio frequencies in the "medium" to "high" frequency (MF-HF or 300 kHz to 30 MHz) bands are most affected. MF includes the AM radio band, while HF includes "shortwave" which is important for aviation communication. Under normal conditions, transmissions in these bands travel in straight line paths with little absorption, until they hit layers in the ionosphere which reflect them, typically at about 100 km altitude. The reflection back downward allows such transmissions to be received well over the horizon from where they were originally broadcast, making reception of "skywave" possible at remote locations. Regions lower than the normal height of the ionosphere can be ionized by solar X-rays and absorb radio waves since they have a lot of collisions with neutral atmospheric constituents. This absorption, plus the shorter range of any reflections that do occur, can cause a radio blackout. A small X-ray event typically does not last more than a few minutes, but the loss of communications is sudden and cannot be predicted. During periods when the Sun is very active, as in Figure 4.11, blackouts can be frequent. As shown for a different event in Figure 4.12, the region affected by a blackout may cover a large part of the daytime hemisphere of the Earth.

An X-ray flare may be followed by a "proton storm". Protons, as discussed above, are complex ensembles but often regarded as "elementary" particles, but high-energy protons are almost never naturally produced on Earth. Protons are an important part of space radiation hazards, discussed mostly in Chapter 8. However, radioactivity on Earth first became apparent with "alpha" particles. The almost immediate consequence of the discovery of X-rays was the discovery of radioactivity, paradoxically involving an almost unknown substance at the time, helium (Figures 4.12, 4.13).

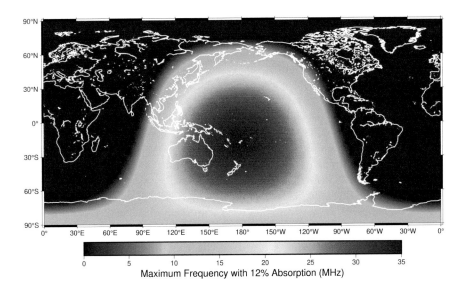

**FIGURE 4.12** Radio blackout due to solar flare on November 7, 2022. Frequencies up to 35 MHz were affected with a factor of more than 1000 absorption at low frequency (author, data from NOAA SWPC, 2022).

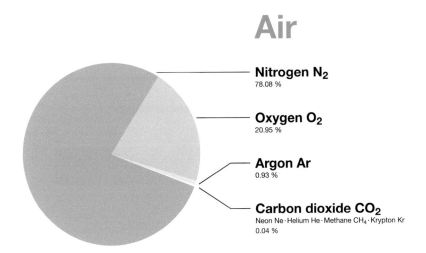

**FIGURE 4.13** Air is 78% nitrogen, 21% oxygen. The 1% argon went uninvestigated through most of the nineteenth century. Dry air has only minute amounts of other gases (Shutterstock).

## 4.3 UNNATURAL NATURAL GASES

Earth's atmosphere is comprised of 78% nitrogen and 21% oxygen on the basis of the number of molecules. It has about 1% of another gas which was not detected by the earliest investigators, and all other gases are present in dry air in only trace amounts.

A further atmospheric constituent, and major greenhouse gas, is water vapor. Early experiments on air were done by Henry Cavendish, the wealthy amateur scientist most active in the late eighteenth century, who also famously investigated gravity as previously noted. He found that, when most of the air in a flask was removed by chemical means, there remained about 1% of the original gas as a small bubble. This is in reference to dry air since water vapor is present in highly variable amounts, and then, as now, is usually excluded from the count. For over 100 years, nobody followed up on this finding. Argon is a non-reactive "noble" gas, with, in fact, the name meaning "lazy" in that it is not active. For work done in 1894, Lord Rayleigh won the Nobel Prize in Physics in 1904 for following up Cavendish's hint. His colleague in this, Sir William Ramsey, won the Nobel Prize in Chemistry the same year for his work on noble gases, which later included the discovery of neon (meaning "new") in 1898. Noble gases do not react, not even with themselves, and so exist as single atoms. Argon is denser than air, so stays near the surface of the Earth rather than rising to the top of the atmosphere and escaping. Neon is less dense than air and thus does escape, making it rare on Earth whereas argon is common, if unnoticed. The reason that these gases are "noble" and unreactive is that their outermost electron shells are full, leading to stability for quantum mechanical reasons.

Ramsey also made a discovery in 1895, the implications of which were mysterious. It was known that the uranium mineral cleveite gave off a significant amount of non-reactive gas when heated or dissolved. Ramsey, of course, was interested in determining whether it might be argon, but causing it to fluoresce showed a spectrum previously never observed on Earth. This was the spectrum of helium (Figure 4.14), discovered in the Sun in 1868 due to a previously unknown bright yellow spectral

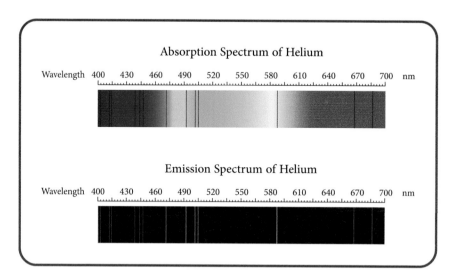

**FIGURE 4.14** Helium spectrum in absorption (top) and in emission (bottom) as it was first discovered in the Sun during an eclipse. The "lines" are shown as a function of wavelength in nm, with the yellow line near 588 nm characteristic of helium (Shutterstock).

emission line observed in prominences during an eclipse in India. In Figure 3.10, from 1869, this line was marked as "?sodium", which is curious since the author, Lockyer, had been involved a lot in research on the new line. The story of who actually first discovered helium is thus confusing, but the term "helium" came into use in 1871, with the name derived from "Sun" in Greek: "*elios*", with the apostrophe indicating an h that is basically silent. Its line is near the "sodium D" double line, so marked in Figure 3.10, which is also visible near the middle of the color spectrum at about 589 nm. Note that there is a large offset in line positions in the two panels of the figure. Helium's low density allows it to escape from Earth, with that and its chemical inertness explaining why it had not been discovered previously. But why should it be in uranium ore? Helium is the simplest noble gas and has a very stable nucleus in addition to a very stable electron cloud consisting of two opposite spin electrons in a spherical cloud near the nucleus. Helium has so much pull between the +2 charge of protons in the nucleus and the −2 charge of the electron cloud that it is the smallest atom. As a nucleus, helium is so strong that it seemed to be an elementary particle and came to be called an alpha ($\alpha$) particle, bearing the first letter of the Greek alphabet since it was the first recognized radiation particle. We now know that this strength is due to helium also having two neutrons in its nucleus, adding no repulsion due to charge but adding more binding due to the strong force. Helium may be simply an element like all others, but played a major role in the discovery of radioactivity. Helium is famously lighter than air and is used in inflating toy balloons. For reasons to be explained later, it is a major constituent of the Sun (and thus of the escaping solar atmosphere known as the solar wind). The connection of uranium to helium soon opened a whole new world to physics.

## 4.4  A GLOWING DISCOVERY

Henri Becquerel, much like the rest of the scientific world, was fascinated by the news from Röntgen about the properties of X-rays. He came from a scientific family which, for three generations, had studied photoluminescence, with academic posts in Paris enabling such study. Both fluorescence and phosphorescence fall under the term "photoluminescence", which is a bit of a redundant term, using both Greek and Latin words about the emission of light. Becquerel noted that Röntgen had pointed out that it appeared that X-rays originated, not from the Crookes tube in general, but more precisely from a fluorescent spot on its glass wall. This possible link between the new rays and the family specialty motivated Becquerel to undertake investigations seeking to see whether phosphorescence also produced X-rays. Phosphorescence can often be activated by light and endure for some time: within the Becquerel family history stood decades of study of mineral and chemical substances that glowed in the dark, usually after activation by being exposed to sunlight. Certain uranium ores were fluorescent (Figure 4.15) and/or phosphorescent.

The key to the hidden world of "inner space" was uranium, almost certainly not known to Becquerel as a source of helium gas, but rather due to having phosphorescent ores. Uranium's feeble connection to the solar system came in that, when identified as an element in 1789, it was named after the planet Uranus, discovered less than

**FIGURE 4.15**  Uranium ore (meta-autunite) under ultraviolet light, from miniera Assuncao, Fereira de Aves, Portugal; 4 cm across (Shutterstock).

a decade beforehand. Its discoverer, Martin Klaproth of Berlin, analyzed ore from the Erzgebirge (near the source of Ritter's chemicals referred to above), and found that uranium could tint glass. Although Klaproth did not separate the metal, uranium is the densest element in that form. It is also not very radioactive; it is used in modern times in armor-penetrating weapons in the form of "depleted uranium", which has had the most radioactive isotopes removed, leaving density as its main attribute. While Earth's magnetism is an indication of its separation into an iron-nickel core and rocky upper layers (called "differentiation"), uranium, despite its density, is concentrated in Earth's crust due to its "lithophile" nature. Lithophile means "rockloving" and in this context, its opposite is "siderophile" or "iron-loving". Lithophile elements are mainly found in the upper layers and siderophiles in the core. As such, uranium is about as abundant as secondary minerals like cobalt and lead and mined in the major mining countries of the world. Of course, such mining now is due to its current importance in energy production. In the nineteenth century, uranium was mined for use in glasses and glazes for pottery, producing some interesting colors. The interest in uranium for the Becquerel family was mostly the phosphorescent nature of some of its compounds and crystals. Röntgen's mention of glass fluorescence in the walls of a Crookes tube stimulated Becquerel to investigate whether fluorescent or phosphorescent minerals or crystals in general could produce X-rays.

Early in 1896, Becquerel communicated to the Academie des Sciences that a compound of potassium sulfate (a fertilizer) and uranium sulfate in crystals emitted radiation that seemed like X-rays after being exposed to sunlight. He noted that another researcher had found the same effect with "commercial phosphorescent calcium sulfate", which could be gypsum treated in some way, which supported his exciting new finding that sunlight could apparently induce phosphorescent materials to produce X-rays. The other reports could not be reproduced, so he continued his own work with uranium salts. He noted that the uranium compound showed phosphorescent emission for only a short period after light exposure, but that the inferred X-rays persisted. For reasons that are still not clear, Becquerel developed photographic plates, used to record the penetrating radiation, when his uranium compound had *not* been exposed to sunlight. He got the same result as when exposed to sunlight directly, or under various contrived circumstances like reflection from a mirror first. Basically, he would conclude that it was some *intrinsic* property of the uranium salts that made a penetrating radiation, and not their phosphorescence. This he backed up by experiments with non-fluorescent compounds involving uranium, moving on to find the greatest effect from pure uranium metal. He also experimented on the penetrating power by placing obects between the uranium and the photographic plates (Figure 4.16). Becquerel wrote of "radiations actives" and the rays came to be called "uranium rays", later changed to "Becquerel rays" by others when it was realized that another natural element, thorium, also emitted them. Becquerel applied the techniques of physics to the rays, finding that there were three sorts, which ultimately were called alpha, beta, and gamma ($\alpha$, $\beta$, $\gamma$) somewhat unimaginatively after the first three Greek letters (Figure 4.17). Ultimately it was another scientist active in Paris at about the same time who was credited with originating the term "radioactivity". She was a doctoral student (although not with Becquerel as supervisor) and shared the Nobel prize with him (along with her husband) in 1903. Her name was Marie Curie, née Skłodowska, and she had come to France from Poland (then part of the Russian Empire) twelve years earlier to study.

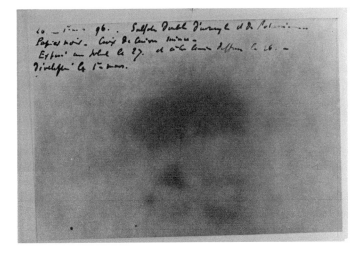

FIGURE 4.16 Becquerel test of emission from uranium salts showing their partial penetration of a copper cross (Becquerel, 1896).

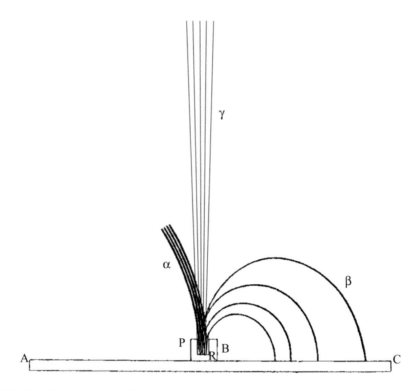

**FIGURE 4.17** Figure from Madame Curie's PhD thesis illustrating the 1903 concept of α, β, and γ (alpha, beta, and gamma) ray deflection in a magnetic field. Radium (R) is in a lead cup (P) and radiates these high-energy emissions either directly or through daughter products (Curie, 1903, retouched).

At this time, it was generally believed that matter was divided into atoms, but nothing was known about their structure. It was, however, possible to infer that cathode rays (now called electrons) were much less massive than atoms. Faraday, in 1833, had quantified the process of electrolysis, which allowed the movement of atoms through a chemical solution under the influence of an electrical potential. An electrical current flowed, which could be measured, and mass was transferred to an object, for example, a brass piece coated with a thin layer of silver. Another electrolysis process allowed the separation of hydrochloric acid (HCl) into measurable quantities of the poisonous gas chlorine and the explosive gas hydrogen. Since the atoms of both moved through the liquid and came to electrodes at either side, they were called *ions* from the Greek word for "moving". We use this word still for atoms bearing a net charge, including free positive ions in space. Since the current and mass of hydrogen could be measured, one could calculate that, for hydrogen ions, the ratio of charge to mass (in present SI units) is $9.57 \times 10^7$ C/kg. The coulomb (C) is itself a rather large charge but more conveniently recalled as the sum of elementary charges ($6.24 \times 10^{18}$ of them) flowing in a circuit each second when one ampere (A) flows in it. The amount of charge in a given mass of protons is enormous compared to the very tiny amount of *unbalanced* charge normally present in matter surrounding us. For example, a large melon is mainly made of water and carbohydrates, each with protons. If the electrons accompanying the roughly 1 kg of protons could be removed, the protons would repel each other with enormous force. Not only would they go to the surface of the modified melon as they tried to repel each other as far as possible, but even there they would have an energy of about 100 billion megatons equivalent: two billion times larger than the largest nuclear bomb ever detonated. Even if we could take away one coulomb by flowing electrons in a modest electric circuit, the remaining protons in a melon would have the energy equivalent of 1,000 liters of gasoline. These examples illustrate that charged particles are in almost perfect balance in our everyday lives: if they were not, large amounts of energy would be exchanged with charged bodies. When current flows in a wire, the wire itself remains almost perfectly electrically neutral as electrons swim past a structure of metal (often copper) nuclei. What we perceive as a "static" electrical shock carries only millijoules (mJ) of energy due to a tiny charge imbalance. Taking the fruit analogy a bit further, these mJ could raise an apple by only a few mm near the surface of the Earth (Figure 4.18). The whole electrical industry that grew in the nineteenth century viewed electrical charge in chemistry as borne on particles of about the mass of a proton, but did not know what carried charge in electrical circuits. When the electron was discovered, it was found to have a mass 1836 times smaller than that of a proton. If protons already

**FIGURE 4.18**  If Newton lifted the apple 1 m, it would take about 1 J (joule) of energy. The gravitational force on an apple is about one N (newton). Lifting 1 mm uses one mJ (Shutterstock).

carried an incredible amount of mass per kg by comparison to everyday objects, the electron carried far more. In fact, $(e/m)_e = 1.76 \times 10^{11}$ C/kg. This quantity was precisely determined by American Robert Millikan as reported in 1913, using studies of tiny oil drops obeying Stokes' Law in the fluid of air.

The apparatus described schematically in Figure 4.17 by Mme. Curie had been built by Becquerel a few years prior to her thesis being written. The Nobel Prize in Physics in 1903 was awarded half to Becquerel and one quarter to Pierre Curie. Marie's share of the prize was "only" one quarter, but having won the Nobel Prize likely helped in getting a PhD the following year!

The stability of matter had been a basic concept developed to a sophisticated level by the turn of the twentieth century. Although, of course, matter changes form, how that happens had been systematized through chemistry, with certain very basic rules established over the course of the nineteenth century. Two of the most fundamental principles were that mass and energy did not appear or disappear in the course of chemical reactions. Electric charge could also be accounted for: if an ion went to one side of an electrochemical apparatus, its charge equivalent went to the other side. These principles were so well established that they were called conservation rules: mass, energy, and charge were conserved in chemical reactions. Radioactivity violated all the rules! After the initial discoveries, much progress was made by a person with an unusual background: a farm boy from New Zealand.

## 4.5  SYSTEMIZATION OF NUCLEAR PHYSICS

Much as Marie Curie had sought educational opportunities outside of her homeland, Ernest Rutherford came from New Zealand to the University of Cambridge in England in 1895. Although he already had a master's degree from New Zealand, he graduated with a "BA Research Degree", and moved on to McGill University in Canada in 1898. He was awarded a "DSc" degree in 1901 from the University of New Zealand, now disaggregated. In 1902, Rutherford and colleague Frederick Soddy noted that it was rather remarkable that helium seemed to be present in uranium ores despite them not allowing the gas to diffuse through them, meaning it did not come in from elsewhere. The alpha ($\alpha$) particles emitted from radium were shown in 1903 by Ramsay, who had isolated argon, and Soddy, who visited him in England with a precious radium sample from Rutherford's lab, to be helium nuclei. In a very clever experiment by Rutherford several years later, the direct "real-time" production of helium nuclei was confirmed. An initially empty (vacuum) chamber separated from the radium by a *very* thin glass partition (since alpha particles do not have much penetrating power) began, over time, to fill with helium as evidenced by an electric discharge showing the its yellow light. Mass was appearing in a vacuum! It was also shown that radium produced a gas, initially known as "emanation", that was much more radioactive than the original metallic radium itself. This gas is now known as radon. Being a heavy gas that does not react chemically, it can float around freely and accumulate in enclosed spaces to create a health hazard in regions of the world that have high natural uranium content in their soils (uranium being at the top of the reaction chain that gives radium and radon).

Another aspect of radium that became apparent once it was separated by chemical means from uranium ore was that it seemed to violate the energy conservation rule: with no apparent change, radium ore was warm to the touch (of course, touching it turned out not to be a very good idea). Energy in the form of heat appeared to be coming out of nowhere! Finally, it was soon clear to Becquerel that the other radiation, beta (β) rays, were charged. In a setup very similar to that given schematically by Mme. Curie (Figure 4.17) and with an elegant simple analysis of the expected path of charged particles, he showed that beta rays were negatively charged, bent by a magnetic field. Although the requirements of charge neutrality meant that a small current flowed through the supporting apparatus to make up for departing electrons (which beta rays were soon shown to be), electrified particles were flying off the radium sample, without being replaced by a chemical reaction. As shown schematically by Mme. Curie, alpha radiation in the same magnetic field did not undergo as much bending, but it was eventually established, in devices with stronger magnetic fields, that they did bend in the opposite direction, meaning that they bore a positive charge. The easier deviation of beta rays led them in more refined experiments to be shown to be more energetic versions of the well-known "cathode rays", which, in the form of "corpuscles", were the first subatomic particles to be identified, by J. J. Thompson at Cambridge near the turn of the century. These are of course now known as electrons. Gamma (γ) rays did not respond to even the strongest magnetic bending fields and are now known to be a form of electromagnetic radiation of shorter wavelength than X-rays.

The concept of conservation of mass leads to the only equation which appears explicitly in this book: $E = mc^2$. This "famous" equation relates mass to energy, with the proportionality being a very large number. A small amount of mass lost converts into a large amount of energy since, in SI units, $c^2 = 8.99 \times 10^{16} \approx 9 \times 10^{16}$. Unfortunately, the single most dramatic and destructive illustration of this conversion is in nuclear weapons. That dropped on Hiroshima, Japan, in 1945, converted about 0.7 g of matter, about the mass of a metal paperclip, to energy. Since the SI mass unit is the kg, this is $0.7 \times 10^{-3}$, so $E = (0.7 \times 10^{-3})(9 \times 10^{16}) = 6 \times 10^{13}$ in SI units, i.e., joules. One kiloton of energy (equivalent to blowing up one ton of TNT) is 4.2 TJ or $4.2 \times 10^{12}$ J, so simple division shows that the Hiroshima bomb released about 15 kilotons of energy. Current nuclear submarines carry warheads of about 30 times this energy, with a destructive force hard to imagine. Compared with the mass of reactive ("fissile") material in a warhead, the actual mass changed to energy is small. This is true even in individual steps in a decay as pictured by Mme. Curie, but it is nevertheless convenient to talk in energy equivalents of mass, basically, units arising from the rearrangement $m = \dfrac{E}{c^2}$. This both provides us with conveniently sized units, and allows a direct association of mass conversion and energy released. Electrons are deviated more than alpha particles by magnetic fields since they have half as much charge (but negative) but about 1/8,000 the mass. The speed also enters into the bending force in that faster particles feel more force, but mostly we note that electrons have a very small mass compared with particles found in the nucleus. This mass in SI units is $9 \times 10^{-31}$ kg.

It is quite clear that a more convenient mass unit would be helpful, but energy units also need some attention when we speak of the subatomic domain (as will be discussed in the chapter about the solar wind). Moving one electron through a potential difference of 1 V (volt) will cause it to gain $1.6 \times 10^{-19}$ J. This type of energy gain is common in small circuits like those of flashlights or LED indicators which can be powered by batteries totalling a few volts "electromotive force". Of course, to do perceptible work, like making light, many electrons must be moved. The number of single charges needed to make the common value of 1 A in such a circuit is in fact one divided by the electron charge, which is, as mentioned above, borne on $6.24 \times 10^{18}$ electrons. If we regard such a tiny amount of energy as typical and useful in discussing everyday life (which includes all of chemistry, based on the electron changes in atoms which typically involve electrons moving through potentials measured in low numbers expressed in volts), then it will not surprise that numbers involving nuclear mass change end up large even if this "natural" unit is used. To give a more precise value, we regard $1.602 \times 10^{-19}$ joules as a new unit called the "electron volt" (eV), but mass destruction even of electrons involves about a million of these energy units. The electron mass being $9.1 \times 10^{-31}$ kg, its energy equivalent is this value times $c^2, \approx 9 \times 10^{16}$ to get $8.2 \times 10^{-14}$ J. In turn, dividing by one eV to express it in those units, we get $5.11 \times 10^5$ eV as the mass energy of an electron. This is more conveniently expressed as .511 MeV, about half a million electron volts. Since this is an energy, we sometimes speak of the mass itself as being this energy divided by $c^2$, so that the electron mass is 0.511 MeV/$c^2$. While this is more correct, it is fairly common just to use the energy-equivalent values when speaking of masses, i.e., in this case, just say 0.511 MeV. For clarity, we will avoid this possibly misleading shorthand below: MeV will be an energy unit and MeV/$c^2$ a mass unit. In very loose talk, we regard the mass of an electron as being about half an MeV/$c^2$ and nucleons (either proton or neutron, they have about the same mass), which are about 1836 times more massive, as having masses of about 1 GeV/$c^2$.

The masses just discussed are those associated with the mass energy of the particles themselves. Total energy can include also that due to being in fields (potential energy) and from motion (kinetic energy). The equation $E = mc^2$ is associated with the "special" theory of relativity, which, in Einstein's original conception, tied together light and motion, or more directly, electrodynamics and motion. The general theory of relativity, which was discussed in Chapter 2 as an invisible "force", deals with gravity. They share a common way of depicting space in four dimensions, but, in special relativity, space is "flat". Effects of special relativity generally involve expressions involving the speed of a body over the speed of light, squared, as can be deduced by extending our discussion of light clocks a bit further. In parallel with the idea that mass destruction, being proportional to $c^2$, makes values very large, *division* by $c^2$ makes special relativistic effects small in the solar system. Those effects include apparent shrinkage of objects in the direction of motion, changes in time as we saw in a moving frame as opposed to that in the frame regarded as stationary, and increases in apparent mass of moving bodies. The mass energy we have been discussing is often called the "rest mass". We noted that chemists of the nineteenth century arrived at the conclusion that masses do not change in chemical reactions.

Are nuclear reactions different in some way? No, there is indeed mass change in chemical reactions too, but far too small to have been detected by even the best analytical chemists of the nineteenth century. This is also related to how small relativistic effects are in everyday life, at least as far as the electrons involved in chemical reactions are concerned. These are the outermost electrons, not moving very fast. The inner electrons in heavy elements can have major relativistic corrections needed. A further result of special relativity and the equivalence of mass and energy is that total annihilation of matter is possible. Due to other quantities that must be conserved but are beyond the scope of this discussion, this generally is through interaction with antimatter. In the "outer" invisible part of the Solar System, antimatter is not widely found, although there is antimatter inside the Sun, with any given particle of it not existing for very long.

More specifically, in the Sun's core, where fusion reactions take place, one part of the reaction chain generating solar energy involves annihilation of positrons produced by rapid nuclear decay. The positron is the antiparticle of the electron, and in the solar interior there are abundant electrons, so positrons are rapidly annihilated. The "lucky" electron one links up with, along with the positron itself, become "pure energy" in the form of gamma rays, with the total energy that of the original particles (which is mostly rest mass energy). The gamma rays are not directly observed since they are deep within the Sun and simply heat the local matter. However, we should not regard antimatter annihilation as an energy source found only in science fiction, since about 10% of the Sun's energy output channels through this form of matter destruction.

## 4.6 DECAY CHAINS

A brief look at the disintegration chain of radium will show us some aspects of radiation in space, which is considered in a later chapter, but we need to be guided by an advance view of what is *not* there. Radiation in space is dominated by protons and neutrons, and these are not common features of "radioactivity" as discovered by the pioneering physicsts. These nucleons are important components of space radioactivity, so why do they not have a relation to radioactive elements? We need to look in more detail at decay or disintegration chains that do characterize Earth-bound radioactivity. They are easier to understand than chemical or nuclear reactions, which involve two or more "ingredients" interacting with each other, possibly operating in both forward and reverse directions at the same time. At any step in a decay chain, only one nucleus needs to be considered and the reactions proceed in only one direction.

The timescales in a decay chain determine which products survive and in what proportion. These timescales are commonly given as *half-life*, the time after which half of the original material remains after the rest has decayed. It can hardly be surprising, in view of Mme. Curie's diagram, that most of a decay chain consists of emission of the particles depicted, alphas and betas, with the non-deviated gamma rays being electromagnetic radiation. The effects of emission of these radiations differ in what they carry away: alphas are helium nuclei consisting of two protons and two neutrons, while betas are electrons. Gammas are not particles in the usual sense, although of course wave-particle duality gives them some particle-like aspects. They are energetic enough that single photons can be detected, but they have no rest mass. Nuclei, being made of neutrons and protons, have masses that are proportional to their *(atomic) mass number*, basically being that number times the *atomic mass unit*, which in turn is close in mass to that of a proton. However, the chemical properties are essentially all due to the charge in the nucleus, which is the number of protons, or *atomic number*. Perhaps a bit confusingly, the atomic number is given the symbol Z, while the mass number is designated A. Historically, this comes from use of the German word "*Zahl*" to denote the number of protons, while it seems A is for "atomic".

The powerful electric force from protons in the nucleus attracts electrons, which arrange around it to neutralize its charge, but are subject to quantum mechanical rules. If we add to that the statement that they may be affected by other atoms being nearby, we have summarized the entire field of chemistry in two sentences! Chemical properties as determined by the number of electrons and their arrangement are summarized in the *periodic table*. They hardly change if a nucleus has different numbers of neutrons. As a result, chemists place objects that physicists would regard as different at the same place in their most important scheme of arrangement. From the Greek, "being in the same" (*isos*) and "place" (*topos*) can be fabricated into the word *isotope*, which is what variants of an element having different numbers of neutrons are called. The role of neutrons is to add to the nuclear "glue" involving the *strong nuclear force* (in its secondary manifestation) to add to that of protons in holding the nucleus together. As mentioned above, the strong force as it operates in binding nucleons to each other is similar to the forces holding molecules together in the sense that it is indirect. Molecules can involve atoms which have no net charge. *Bonds* allow them to form low-energy configurations even in cases where the constituents are not charged. This aspect of the strong force, much like interatomic forces allowing certain molecules to form, has a very short range. The electromagnetic force, on the other hand, falls off with the inverse square of distance (as does Newtonian gravity), but extends far.

If we consider the simple case of bringing two protons together, their repulsion will always dominate. Even if they are made to come close, the nuclear attractive force will not ever become large enough to dominate. As a result, the element with atomic number 2, but no neutrons, which would be a form of helium, is not stable at all. If one can contrive the situation to get a neutron into the mix (and, of course, bringing three particles together is much less likely than just two) then a stable form of helium, helium-3, can arise. The three particles are then regarded as "bound". There is energy emitted in whatever reaction brings nuclei into a bound state, and that energy would have to be supplied again to unbind them. Thus, we regard the energy of a bound nucleus as being negative, much as electron energies are negative when they are bound to a nucleus. Similarly, the energies of orbiting bodies are negative with respect to what their energy would be if very far from each other. Bound systems can be regarded as being in an "energy well". The "usual" form of helium has two protons and two neutrons, and they hold together very tightly, so this form is found most frequently, and indeed the alpha particles emitted from nuclei are in this very tightly bound form. Nevertheless, helium-3 is also a stable isotope, although easier to break apart than helium-4. To distinguish such isotopes, standard notation puts the mass number as a preceding superscript and the atomic number as a preceding subscript, along with the symbol (a letter or combination) for the element, so that helium-three and "standard" helium are respectively $^3_2\text{He}$ and $^4_2\text{He}$. This is just a bit redundant in that the atomic number already specifies the element, but it is also useful to think of names, not just numbers. We will routinely refer to isotopes by name and mass number, or symbol and mass number, e.g., helium-3 or He-3, and sometimes $^3_2\text{He}$, if it is also handy to be reminded of the charge.

Alpha emission decreases the atomic number by two by removal of two protons, and the mass number by four since a total of four nucleons are removed. See, for example, radium decaying to radon at the top right of Figure 4.19. Once the Curies isolated radium as a chemically separable element, it would have been decaying to radon gas with the fairly long half-life of 1,600 years. The difficulty of such extraction is why Marie Curie received the Nobel Prize in Chemistry in 1911 to supplement the part of the Nobel Prize in Physics awarded in 1903. Radium-226 is essentially the only natural radium they could have extracted from uranium ores, due to its long half-life. Other isotopes of radium have very short half-lives, so they decay rapidly and are present in ore in even smaller amounts than radium-226. The radon-222 gas it decays to would mostly have been trapped in the radium metal and in turn decayed with a half-life of 3.86 days to form polonium (named for Marie Curie's homeland). With a short half-life of only 3.1 minutes, this in turn decays to the radioactive Pb-210 form of lead. Its decay is not by alpha emission but rather by beta emission. Such removal of the negative charge of an electron *increases* the atomic number by one, but since electrons are very light, does not change the mass number. Note, however, that neutron number is on the vertical axis of the figure, and that does change by −1. Both alpha and beta decays are visible in the entire radium decay chain shown. Radium, in turn, comes through several decay steps from uranium-238, with that upper portion of the decay shown in Chapter 8. Whereas radium-222 has a half-life of 1,600 years, the half-life of uranium-238, the most common isotope of uranium, is 4.5 billion years, about the age of the Solar System, so that it may be considered to be primordial, that is, present since the formation of the Earth. Gamma emission changes neither atomic number nor mass number: a gamma emitter remains the same isotope much as a radiating atom remains still that type of atom (since photons carry no charge). As such there are no gamma emission steps *shown* in the decay chain, but the decays can leave nuclei in states with excess energy that can be emitted as gamma radiation. For example, the decay of radium to radon can leave the latter in an excited state that relaxes by giving off a gamma ray photon of energy 186 keV. Figure 4.19 therefore is an accurate picture of what comes out from sources that were initially radium: alpha, beta, and gamma from various elements arising from the decay chain. The alpha emissions, shooting out massive, energetic particles, take away most of the energy given off in the decay. It is useful to define the neutron number N to go with the effective "proton" number Z and make isotope charts like

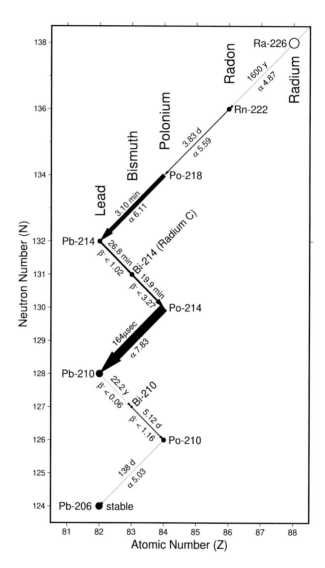

**FIGURE 4.19** Radium decay scheme. Arrow width indicates how fast decays take place, as also shown by the half-life above the line. Particle energies in MeV are shown below the lines. A large line going into an isotope with a small line out will result in buildup of that nucleus, and vice versa. The number of protons (Z) and of neutrons (N) characterize the isotopes shown (author).

Figure 4.19 in terms of these two numbers as shown (sometimes A and Z are used). Then decays, and to some extent reactions, can be shown as progressions on a chart much like they would be on a board game.

   In the historical development, there was a point in the early twentieth century when the basic constituents of atoms were thought to be electrons and alpha particles.

This view prevailed for a while, but researchers, with Rutherford leading the way, used alpha particles as projectiles to further explore the atom. This approach led to a startling discovery in 1909. Although co-workers later recounted that it was basically on a whim, Rutherford bombarded a sheet of gold with alpha particles. It must have been more than a whim, since data had to be accumulated laboriously by counting the individual flashes of alpha particles on a luminescent screen. Most alpha particles passed straight through, but some were deflected through large angles (Figure 4.20). Rutherford was famously quoted later in his life reflecting that this was "as though you had fired a fifteen-inch shell at a piece of tissue paper and it had bounced back at you". The usually observed small deflections were consistent with a model with positive and negative charges smeared out, so that no net electric force would be felt by the positively charged alpha particles in traversing the gold sheet, allowing them to pass through with little deviation. Large deviations are only possible if there is a large force to bend the particles and, in the extreme case, reverse their direction of motion. This force could only arise if there was a concentration of positive charge that was very small in size. This led to the concept of the atomic nucleus. Furthermore, the force being electric (gravity can be neglected and no other forces were known in 1909; at the small depth of penetration, nuclear forces are not effective), the exact same laws of orbital mechanics as for planets could be used, but

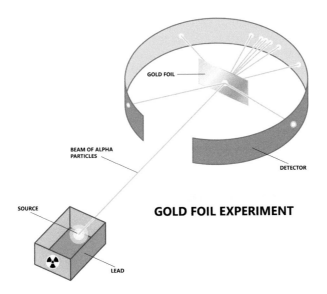

**FIGURE 4.20** In Rutherford's gold foil experiment, alpha particles from a source came out through a hole in a lead box to hit a very thin gold foil. While the vast majority passed almost directly through to hit a fluorescent screen, some were deviated to angles of 90° or more. Most passing directly through indicated that the scattering gold atoms were mostly empty space, while the few large deviations showed that a massive, positively charged nucleus must be at their centers. The scattering is a 3-D process, but it was only practical to make and observe a screen as shown (Shutterstock).

with a repulsive force. There arose hyperbolic orbits (see Figure 9.5), which also can arise with attractive forces if the relative speed is high (unbound orbits). With repulsive forces, only hyperbolic orbits are possible, but the mathematical form is the same as Newton discovered for inverse square law gravity. The accuracy of the solutions, which are at a limit in which quantum effects are minimal, showed that near the nucleus was empty space. The nucleus itself was very small (ca. $10^{-15}$ m diameter) and isolated in the center of an atom. The ratio of the size of a nucleus to that of an atom (about $10^{-10}$ m across) is about the same as that of a small seed to that of a football field.

One aspect of decay chains is that different elements, with different chemical properties, are involved. As mentioned, Marie Curie was awarded the Nobel Prize in Chemistry in 1911. She remains the only person to have received the Nobel Prize in two different sciences. Radium was a good starting point for other investigations since it could be isolated from ore and concentrated to give a high activity level. Various researchers realized that one of its products was a gas, which was found by realizing that opening windows in labs reduced their radiation level! Mentioned above in relation to noble gases, this gas, radon, has no stable isotopes, and can be a concern in closed spaces in regions where building materials contain uranium. Unlike the noble gas helium, it is denser than air and thus goes down into below-ground spaces. As a noble gas, it does not bond with other materials, but its radiation products, such as polonium, bismuth, and lead, are radioactive heavy metals which can. This new world of radioactivity features a colorless, odorless gas that changes into tiny metal particles! Since there are numerous radioactive elements in decay chains, activity profiles in a sample change through time. The final stage in the change shown is the decay of polonium-210 to stable lead with a half-life of 138 days. This makes this isotope of polonium particularly poisonous because, if introduced into the body, it can irradiate internal organs with alpha particles over a long period of time. Alpha particles are not dangerous external to the body since they are stopped by the outer (dead) skin with little effect, but if allowed to interact with live cells directly, they have catastrophic health effects. Rutherford was particularly enthusiastic about the alpha particles emitted by polonium-214, with a high energy of 7.83 MeV, emitted at a high rate. With these, atoms could be bombarded in an attempt to get close to the nucleus, now that it was known to exist. This approach brought a double surprise.

It was clear that a nucleus as a small body of large positive charge should repel its own constituents very strongly. This led Rutherford to propose that a new force must exist to overcome this repulsion and hold nuclei together: the first inkling of the strong nuclear force. In his 1913 book, "Radioactive Substances And Their Radiations", he wrote that:

> No doubt the positively charged centre of the atom is a complicated system in movement, consisting in part of charged helium and hydrogen atoms. It would appear as if the positively charged atoms of matter attract one another at very small distances, for otherwise it is difficult to see how the component parts at the centre are held together.

Even at this time he did not use the word "nucleus", but that soon came into use for the small, dense, charged center of atoms. The idea that hydrogen atoms might be in the nucleus went back to the idea of English chemist William Prout, at the beginning of the nineteenth century, that single hydrogen atoms might build up to form the heavier atoms. With only alpha particles, one would build up charge by twos and mass by four atomic mass units, so some protons needed to be tossed in to make whichever element was in question. Rutherford took his experiments further to try and probe nuclear structure, including for elements in gaseous form. For this, he used a scintillating screen far enough away not to see alpha particles striking it, but with nitrogen he noted an "anomalous effect", published in 1919. Flashes differing in character from those produced by alphas were deduced to be from hydrogen, making Rutherford conclude that hydrogen nuclei, which he later dubbed "protons", came out from nitrogen nuclei. Since the latter had a mass of 14 atomic mass units, he deduced them to be formed of three alpha particles and two H nuclei. Although he had induced the first artificial nuclear reaction, by observing only one product (the protons), he did not correctly deduce what the reaction was. He concluded that "the nitrogen atom is disintegrated…, and that the hydrogen atom that is liberated formed a constituent part of the nitrogen nucleus". Rutherford in 1919 had moved to Cambridge after a dozen years at Manchester University and directed work there by Patrick Blackett, with the more sophisticated apparatus of the "cloud chamber", which showed reacting particles' paths. From thousands of photographic plates, mostly in stereo, a few showed the paths of particles before and after reactions. In these, not only the incident alpha particle and the exiting proton could be seen, but also a nucleus moving out to make a "star". The nucleus remained and had *not* been "disintegrated". Rather than a disintegration, similar to an induced decay, a more complex reaction had taken place.

The original nucleus, originally at rest, could not be seen, but from the paths of the other particles, their energy and mass could be determined. The product nucleus in the case of alpha particles striking nitrogen was oxygen-17. This form of oxygen has eight protons, one up from nitrogen, and nine neutrons rather than the eight that characterize the far more common isotope, oxygen-16. It is nevertheless stable: Blackett stated that "If it is stable, it should exist on the earth…in such small quantities as to escape detection". Later measurements showed it to be present as about 0.04% of all oxygen at the Earth's surface. We can write the reaction involved as $^{14}_{7}N + {}^{4}_{2}He \rightarrow [{}^{18}_{9}F]^{*} \rightarrow {}^{1}_{1}H + {}^{17}_{8}O$, in which the first step resembles a reverse of alpha decay, increasing the mass of nitrogen by four and its charge by two, consistent with the absorption rather than the release of an alpha particle. However, the intermediate state of fluorine (F) is in an excited state denoted by an asterisk, and decays effectively immediately by release of a proton, transforming down by one mass and one charge number. Fluorine-18 does exist, but, if not in this excited state, decays by release of a positron in about a two-hour timeframe, to form stable oxygen-18. The net result was that rather than disintegrating, the nitrogen nucleus was changed into an oxygen nucleus by the impact of an alpha particle, with one proton released. Blackett's work was published in 1925, which is regarded as the true date of publication of the discovery of nuclear reactions. Since he had analyzed about 23,000 photographs, with about 270,000 alpha particle tracks over the preceding 5 or so years, it is rather hard to nail down an exact discovery date!

## 4.7 A BIG NOTHING

During most of the 1920s, the theory of "intranuclear electrons" was important in nuclear physics. In logical fashion, since Rutherford thought of nuclei as having in them the things that came out of them, which seemed to be electrons, alpha particles, and now protons, it seemed that those were the constituents of nuclei. By combining these as needed, any combination of mass and charge could be made. This was a necessary feature of a theory of nuclei, since isotopes had been discovered with the development of the mass spectrometer (Figure 4.21), which has a heritage in instruments now used to observe atoms in space. For example, the hypothesized oxygen-17 could be shown to be present as a small signal of higher mass if an environmental oxygen sample was analyzed. Since electrons are about 0.05% the mass of a proton, they hardly counted in summing the mass of nuclear constituents, but, if present, could adjust the charge. As a result, Rutherford thought that alpha particles might be present to give four mass units and two of charge, if they had two electrons tightly

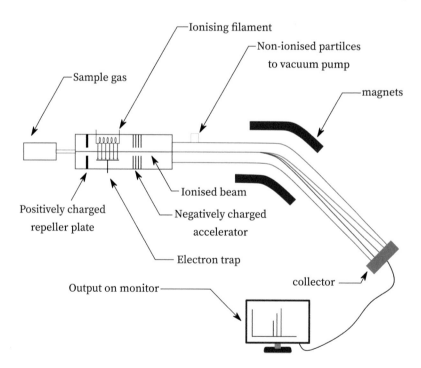

**FIGURE 4.21** Mass spectrometer. Given the same energy in the electric section at left, ions bend in the magnetic field more if of low mass. Much as a prism disperses light, ions are spread out by mass on the collector (Shutterstock).

bound to them. A handier concept for adjusting the mass by one without affecting the charge was to have protons with electrons tightly bound to them. Rutherford did know from scattering experiments that the nucleus had a diameter of only about $10^{-15}$ m, so "tightly bound" means *very* tightly bound. In fact, in 1920, Rutherford referred to these hypothetical proton-electron neutral particles as "neutrons", which of course is not what we now mean by that term. Rutherford also used this binding concept to explain alpha decay. Within the nucleus, alpha particles would have electrons tightly bound to them, but if they got out of a nucleus, with some interaction that stripped the electrons, they would be repelled vigorously as positive particles, gaining the high energies observed. Rutherford's reasoning fell afoul of developments in quantum mechanics, which led to the discovery of the final nucleon, the neutron in our current sense.

In 1927, theoretical work by Werner Heisenberg in Germany showed that having electrons tightly confined in the nucleus was an untenable concept due to the "uncertainty principle". Basically, the wavelike aspects of particles become important on a small scale, and the smaller their mass the larger the scale at which this effect dominates. Although specifying where heavy nuclei lie gives little problem, the much lighter electrons cannot be "pinned down" so easily. Knowing their position to be in a nucleus would imply a very high momentum for them. Since energy is proportional to the square of momentum (for low energies), intranuclear electrons would have fantastically high energies, certainly more than those of electrons observed in beta decay. Another development in quantum wave theory was the concept of "tunneling". If the nuclear force held alpha particles in the nucleus, how could they get out? In the classical view (Figure 4.22, top) the nuclear force would have held them in and made a "barrier" impossible to get over for alpha particles of the observed energy. Russian physicist George Gamow, working in Germany in 1928, noted that, in wave mechanics, the wave function probability for existence of an alpha particle "penetrated" or "tunneled through" the barrier. Even with a simple square barrier, he could do calculations relating the energies and half-lives for various alpha emitters with astounding accuracy over a huge range of half-life values. At the end of the decade, the concept of intranuclear electrons was finished.

Fortunately, experimentation proceeded apace, with groups in Germany and France noting first, that there was a strong emission when the light element beryllium was bombarded with alpha particles, and second, that this emission, which was uncharged, caused protons to be ejected from hydrogen-rich substances. Working in Rutherford's Cavendish Laboratory, James Chadwick found that the only consistent explanation of these two facts was the emission of "a proton and an electron in close combination, the "neutron" discussed by Rutherford", as published in 1932. He correctly deduced that the particles came from the overall reaction $^{9}_{4}\text{Be} + ^{4}_{2}\text{He} \rightarrow [\,^{13}_{6}\text{C}\,]^{*} \rightarrow \,^{12}_{6}\text{C} + ^{1}_{0}\text{n}$, where the last "element" is the neutron. Its mass could be deduced from the puzzling effect of proton ejection. Gamma rays, also uncharged, would not be able to transfer enough momentum to protons in a secondary target (like paraffin) to eject them. Only a particle of about the same mass could do so, and being about the same mass as a proton seemed to fit well the bound

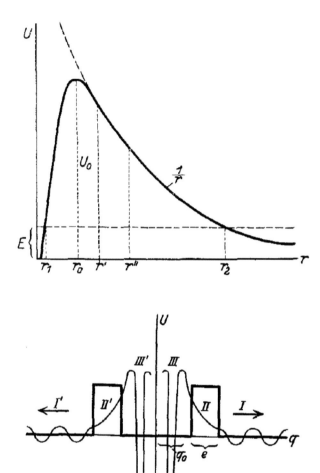

**FIGURE 4.22** Gamow explanation of alpha decay, impossible in the classic picture (top) but possible in wave mechanical barrier penetration (bottom) (Montage from Gamow, 1928).

proton/electron model. Although neutrons can themselves have beta decay, emitting an electron, they are no longer thought of in the untenable Rutherford form, but like the protons of Figure 4.1 have a complex internal form that can result in this without an electron having been initially present.

Chadwick was awarded the Nobel Prize in 1935, a short wait for work published in 1932. Neutrons offered a perfect way to explain the structure of all isotopes, since they could add one unit of mass "as needed" and no charge. Quite quickly, the deficiencies of the bound proton/electron model (only some of which were listed above) were realized and the neutron taken as a nucleon, of mass slightly greater than the proton, unstable outside nuclei and even inside some of them. Indeed, the discovery of the neutron allowed much of the subsequent progress of nuclear physics, especially when its role in allowing fission chain reactions was realized.

In the modern view of particle physics, which cannot be explored in detail here, although neutrons are nucleons, they (like protons) are not fundamental particles. From a practical standpoint, neutrons are not necessarily stable. The mass of the neutron is 939.57 MeV/c$^2$, while the proton mass is 938.28 MeV/c$^2$, and the decay is by beta decay, much as shown at some steps of the decay chains discussed, so an electron of mass 0.511 MeV/c$^2$ is also involved. This leaves 0.36 MeV of energy which can be released in such a decay, so it is energetically favorable. Free neutrons decay with a half-life of about 15 minutes. Beta decay, including that of neutrons, involves the weak nuclear force and will be revisited to explain a mysterious invisible radiation from the Sun that permeates the Solar System. Neutrons themselves are not present in large numbers in space, due to their short half-life, although they may arise through nuclear processes and even pose a hazard. Stable protons and electrons dominate the composition of the non-planetary parts of the Solar System, with helium nuclei the next most common "particle" constituents. We will revisit these facts in later chapters, giving the reasons that space radiation differs significantly from the type found near the turn of the twentieth century.

## 4.8 SUMMARY

The previous chapter dealt with the properties of electromagnetic radiation, and this one mostly with what are regarded as "particles". In modern physics, the theory of quantum electrodynamics is remarkably complete and successful, describing electromagnetic radiation and its simpler interactions. For particles, more complex theories are needed for a full description. However, in the remainder of our survey, it is adequate to regard matter as being made mostly of protons, neutrons, and electrons. There was a long path to get to even that point, and surprisingly, the initial study of radioactivity has little to do with the types of "radiation" detected in space. A key link is alpha particles, which are helium nuclei, but the origin of these in space is entirely different from their source on Earth in the minerals that gave rise to our knowledge of nuclear structure. X-rays were important in getting this discovery process going, so, arising out of the last chapter as a form of electromagnetic radiation, we have discussed their important role in the radiation environment of space. That will be explored in more detail later, and we will also explore the origins of all matter in the Solar System. First, we will detour slightly to look at some aspects of small bodies of the Solar System.

# 5 Beneath Our Notice

Ancient peoples looked at the Solar System and noted its brighter members. Although moving and classed as planets, most visible at night were starlike in appearance, if often brighter than the brightest stars. Stars themselves were placed into six ranked categories of brightness, from first to sixth. The brightest, or first magnitude stars, we now know to be roughly one hundred times brighter than the faintest. The eye, like some other human senses, has a *logarithmic* response, which permits a large range of sensitivity. Factors of brightness, rather than absolute or linear brightness, are detected by the eye. The ranking from brightest down, and logarithmic response, have left their heritage in the astronomical *magnitude system* still used today to measure brightness in the sky. Among objects seen in the night sky, the Moon is the brightest, attaining a magnitude of about −13, the second brightest being Venus at −4.4. The magnitude of the brightest star, Sirius, is −1.3. The summer star Vega has a magnitude of 0.0, so nominally bright stars have a magnitude of 0. Each magnitude *lower* is a factor of 2.512 brighter, so if we regard Venus as being −3 when near but not quite at maximum brightness, the 10-fold magnitude difference with respect to the Moon means that the latter, when full, is 10,000 times brighter ($2.512^{10}$). On the other hand, the faintest stars visible on a moonless night from a dark location have magnitude 6, so are about $2.512^6 = 251$ times fainter than the typical brightest stars also visible. Besides the classical planets detected by ancient peoples, two further wandering objects can be seen by the unaided eye but never were noticed. These are the planet Uranus and the asteroid 4 Vesta, the 4 designating the order of discovery among early minor planets. Figure 5.1 shows both objects visible in the same general direction in the year 2031. This scene also features the planet Saturn (magnitude 0.15) near the bright star Aldebaran (0.85, slightly fainter), which is in the V-shaped star cluster, the Hyades. This is the "head of the Bull" (Taurus), with Aldebaran being the "bull's eye", although actually its only relation to the cluster of stars is being lined up with them.

Uranus attains a magnitude of 5.6, so is brighter than the typical "faintest" star. It is near this magnitude in the figure (at upper left) and under a dark sky would look to be a star among many other faint ones. The asteroid Vesta can be even brighter at 5.1, but, in this figure, it has a magnitude of 8.1, well below the threshold of visibility. Uranus, being in the outer Solar System, is already remote and its distance from Earth does not change by much relatively. We are usually seeing it almost fully illuminated by the Sun as we look outward. It thus does not change brightness much when in the night sky. Nearby planets and asteroids vary their distance and viewing angle a lot, thus varying a lot in brightness. As here, Vesta is usually too faint to see and not often bright enough to be visible. In a dark night sky, the human eye can detect over 2,000 stars, and Vesta and Uranus, near the detection limit, went unnoticed among the myriad faint stars which do not usually figure in the outlines of constellations. Neither was very noticeable, so

DOI: 10.1201/9781003451433-5

FIGURE 5.1    Stars in the constellation Taurus as seen on March 19, 2031, with Uranus barely visible near the top. The path of Ceres in 1801 is shown in red (Stellarium, modified with overlay).

despite being visible to the unaided eye, Uranus was found only in 1781, Vesta only in 1807, both with telescopes. A new planet and several asteroids being found unleashed a new era in the discovery of celestial objects which had previously been unnoticed.

As noted in Chapter 2, dwarf planet 1 Ceres was the first "asteroid" discovered and is much larger than Vesta. By coincidence, it was in almost exactly the same position as Vesta in the figure when discovered and had a similar magnitude of 7.7. Ceres was discovered (in Figure 5.1 its path in 1801 is marked in red) due to systematic surveying with a telescope and never attains naked eye brightness. Despite its smaller size, Vesta has a high reflectivity or *albedo* of 0.4, reflecting 40% of light striking it. Ceres is more typical of asteroidal bodies in having an albedo of 0.1 (in fact, most are yet darker), so it never gets bright enough for the unaided eye to see.

Ancient peoples knew of comets, which can on rare occasions become brighter than Venus and have a tail covering far more area of the sky than does the Moon. They did not know whether comets were part of the atmosphere or celestial objects. There may have been a hesitation to accept comets as being celestial since they showed a type of change that was not thought to belong to the perfect celestial domain. They were ephemeral in time, rarely being seen for more than a month, and changed in brightness and shape. The Moon seems to have been forgiven for its phases, but other celestial bodies were not thought to change, so a comet was thought of like a cloud, changing within our atmosphere. Another sky phenomenon with a celestial connection was *meteors*. The connection to the word *meteorology* makes a link to the atmosphere, and, in this case, the origins of the phenomenon are in deep space, while the phenomenon itself is atmospheric. Some cultures did realize that stones sometimes fell from the skies, associated with very bright meteors called *fireballs*. Such stones are called *meteorites*. Western civilization did not rapidly make this connection. For example, American President Thomas Jefferson, a noted amateur naturalist from Virginia, supposedly said that he could more readily believe that Yankee professors would lie (about meteorite origins) than that stones could fall from the sky.

The ancients did not comment on light cones visible at certain times of year long after sunset or before sunrise, likely thinking them to be part of the rising or setting atmospheric phenomena. Extending along the ecliptic, these are called *zodiacal* light. Nor did the ancients apparently ever notice the small blur we now call *gegenschein*, at the antisolar point. These phenomena are located along the zodiac, or the apparent path of the planets, and we now know that they are manifestations of dust in space. From our modern perspective, we know in a general way that the Solar System formed from a cloud of gas and dust. Yet there is little gas and dust in it today. What happened to it?

These diverse topics, most seemingly unrelated, have a connection to minor bodies of the Solar System. In most cases, gravity is the present dominant force on them, but since the Solar System is very old, some very subtle effects have a surprisingly large influence.

## 5.1 ANCIENT ROCK

Earth is famed for its geological activity. Earthquakes and volcanic eruptions are dramatic events frequently in the news. Weather also is newsworthy when, as frequently happens, it affects human activities. The hydrologic cycle and effects of winds and water flow cause weathering, a change that is slow on the timescale of a human life. On a yet-longer scale, Earth's surface itself is processed by continental drift and volcanism, with the ocean floors constantly renewed by eruption (Figure 5.2) and in other places sucked into the depths. This high level of activity makes Earth a terrible place from which to study its own ancient history. So much of the evidence has disappeared!

Although some geological activity happens quickly, most such processes are slow relative to human lifetimes. Studies of geology in the late eighteenth and early nineteenth centuries brought about the view that slow, uniform processes produced most of the stacked sequences of rock exposed as sedimentary layers. In some places,

FIGURE 5.2    Kilauea volcano lava entering the ocean at Kapoho Bay, Hawai'i on June 4, 2018. Six months earlier, the author had rented a house visible at right and eventually destroyed, a rare instance of geological processes proceeding on a human time scale (US Geological Survey, Public domain).

these show regular layering over hundreds of meters, while a reasonable rate of accumulation, based on present mechanisms, is only on the order of a millimeter per year. One high cliff might evidence hundreds of thousands of years of accumulation, putting to rest calculations based on the Bible that claimed an age for the Earth of only a few thousand years. Such stratigraphic methods are based on the concept of *uniformitarianism*, that the present is the key to the past. Common slow processes of growth and decay left evidence from the past directly leading to concepts of very long timescales in geology. It was, however, in the twentieth century that scientific dating took a huge leap forward after the discovery of radioactivity. By studying how much remains of a decaying substance as compared with the daughter products arising from that decay, one can accurately determine the ages of rocks, be they of inorganic or organic origin. It has been established that Earth is about 4.5 billion (4.5 $\times 10^9$) years old so that the decay of uranium-238 to lead is a good means for dating older rocks. The lower part of this decay chain was shown in the previous chapter, and it will be revisited in the final chapter.

For relatively recent remains of once-living things or beings, the carbon-14 method is famed. Most carbon involved in life is of the most common form, carbon-12. Elements have their chemical properties determined by the charge in the nucleus, which comes from protons, their number thus being the *atomic number*. Chemists usually remember the atomic number associated with the element's *chemical symbol*, which in the case of carbon is C. As a reminder, the atomic number may, a bit redundantly, be specified at the lower left of the element symbol thus: $_6$C. The three important "heavy" elements, carbon, nitrogen, and oxygen, not only have an increase of one proton in sequence, thus are respectively $_6C$, $_7N$, and $_8O$, but also are most stable if they have an equal number of protons and neutrons. Neutrons add mass very similar to that of protons to an atom, but no charge. Charge determines the interaction of a nucleus with electrons to form atoms, so neutrons have very little influence on chemistry. Their role may mainly be thought of as providing more nuclear force to hold together a nucleus that really would like to fly apart since all of the positively charged protons would like to repel each other. Conversely, nuclei with too many neutrons tend not to be stable, as is the case for carbon-14. Neutrons being nearly identical in mass to protons, the *atomic mass* is the sum of their numbers, and is written at the upper left of the symbol. We deem elements with varying numbers of neutrons to be *isotopes* and write the mass number at the upper left. The most stable forms of "CNO" elements are thus, with equal numbers of protons and neutrons, $^{12}_{6}C$, $^{14}_{7}N$, and $^{16}_{8}O$. The number of neutrons, sometimes, as in Figure 4.19, called $N$, is the difference between atomic mass number and atomic number. A couple of minor points should be made before proceeding. In compounds, i.e., reacted substances, the number of a given element in a molecule is written at the lower right. Among "CNO", nitrogen and oxygen are gases at room temperature and pressure, but with a pair of atoms: respectively, they exist as $N_2$ and $O_2$ molecules. *Compounds* between different atoms may be written as, for example, $CO_2$, with two oxygen atoms bound to one of carbon (a 1 is not usually written in compounds), or $H_2O$, with two hydrogen atoms attached to an oxygen atom. Furthermore, the atomic mass may reflect several isotopes mixed in some proportion in nature. In cases where two or several isotopes are common, the mass number of an element as found in nature may not be nearly an integer. This caused some confusion in the history of chemistry but is now readily understood. When chemical substances form compounds under the conditions prevailing near the surface of the Earth, it is usually from the neutral state, with the compound having a lower energy than the original atoms.

Returning to carbon, with several isotopes, carbon-14 or $^{14}_{6}C$ is the most useful for *radiocarbon* dating. The mass number being two higher than that of the more common $^{12}_{6}C$, it contains two extra neutrons. It is formed in the upper atmosphere by cosmic ray interactions. Cosmic rays will be described in a later chapter; suffice it to say here that they cause nuclear reactions (at a relatively slow rate) in the upper atmosphere and near the surface of Earth, which release neutrons. Being neutral, neutrons are able to easily penetrate the nucleus of the most common element in air, nitrogen, causing the release of a proton and formation of carbon, which has a charge one less than that of nitrogen ($^{14}_{7}N + ^{1}_{0}n \rightarrow ^{14}_{6}C + ^{1}_{1}H$). Carbon-14 is radioactive due to having two extra neutrons as compared to the stable form of carbon, decaying back to nitrogen by emission of an electron from the nucleus (beta decay: $^{14}_{6}C \rightarrow ^{14}_{7}N + e^- + \bar{\nu}_e$). However, it is chemically nearly identical to stable carbon. It thus fully participates in the chemistry of life, in which carbon plays an essential role, once mixed by atmospheric processes into the biosphere. However, once a living object, be it animal or plant, stops interacting with the biosphere by dying, carbon-14 is no longer refreshed and starts to decay. On a timescale suited for studying organic historical artifacts (such as wood), it decays: in only about 5,730 years (see Figure 5.3), half of it is no

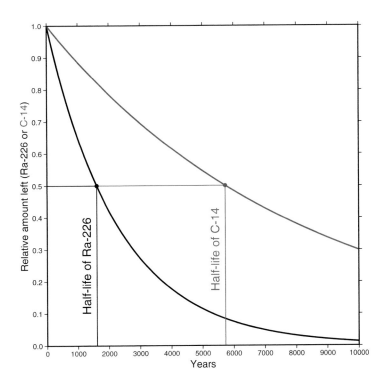

FIGURE 5.3    Decay of Ra-226 (black) and C-14 over 10,000 years (author).

longer present. This figure graphically presents the half-life mentioned before, in which half of any initially present radioactive material decays. If we compare an older sample from an organism with a modern one, the older one will have a lesser ratio of carbon-14 to carbon-12, since the former decayed away in it, while the latter did not. After another half-life of 5,730 years, half again of what was left after the first 5,730 years remains. And so it goes, so that the age of a sample may be determined by the amount of carbon-14 remaining in it, as detected by its distinctive beta signature. There is not a large proportion of carbon-14 in the first place, and, eventually, the amount in a very old sample becomes undetectable. Radiocarbon dating can be used only for rather young objects compared with the long geological timescale. It can be used for most of the time during which humans have been present on Earth, which is insignificant on the timescale of the Solar System. For dating the Earth itself, uranium's most common isotope, U-238, is more useful since it has a half-life comparable to the age of the planet.

The concept of "long geological timescale" arose out of parallel developments in astronomy, biology, and geology. As modern science developed, and especially in the nineteenth century, the concept of a very ancient Earth started to take hold. Ancient forms of life were found as fossils in deep sedimentary layers that clearly, from the point of view of uniformitarianism, took a long time to accumulate (Figure 5.4).

**FIGURE 5.4** Layered rocks on the Gulf of St. Lawrence in eastern Canada. Presently at sea level, these rocks were deposited in deeper waters about 375 million years ago in the Paleozoic Era. Studying fossils and correlating such layers played an important role in the realization of the great age of the Earth by geologists of the nineteenth century (author photo).

In addition to the mere deposition of a large amount of material, changes in the fossils due to evolutionary processes (another parallel development) were clear. Mountaintops could hold ancient marine fossils showing that they had been pushed up from the sea, clearly a slow process. The depth of ancient time started to become apparent, but how could this be quantified? Puzzles also arose. As mentioned in the previous chapter, nuclear burning in the Sun does not make it hot, rather it makes it long-lived. The Sun would be equally hot without nuclear fire. However, since it is losing energy at a prodigious rate, it would not remain stable. The *Kelvin time* of the Sun indicated that it could draw on slow gravitational collapse to remain alight for only about 10,000 years. This concept was named for Lord Kelvin, who was active in many domains and whose name is now the unit (kelvin) of absolute temperature. We will see later that the Sun's timescale could be "pushed" into the millions of years, but that is still thousands of times shorter than other lines of approach gave as an age for the Earth. By the end of the nineteenth century, all indications were that Earth was much older than the Kelvin time and that the Sun, at least as old, needed some other source of power than gravitational collapse. Conveniently, this is when radioactivity was discovered by Becquerel as outlined in the previous chapter.

Although the first radioactive element known was uranium (previously mostly used as an ingredient in ceramic glazes), its decay products included radium, which was comparatively short-lived, and which released huge amounts of energy per gram as compared with any chemical reaction. With a half-life of about 1,600 years, the most common isotope of radium releases this energy more slowly than many familiar chemical reactions (consider a flame), but for a very long time. The power output is less, but over time, the total energy released is far more. Clearly, nuclear processes could power the Sun, although the basic details did not get sorted out until the mid-twentieth century. As will be discussed later, those details include that nuclear reactions are the basic source of stellar energy, not simple decay. The existence of radioactive decay gave the ability to quantitatively date rocks on Earth in a manner similar to that mentioned for carbon dating, although using decays with much longer half-lives appropriate to the long timescales of geological as opposed to biological processes. The upper part of the uranium-238 decay will be discussed in Chapter 9, but uranium is plentiful and decays ultimately to lead, which has many stable isotopes. Since uranium turns into lead, the less uranium and more lead a rock contains, the older it is. Instead of the timescales of only thousands of years involved in carbon-14 decay, some isotopes of uranium have a half-life in the billions of years. When techniques were properly calibrated and the details of decay chains taken into account, the oldest rocks on Earth were found to be about four billion years old. The Sun must, by inference, also be at least that old.

From energy sources known in the early twentieth century, the Sun would not have been able to have put out energy at its present rate for as long as the age of Earth. This problem was not resolved directly by radioactivity. Barring the interior being much different from the surface layers (and such things as its known mean density precluded this), there were not enough radioactive materials in it to power it by decay. In the 1930s, nuclear reactions that released energy through splitting the atom (fission) were discovered. Sadly, the first practical application of this release was in war, and, by the early 1950s, it had also been demonstrated, although not on civilian populations, that fusion, the reaction of light elements to form heavier ones, could be even more powerful. By the 1950s, it was worked out how fusion could power the Sun based on the fact that it had plentiful hydrogen. Some concerns and details of this process will be detailed in a later chapter that discusses the *neutrino* by-product of these reactions, streaming invisibly through the Solar System. The age problem of the Sun was thus resolved since its plentiful hydrogen and the large energy release per atom in fusion could explain it retaining its present form for billions of years.

Developments in nuclear physics tied the Earth, the Sun, and the stars together in a consistent picture of great age. What could be said of the only material on Earth truly directly tied to the skies, meteorites? Surprisingly, almost all of them could be very precisely dated to exactly the same age, tightly clustered around 4.55 billion years, fully hundreds of millions of years older than any Earth rock. In turn, however, the question came to be asked as to where meteorites come from.

## 5.2 ROCKS FROM SPACE

Meteors and meteorites were discussed in ancient Western culture; however, by the time that they had passed through the Aristotelian worldview, their true nature had been obscured. The argument may be made that other cultures had proper interpretations, allowing them to conclude that stones actually did fall from the sky. In the West, the evidence was fragmentary and the approaches diverse. The Roman philosopher Lucius Annaeus Seneca discussed many "Natural Questions" in a surprisingly modern way in his book of that name written in 63 CE. Very near the beginning, he discusses "fires" in the sky which he relates to lightning. Of bigger fires or "long stars", he mentions that they may do damage on Earth, but again this is explained by an analogy to lightning. He appears to use his own observations, Aristotle, and likely secondhand accounts to make a consistent explanation, and clearly had seen bright meteors or fireballs, but likely never witnessed a meteorite fall. His near contemporary, Gaius Plinius Secundus (usually known as Pliny the Elder) also had a modern bent in his research as reflected in his attempt to document the eruption of the volcano Vesuvius as it destroyed Pompeii in 79 CE, which led to his death. His "Natural History", like Seneca's work, used his own but also other sources. One of them is extrapolated work of Anaxagoras, the Greek philosopher who lived about 500 years before his time. Anaxagoras had knowledge of a large meteorite having fallen during the daytime in Aegospotami ("Goat River") in what is now Turkey. From the date given in his writing, the event took place about 467 BCE. The story gets a little murky in that Anaxagoras was said to have predicted it. While Anaxagoras may plausibly be credited with deducing the cause of eclipses, prediction of a meteor fall was then certainly not possible. It has only very recently been possible to discover meteoroids in space and predict their impacts. In any case, Anaxagoras was greatly influenced by the event, to the extent of making a connection between Earth and the heavens, claiming that based on this evidence, things like the Sun, Moon, and stars must be made of stone like the meteorite. This meteorite fell during the day, so he associated it with the Sun. Its light, and that of other bodies, was inferred to be due to them being very hot. In view of the Aristotelian separation that dominated ancient Greek thinking, this heretical idea did not persist and indeed was viewed as ridiculous. Pliny, however, knew of revered meteorites, and had even seen one, and thus claimed that the fall of stones from heaven could not be doubted. Despite the importance of his "Natural History", the Aristotelian view held that if stones fell from the sky, it must be because the wind had picked them up from their natural place on the surface of the Earth, to which they returned. Comets were implicated as the cause of the winds, and the clear view of Pliny was obscured. The term "*meteoros*" generally meant "things in the sky" to yet earlier writers, and both comets and meteors were viewed as what we now would refer to as "meteorological" phenomena. "Meteoritical" denotes things to do with meteors or meteorites in modern usage. The term "meteor" in our modern sense only came into use in the fifteenth century, and the modern view of falling rocks is yet more recent and had a difficult birth.

The concept of meteorites as stones from the sky was difficult for scientists to accept, no less than Newton having claimed that interplanetary space was a void. A "Great Meteor", seen in England in 1783 (Figure 5.5), presaged by about a decade the realization that this could not be the case. Despite William Herschel having recently been appointed as the "King's Astronomer" and the event being observed at Windsor Castle, he does not seem to have presented an opinion on the matter. This

FIGURE 5.5 The Great Meteor of August 18, 1783, as seen from Windsor Castle, UK. Small fragments following the main body are shown here as one tail (Sandby, 1783: public domain, cropped).

possible lack of interest by an astronomer in what was then regarded as an atmospheric phenomenon was made up by his grandson, A. S. Herschel, who much later made important contributions to meteoritic science. Another scientist from what is now Germany, Ernst Chladni, possibly not knowing of this event but using earlier records, claimed in 1794 that iron masses could fall, and had fallen, from the sky. However, it was only after a daylight event in l'Aigle (Normandy) in France in 1803 that the investigative work of Parisian Jean-Baptiste Biot resulted in a widespread acceptance of the concept of iron or stones falling. As mentioned above, Thomas Jefferson as president of the US was said to claim that he could more easily believe that Yankee professors would lie about stones from the sky than that they could actually exist. Although Jefferson was a naturalist of some scientific renown, this

## METEORIC STONES.

THIS phenomenon in nature is calculated to awaken anxiety, excite wonder, and perhaps occasionally to produce apprehension. No satisfactory solution of the secondary cause of their existence has yet been given. We shall not attempt one. Their existence, and the circumstances of their fall, in the succeeding accounts, selected from many others which might have been given, prove that only a present and all-powerful Deity can protect us from the danger of sudden death, in the bright sunshine or in the lowering thunder tempest.

THAT meteorites do really fall from the upper regions of the air to the earth, can no longer be doubted, unless we are determined to reject the evidence of human testimony. These bodies have a peculiar aspect, and peculiar characters, which belong to no native rocks or stones with which we are acquainted. Their fall is usually accompanied by a luminous meteor, which is seldom visible

FIGURE 5.6   The first description in English of meteorites and their hazards, in 1823 (*The Monitor*, Vol. I, September, 1823: public domain).

comment appears to be apocryphal, reflecting the distrust of a Virginian southerner for northern liberals, thus possibly showing that some things never change. Since the writings of Chladni and Biot were in French and German, the first use of the word "meteorite" in English appears to be in 1823 in the former Boston publication "*The Monitor*" (Figure 5.6), which contained a mixture of religious and scientific articles. It had the wisdom to also state that the cause of the existence of such a phenomenon could not be given. It was only nearly 200 years later, in the decade of 2010, that a full and satisfactory explanation of how most meteorites come from the asteroid belt to Earth was developed, as will be recounted below.

## 5.3  MINOR PLANETS

By the turn of the nineteenth century, the ability to calculate the orbits of the known planets was well established, largely through the work of "*mécaniciens*" in France, as detailed in Chapter 2. A mechanical system with the gravity of the Sun as the dominant central force, yet the fine details well accounted for by interactions among the planets (dominated in turn by the largest, Jupiter), worked well mathematically. However, the search for a larger pattern remained of great interest. Kepler, obsessed with metaphysics, had sought patterns in music and geometry. Two hundred years later, such a pattern had not been verified. Even today, a basic search for patterns and their origin motivates astronomers. Our powerful modern observational techniques have found many planetary systems, while advanced computing allows us to simulate myriad varieties. Yet an overall pattern eludes us. As mentioned above, by the late

1700s, the preferred pattern for the Solar System as then known was called Bode's Law. It no longer has sufficient interest to even motivate giving details, beyond that it dealt with the spacing of the planets and pointed out that the zone between Mars and Jupiter was empty and perhaps should contain one. Telescopes and organized search methods were by then well enough advanced that several groups of astronomers were searching for a planet in this zone. On January 1, 1801, which technically was the first day of the nineteenth century, Giuseppe Piazzi, observing in Sicily, found a candidate object, which was promptly lost in the Sun's glare with few observations taken. The approximate path is shown in Figure 5.1 in red, from actual observations from discovery until February 11, 1801, when it entered evening twilight. The difficulty of determining an orbit from the short arc of observations shown can be imagined, and with the small aperture and field of view of telescopes then available, the object could only be "recovered" once it entered the morning sky if an accurate position was known. The observations, such as they were, made their way to the German mathematician Carl Friedrich Gauss, who applied statistical techniques much different from those of classical mechanics to predict where the object would be able to be seen when it later moved out of the Sun's glare. Gauss was so secretive that we still do not know exactly what he did, but it likely resembled curve-fitting methods that are now associated with his name. In any case, a few months later the new "Planet" was found near the position Gauss specified. Named Ceres after the patron goddess of Sicily, it became the prototypical asteroid, since several others were discovered soon after. The largest (although not the brightest) asteroid, Ceres is now regarded as a dwarf planet and was recently visited by the Dawn spacecraft, which is described in Chapter 9 from a technical point of view.

The term "asteroid" came about when it was realized that none of these bodies showed an appreciable disk. The planet Uranus, discovered (Chapter 2) by Anglo-German amateur astronomer Herschel in 1781, in any but the smallest telescopes could be seen to have a round disk. Indeed, many observers searched out comets, with the telltale sign of such a discovery being a fuzzy appearance, not point-like as a star would be. The Messier catalog much beloved of amateur astronomers was initially made to indicate objects that did not look like stars but were also not comets: now it has a fine selection of galaxies and nebulae to observe. Asteroids, although much closer than Uranus or most comets upon discovery, are so much smaller that they appeared starlike even under the high magnification. The "*aster*", meaning "star", part of their name, derives from this. Herschel, whose rise after the discovery of Uranus was, shall we say, astronomical, turned his attention, and the best telescopes in the world, to the problem of what the new objects could be, especially after the rapid discovery of a second one (2 Pallas) by Olbers in Bremen. In 1802, he published studies in which he compared their appearance as seen in his large telescope to that of a small disk some distance from his telescope, and perhaps unsurprisingly came up with inaccurate diameters. Nonetheless, they did appear to have disks, but no coma like a comet. Herschel concluded from this and their orbital properties that they were of a different nature than planets, and made up a name he rather liked: "From this, their asteroidical appearance, if I may use that expression, therefore, I shall take my name, and call them *Asteroids*". Most asteroid discoverers, perhaps in the vain hope of matching Herschel as planet discoverers, used the term "Planet", or perhaps begrudgingly "minor planet", for their discoveries. Despite being coined in 1802, the term "asteroid" came into common use only in the twentieth century.

Where Bode's law predicted a planet, by the end of the nineteenth century hundreds of small bodies were known. The force of Bode's idea was strong enough that for a while the idea that asteroids might be not the predicted planet, but its exploded remains, was in vogue. We now know that the total mass of the asteroid belt is less than 0.1% that of Earth. Although the primordial density may have been much larger, there is no evidence of a truly planet-size object ever having been in the "predicted" position. Although the asteroid belt has a low mass density, there are many asteroids. The rate of discovery greatly picked up upon the application of photographic techniques, and with ever larger telescopes, they started being found even on images taken for different purposes, meriting the term "vermin of the skies". Now, hundreds of thousands of asteroids are known, not all having orbits with the limits of the orbits of Mars and Jupiter, as shown in Figure 5.7. The basic approach to asteroid discovery with digital images is to look for change. If one re-examines Figure 5.1, the red dots for Ceres are in a different position each night. Most other things in the image are not, and some like Saturn have known motion. So, if one compares images, the things that change and have not been noticed before are usually new asteroids.

In contrast to comets, known since ancient times, asteroids do not generate large amounts of gas even though they are in zones of the Solar System in which they would be warm enough to do so if they had ice in them. Comets indeed have

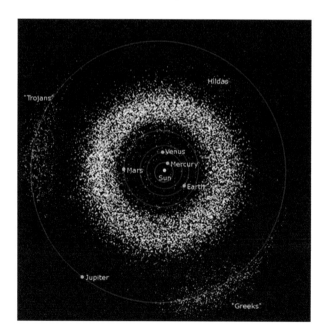

FIGURE 5.7    Location of known asteroids on July 6, 2006 (Wikipedia-Inner, 2006: public domain).

a nucleus which may be the size of an asteroid, but made of material susceptible to evaporation into space to take up a huge volume and reflect light or fluoresce. Already in 1802, Herschel commented that asteroids might be comet nuclei that had lost their volatiles, and some small proportion of them may indeed be. Most asteroids are instead rocky bodies, as verified by the numerous spacecraft that have recently visited them. Since asteroids are very small, they have little mass, both as individuals and in the aggregate. Unlike what Bode's Law implied, there is not nearly the mass of a major planet in the asteroid zone: it is an under-dense region of the Solar System. Also, the distances between asteroids and their small size mean that, in a far cry from popular depictions in movies, one would not see other asteroids even if standing on one (unless it happened to have a satellite asteroid). The asteroid "belt" hardly has any material in it. Nevertheless, it is the main source of meteorites. Minor impacts are enough to dislodge material from asteroid surfaces and place it on paths that intersect Earth to fall as meteorites. We do not feel that such "direct" delivery is the main way in which meteorites come to Earth, as will be seen below. In some very rare cases, however, catastrophic impacts on larger bodies can send meteorites from them to Earth: having lunar samples, we find that some rare meteorites are almost identical to them, and surely came from the Moon, ejected in crater-forming events. We also know through gas analysis that certain anomalous meteorites are from Mars. These are in fact our only samples from Mars at present. These two very rare groups are also the youngest meteorites, ejected at random times by impacts. As mentioned above, the vast bulk of other meteorites all have almost exactly the same age, about 4.55 billion years old.

The meteorite evidence leads to two main conclusions. Since they are about 4.55 billion years old, the Solar System is not older than that. Since their ages are tightly bunched, the formation of the solid bodies of the Solar System was, relatively speaking, very rapid. To understand the present nearly empty Invisible Solar System, we must briefly consider the time of its formation, when it was far from invisible. For that time travel, we start with analogous systems now in the process of forming.

## 5.4 OTHER STAR SYSTEMS AS A CLUE TO SOLAR SYSTEM FORMATION

HL Tauri is a faint star (visual magnitude 14.5) located at the top of the "V" shape of the Hyades cluster shown in Figure 5.1. The nearby bright star Aldebaran is about 20 parsecs (pc) away from us, the cluster itself is about 50 pc, and HL Tau is about 150 pc away, thus not part of the cluster. Much like the Solar System bodies in the same figure, these objects are not physically related but just lined up. Nevertheless, the stars in this part of the sky are on the edge of the Milky Way, where gas and dust are sufficiently dense to allow star formation at the present time. Although HL Tauri is likely no more than one million years old, the structures visible in Figure 5.8 appear flattened and in zones of rings of dense material with gaps between them. Some clumping is evident in some rings, suggesting the formation of planets. The image was taken by the Atacama Large Millimeter/submillimeter Array, using wavelengths

**FIGURE 5.8**   ALMA image of the protoplanetary disk around HL Tauri (ESO/NAOJ/ NRAO, 2014: CC BY4.0).

between about 1 and 3 mm. The array of dishes is at high altitude in a very dry region, allowing these wavelengths to get through the atmosphere, and by the use of many radio dishes over a large area effectively has a very large aperture compared to the short wavelength. In addition to very high angular resolution, mm wavelengths penetrate intervening gas and dust, which is essential in star-forming regions where such obscuring matter is prevalent. In this image, the second gap out is about 30 AU from the star, comparable to the distance of Neptune from the Sun. Millimeter wavelengths allow a "spectral index" to be calculated, so, even though the peak of the spectrum is not in the range of ALMA's detection, temperatures can be estimated, and decrease from about 300 K to 10 K from center to edge. The flattened structure, presence of gaps and clumping, and temperatures are thought to be very similar to those in the early Solar System. Of course, as already emphasized on page 1, our Solar System is basically transparent and no longer has a large amount of dust. We now examine its inferred history, including past and present "cleanup" mechanisms.

## 5.5 THE MESSY WORKSHOP

The dusty, messy formation of the Solar System set the stage for the cleaned-out Invisible Solar System we now live in. By analogy to protoplanetary systems, it was once filled with dust and gas, not the vast voids now separating the few planets. To understand why, we must briefly review the entire history of the Universe. Radioactive dating applied to meteorites shows that the solid materials in the Solar

System date from a little over 4.5 billion years ago, but the last 50 years of astrophysics have firmly established that the Universe itself is about 13.8 billion years old. The Sun was definitely not in the first group of stars to form. We defer discussion of the initial formation of elements until Chapter 8, but the previous chapter introduced the concept of nuclear reactions. The early universe of the "Big Bang" was hot and dense, rapidly cooling, and had a stage lasting about one minute during which neutrons and protons rapidly reacted to form helium and very little else. This stage was over at the end of the "first three minutes", and it was at least one hundred million years before stars began to form. Those stars must have been quite different from the Sun, and definitely could not have had planets similar to ours, for the simple fact that heavy elements needed to form planets were not formed in the Big Bang. More complex reactions, to build up heavier nuclei, could not take place in the rapidly cooling early universe. Much later, the earliest stars formed, with gravity pulling gas together until it got hot enough to have nuclear reactions. In present star formation, both atoms of heavy elements themselves, and aggregates such as the dust visible in Figure 5.8 and Figure 5.9 radiate energy from dense regions and thus cool them. The early stars could not cool effectively, lacking these heavy elements, so are inferred to have been larger than those presently forming. In the present universe, massive stars fuse hydrogen during most of their lifetimes (known as the "main sequence"), using catalytic cycles involving carbon, nitrogen, and oxygen (CNO). The first stars, almost entirely devoid of heavy elements, made helium from hydrogen (neutrons not being abundant after being used up in the Big Bang, and also being unstable as individual particles). Further burning processes, now characterizing the late life stages of "red giant" stars, eventually formed the heavier elements. For example, Aldebaran in Figure 5.1 is a

**FIGURE 5.9**   Hubble (left) and JWST (right) images of the Eagle Nebula star-forming region "The Pillars of Creation". The ALMA field of view is tiny compared to these images (NASA, ESA, Hubble Heritage Project (STScI, AURA): public domain).

red giant, which, having exhausted hydrogen in its core, is now burning it in a shell, and may have started fusing helium. Once helium burning starts in the more massive red giants, a sequence may follow that eventually produces the most common heavy elements if they are massive enough. These elements are recycled into gas and dust clouds in the final stages of stellar evolution, which usually includes the dynamical process of shedding outer layers (with mixed-in heavy elements since such stars have active convection to mix them) as a planetary nebula such as the Helix nebula shown in Figure 5.10. The keen planet and comet hunters mentioned earlier in the chapter gave this name since their disk-like appearance mimicked that of planets, yet they were "nebulous" or cloud-like. By the time that the Solar System formed at about the two-thirds mark in the age of the universe, those heavier elements and many more had been made in previous generations of stars.

The early generations of stars, especially the more massive among them, thus acted as "cauldrons in the cosmos" to brew up heavy elements. The most massive stars not only generate even heavier elements, but they also may spread them in a more spectacular fashion through supernova explosions. Very recent results suggest that the very heaviest elements, such as gold, may be generated in the collisions of neutron stars, which are a yet more extreme endpoint of stellar evolution than the white dwarfs characterizing the centers of the more common planetary nebulae. Totally ignoring the rules of chemistry, astronomers refer to every element heavier than helium as a "metal". Even now, low-mass stars of early generations are observed to be "metal-poor". On the other hand, recently formed stars have much recycled material from earlier generations, and thus are "metal-rich". Most of these heavier elements come from the most massive, short-lived stars of earlier times. Heavy

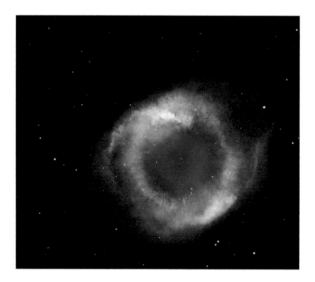

**FIGURE 5.10**   The Helix, a planetary nebula has expanded from a central red giant star now visible as a remnant white dwarf star (NASA, NOAO, ESA, the Hubble Helix Nebula Team, M. Meixner (STScI), and T.A. Rector (NRAO): public domain).

elements such as those present in dust in regions such as the Pillars of Creation shown in Figure 5.9 and also in our own Solar System have by now largely aggregated into planets. They are also found in our bodies, which after all arise from a planet. Calcium, aluminum, iron, and nickel, for example, were formed in very massive stars that underwent supernova explosions. Iron is a vital element for complex life, as it transports oxygen (also made in stellar cores) in our bloodstreams, while, of course, carbon and oxygen combine in our bodies to provide vital energy. In the words of Carl Sagan, "We are made of star stuff". Water is, of course, essential for life, and almost all hydrogen on Earth is bound up in it by being attached to oxygen to form $H_2O$. A light gas if not bound in a molecule, hydrogen would otherwise not be found on Earth. A smaller amount of it is bound to carbon, and, of course, those hydrocarbon molecules are essential for life. By mass, the Solar System has about 2% in elements heavier than helium, helium making up 23%, and the remaining 75% being hydrogen. Although our affinity for worlds like our own might make us think that the heavier elements are concentrated in planets, they are but dust compared to the mass of heavy elements in the Sun. Due to the complex structure of the heavier elements' electron configuration, the Sun's spectrum is dominated by absorption lines of heavier elements, as discussed in Chapter 3, and barely reflects its overall composition dominated by hydrogen. This deficit of hydrogen lines also partly reflects the temperature structure of its atmosphere, which is not hot enough to place the electrons of hydrogen and helium into states giving many visible lines.

The overall process of formation of heavier elements from primordial hydrogen and helium is known as *nucleosynthesis*. The original formation of H and He is sometimes referred to as "Big Bang nucleosynthesis" and then the heavier elements are regarded as coming from "stellar" nucleosynthesis. I do not discuss stellar nucleosynthesis in further detail here, but, since most of the content of interplanetary space came originally from the Big Bang, that will be discussed later.

The Sun, about 4.5 billion years old, is about halfway through its life, estimated to total about 10 billion years. However, it also formed after about 10 billion years of accumulation of nuclear burning processes in earlier stars, so that there was plenty of material of the sort that forms dust around. The proto-Sun, with its heavy element content by mass of about 2%, the rest being hydrogen and helium, was surrounded by the proto-solar nebula of similar composition, as we can now image in systems such as HL Tauri (Figure 5.8). Originally quite extended, the solar nebula collapsed, and much like a skater bringing her arms in during a spin, sped up. To continue collapsing, energy had to be gotten rid of. Emission by the numerous atomic transitions of heavy elements, and even by hot dust, helped with this. As the spin increased, the nebular cloud flattened out. Already, a basic element of the present Solar System was present, a flattened form with material spinning around the central condensation that would be the Sun. The next steps were quite chaotic, and, judging from the varied layouts of now known exoplanet systems, different outcomes would have been possible. What is clear is that there was accretion into large planets, possibly more than we now have, while the space between them emptied out. This process may not have been completed in the present asteroid belt, possibly because not enough material was there, or possibly due to the disruptive gravitational influence of nearby Jupiter.

## 5.6  WORLDS IN COLLISION

With planets formed, their orbits may not have been as regular as they now are. There is ample evidence of a period during which large bodies hit each other. Our own Moon is thought to have arisen from a collision of the early Earth with a smaller body, likely about the size of Mars. The material blasted into orbit rapidly re-accreted to form the Moon, originally much closer to our planet. We know that this large collision took place after Earth had formed a core, since not much of that dense material is found in the Moon. In the outer Solar System, the spin axis of Uranus being tipped on its side is not a possible outcome from the semi-orderly collapse of the overall solar nebula, so it was likely struck by another wandering large body. Scattering processes appear to have ejected an outer belt of objects now known as the Kuiper Belt, of which Pluto is one member. We now know that there are much larger Kuiper Belt objects even further from the Sun than Pluto, accounting for its demotion to the status of dwarf planet. Even more dramatic scattering surrounded our entire Solar System with a cloud of comets, the Oort Cloud, and perhaps even ejected some fraction of them from the Solar System entirely. We have recently been visited by a comet from interstellar space, and by another mysterious rocky object, likely both similarly ejected in the formation of other planetary systems.

A period known as the Late Heavy Bombardment left its mark on ancient surfaces in the Solar System, such as that of Earth's Moon (Figure 5.11). The largest clear crater in this image is 108 km in diameter, and craters of many sizes are seen, some very large to the extent that they triggered flooding by lava. Being much smaller than Earth, the Moon cooled quickly and has not had major geological activity late in its

**FIGURE 5.11**  The first US spacecraft image of the Moon, from Ranger 7 on July 31, 1964. Grid marks are artificial. Ranger provided much more detailed views as it approached to impact the surface (NASA/JPL).

existence. The dark, flat areas known as maria on the near side are pools of solidified lava where a lack of large craters attests to a rapid end to the period of heavy bombardment. Earth bears no known evidence of that bombardment due to its continuing geological activity. Much of its surface has been reprocessed many times through continental drift, so that only some old continental cores have rock even approaching the age of the Solar System.

Some recent theories of the early Solar System attribute the Late Heavy Bombardment to a major rearrangement triggered by resonance between Jupiter and Saturn. Such theories are largely explored through numerical modeling with large computers, but it is interesting to recall Kepler's interest in Jupiter-Saturn relations, and the concern with Newton's gravity theory resolved by the explanation of the Great Inequality by Laplace. Clearly, the largest planetary bodies could have had a major influence if they had acted together. Today, however, Jupiter's single influence is found to dominate.

## 5.7 VACUUM CLEANERS

The above description of the formation of the Solar System is necessarily brief. As Figure 5.11 shows, much can be inferred about the past by studying the surfaces of rocky bodies. With our focus on the present day, and on interplanetary space, it is clear that it went from being very dusty to being, and remaining, very transparent. Parents often tell their children that their rooms are not going to clean themselves, but in the case of the Solar System, this is what appears to have happened. Just how can the Solar System be so clean?

The solar upper atmosphere is blown supersonically into space, as will be explored in detail in the next chapter, but it is very tenuous and has little effect on solid bodies, even those as small as dust. This "solar wind" does not "blow" dust, but perhaps surprisingly, solar radiation, mostly in the form of light, can do so. The energy flux from radiation is about one thousand times larger than that of the solar wind, and it can act on dust grains in several ways. Photons carry both angular and linear momentum. This results in *radiation pressure* that can act on dust grains. Because mass trends as the third power of the radius, while cross-sectional area to absorb radiation trends only as the second power, the ratio of area to mass goes up for smaller bodies. This simple picture, a playoff similar to that used in in Chapter 3 to determine the blackbody temperature of planets, breaks down when the wavelength of light becomes comparable to the size of particles interacting with it. In practice, particles of about 0.1 micron ($10^{-7}$ m) diameter are most affected by radiation pressure. In this size range, the effect of radiation pressure is like "negative gravity" since both the light field and gravity fall off as the inverse square. The effect of radiation pressure is highest for particles in this size range since the mass to be accelerated is small compared with the force applied. Ironically, a popular scientific toy, supposedly demonstrating radiation pressure, does *not* do so. A "light mill", also known as a "Crookes radiometer" and named for the inventor of the Crookes tube that was so important in the discovery of electrons, X-rays, and ultimately radioactivity, will indeed spin merrily if placed in sunlight (Figure 5.12). One side of the

FIGURE 5.12   A "light mill" or Crookes radiometer. Purported to show the effects of radiation pressure, it instead shows subtle effects in rarified gas (Timeline, 2005: CC BY-SA 3.0).

vanes that spin is black and the other is reflective. The incorrect concept is that the black side absorbs momentum from light, and the reflective side gets twice the kick by reflecting it. Unfortunately, the effect is too small to produce the motion observed and instead is due to the interaction of remaining gas in the "vacuum" tube holding the vanes. Experimentally, this may be shown by making a very good vacuum, which causes the effect to go away! It remains true that for small grains, which are much smaller than the vanes ($10^{-7}$ m is much smaller than the diameter of a strand of hair), light pressure can overwhelm gravity in the emptiness of interplanetary space, blowing such small grains away.

For larger grains, the effects of absorption of energy to cause them to heat up may be important. They can radiate into space and there may be anisotropy in doing this, so that recoil from emission of radiation may act like a "jet" by having a preferred direction. Even for a perfectly isotropic small body, emission in the forward direction is blue-shifted (due to the Doppler effect) as seen in the frame of the Sun, whereas backward emission is red-shifted. Since blue light carries more momentum than red light, there is a net backward reaction force on the particle. This slows it down in its orbit, making it spiral into the Sun. This is called the Poynting-Robertson effect and is most effective for particles in the millimeter-to-centimeter size range. Interestingly, this may have been one of the only basic physical effects first published in a religious newspaper (in 1903) rather than in a scientific journal. The force is opposed to the direction of motion (green arrow in Figure 5.13) and is strongest near perihelion if the orbit is eccentric. One can think of the slowdown at that point

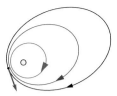

**FIGURE 5.13**   The Poynting-Robertson effect, by which a backward radiation force (green), strongest at perihelion, causes an original orbit (black) to spiral in and become more circular (blue, red) (author).

removing energy and not allowing the particle to climb as far out of the gravitational well as it could without this hindrance. Thus, the orbit gets more circular since there is more decrease in distance at aphelion than at perihelion. It seems counterintuitive, but as the particle falls in, it gains speed as shown by larger arrows on the orbits in the figure. Eventually, the orbit will be nearly circular and the drag force is distributed around it, simply reducing its radius until the particle is evaporated by the Sun. The timescale is about 100 million years for a 1-cm-diameter (gravel-sized) pebble in the asteroid belt. Looking back 4.5 billion years, this is a long enough time to clear out most debris, even though larger particles take longer.

A related effect initially seems quite subtle and is related to the Poynting-Robertson effect. If rotating in the prograde (counterclockwise) direction, an even larger body will warm when in the sunlight, and emit when that warmed area rotates to be in the direction opposite to the motion of the body (which we will consider to always be prograde). There will be a propulsive effect due to the reaction force on emitted infrared radiation. This will expand an object's orbit. If the body rotates in the retrograde sense, the emission will be against the orbital motion, contracting the orbit. For most asteroids, this so-called Yarkovsky force takes so long to act that it is not consequential. It would not be able to move even small ones into the Sun as the Poynting-Robertson effect can for gravel-sized ones. It can, however, cause asteroids to slowly move into zones in which they are affected by resonances, and these can change the orbital characteristics dramatically. We now discussion resonance in the asteroid belt, starting with its history.

## 5.8  KIRKWOOD GAPS

By the mid-nineteenth century, enough asteroids had been discovered (after Ceres in 1801) and their orbits were well enough determined that a statistical approach to their study could be taken. The American astronomer Daniel Kirkwood noted that, although asteroids were randomly distributed in distance from the Sun at any given time, their semimajor axes were not random. He published these results in a chapter in his book "Meteoritic Astronomy: A Treatise on Shooting-Stars, Fire-balls, and Aerolites" in 1867. Even with the small number of objects available for study at the time, 91, of which he excluded 5 at the ends of the distribution, he felt that he could infer that certain values of semimajor axis did not have objects associated with them.

Figure 5.14 shows that the evidence was sparse, but Kirkwood correctly pointed out that Kepler's third law gives a one-on-one relationship between semimajor axis and period, meaning that certain periods were avoided. Asteroids in the main belt, being inside the orbit of Jupiter, orbit with shorter periods, and he suggested that those with small number ratios, of which he cited 1/2, 1/3, 2/5, 2/7, 3/7, and 4/9, to the period of Jupiter, should be lacking. Small number ratios suggest a resonance, with periodic small but similar pushes building up to a large effect over time.

Kirkwood was inspired by, and suggested that an analogous effect took place in, the rings of Saturn. These rings (Figure 3.14, barely visible edge-on in Figure 5.15), which are the delight of many first-time telescopic viewers, had defied explanation since their discovery by Galileo in 1610. The Italian-born astronomer Giovanni Cassini used improved instruments at the Paris Observatory, funded by the Sun King, Louis XIV, to discover four moons of Saturn, perhaps unsurprisingly initially named after the King. He also noted a gap in the rings in 1675 which has borne his name since, clearly visible in Figure 3.14. The NASA Cassini mission named for him orbited Saturn from 2004 until 2017, allowing a thorough investigation. While the rings are a beautiful thing to behold even with a small telescope, they are quite bright. Saturn's obliquity, or tilt of the pole from being perpendicular to the orbit plane, is about 26.7°, comparable to the 23.5° which gives Earth its seasons. Moons tend to lie in the plane of the planet's equator (Earth's Moon is the only major moon not doing so), as do rings. Thus the rings of Saturn have different viewing angles

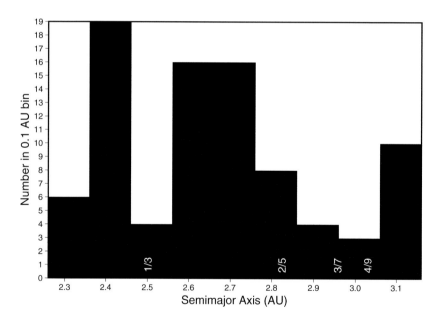

**FIGURE 5.14**   Kirkwood's original 1867 data supporting discovery of gaps in distribution of asteroids by semimajor axis. White figures are periods of asteroids relative to that of Jupiter (author, based on data in Kirkwood, 1867).

FIGURE 5.15 Saturn as seen on September 14, 1789, when the moon Mimas was discovered by William Herschel. Mimas is near the right edge of the thin, near-edge-on ring. Herschel also discovered the next moon to its right, Enceladus (Stellarium).

depending on the Saturn "season". For brief periods, twice in each orbit, they are viewed edge-on, and this is the best time to search for faint objects near Saturn without the glare of the rings. Taking advantage of exactly such a circumstance (Figure 5.15) on September 14, 1789, William Herschel turned his large telescope (4 feet or over a meter in diameter, 40 feet or over 10 meters long) to Saturn and discovered the moon Mimas. He also discovered the moon Enceladus in that year. Herschel generally observed the new moons as "protuberances" on the thin rings. The figure gives an exaggerated view of the brightness of the satellites, so looking for these little bumps on an otherwise nearly linear feature may have taken advantage of how the eye sees detail. He wrote extensively about his observations, which are all the more impressive for having been done at the top of the long ladder needed to use his large telescope. Mimas (Figure 5.16) is only about 400 km in diameter, which is about the minimum size for bodies to become spherical. It does have one very large crater (fittingly named Herschel), giving it the appearance of a "Death Star". The nature of Saturn's rings remained a mystery at this time, Herschel's view that they were solid being little more than an opinion.

Physics was applied to the ring problem by James Clerk Maxwell, later most noted for the theory of electromagnetic waves described in Chapter 2. In 1856, he won the University of Cambridge's distinguished Adams prize for showing that the rings could not be solid and instead must be made of individual solid particles. Later spectroscopic studies showed that they were largely made of water ice and obeyed Kepler's third law as orbiting bodies must. However, the particles are so small as never to have been photographed individually, even by the Cassini orbiter when very near them.

Kirkwood clearly thought of the asteroid belt as being much like Saturn's ring. Indeed, the chapter about gaps in his 1867 book is entitled "The Asteroid Ring Between Mars and Jupiter", and in it he states, "As in the case of the perturbations of Saturn's ring by the interior satellites…Jupiter's influence…would…form gaps in the asteroid 'ring'". He went on to say that, under the influence of Jupiter, an asteroid's

**FIGURE 5.16** Saturn's 400-km-diameter moon Mimas, as seen by the Cassini Saturn orbiter. The large crater Herschel (130 km diameter) gives the appearance of a "Death Star" (NASA/JPL).

orbit would become more eccentric until that zone was "left destitute of matter, like the interval in Saturn's ring". The "distribution of matter...as in the case of Saturn's rings, it would probably break up into a number of concentric annuli." If we seem to have skipped something here, so did Kirkwood. He was clearly referring to the Cassini gap and another one, more recently discovered. The Cassini gap is due to a 1/2 period resonance of particles at that radius with Mimas. However, Kirkwood seemed reticent to publish, and what he mentions in his book rather unclearly, he had not published elsewhere. He did so in the Monthly Notices of the Royal Astronomical Society in 1868, pointing out explicitly the Mimas resonance (and noting, furthermore, that Tethys accounts for a 1/4 resonance). The overall title "On the Nebular Hypothesis, and the Approximate Commensurability of the Planetary Periods" referred to Laplace's hypothesis about the collapse of a proto-solar nebula. This concept had been suggested even earlier than Laplace, by philosopher Immanuel Kant in the mid-eighteenth century. We have discussed above that even young planetary systems are flattened, and clearly thin interstellar material must collapse to form a planetary system. As it does, it likely has angular momentum, at a minimum caused by galactic rotation, but more likely from turbulence on a very large scale within a collapsing cloud complex like those seen in Figure 5.9. This angular momentum will cause an increase in the rate of rotation and a flattening so that, in a sense, it does resemble a "ring", but likely at no point as thin as Saturn's.

Kirkwood's view of the origin of the Solar System is very close to our present one, but in the nineteenth century was out of favor despite the reputation of Laplace. For one thing, it harkened back to the vortex theory of Descartes, which was repudiated by Newton. Kirkwood used the formation of gaps as overall support for the nebular hypothesis. In 1883, Kirkwood revisited his discoveries, seemingly due to competition from Meyer in Switzerland, who published follow-on work acknowledging him, and from a General Parmentier in Paris, who did so without acknowledgement. He refined how sharp the gaps in the asteroid belt were, and reaffirmed that he was the one to identify the physical cause of the divisions in Saturn's ring. Kirkwood seemed to have a reticence about publishing that would not have served him well in the modern "publish or perish" era. When he retired and moved from Indiana to California, aides noted that he had several unfinished manuscripts, which he ordered burned.

The Minor Planet Center in the USA, and several other institutions worldwide, now track asteroids and maintain databases. Discoveries are reported and processed quickly. In 2020, there were about 650,000 asteroids in the MPC database, whose characteristics in the space of semimajor axis and eccentricity ($a$-$e$) are shown in Figure 5.17. The 1/3, 2/5, 3/7 gaps identified by Kirkwood are clear, while his identification of a gap at 4/9 period ratio seems suspect. The eccentricity of an asteroid's orbit is very important in that both it and the semimajor axis determine how close and how far from the Sun, and thus from the inner planets, an asteroid goes in its orbit. Planet-crossing lines in $a$-$e$ space are shown for Earth and Mars in the figure. Any body above the Earth-crossing line can potentially hit or have a gravitational interaction with our planet: a similar consideration applies to those above the Mars-crossing line. It is not a coincidence that the upper limit of the distribution of asteroids in the inner part of the main asteroid belt is near the Mars-crossing line. Also, since Mars has a rather elliptical orbit (e = 0.0935: recall the issues Kepler had due to this), the nominal Mars-crossing line is wider than shown. Asteroids usually start interacting with Mars once they start crossing its orbit. The word "start" is used here since we are about to explore how asteroids move around. However, note that, because planet-crossing asteroids can hit or interact strongly with the planets whose orbits they cross, Figure 5.16 does not show a static situation: the given individual asteroids above the planet-crossing lines are not going to be there for long, at least on astronomical timescales!

Objects coming closer than 1.3 AU from the Sun (about halfway between the Earth- and Mars- crossing lines in Figure 5.17) are considered to be Near-Earth Objects (NEOs). Of these, the vast majority are asteroids. Figure 5.18 shows the logarithm of the numbers in various classes of known NEOs as they were discovered in this century. The total number has risen from approximately 1,000 to over 32,000 in 22 years, with still a high rate of discovery, including for objects greater than 140 m in diameter. Usually, the diameter is estimated from the brightness and distance with

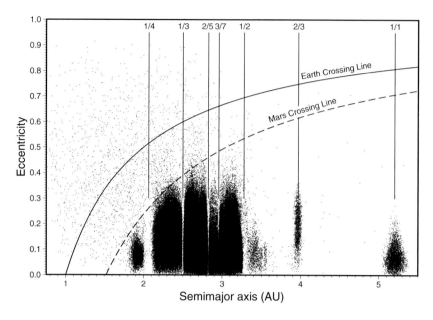

**FIGURE 5.17** Eccentricity and semimajor axis of approximately 650,000 asteroids. Kirkwood gaps and resonances, which cause accumulation of asteroids, are marked by vertical lines with ratio of the period at that semimajor axis to that of Jupiter (author, data from Minor Planet Center).

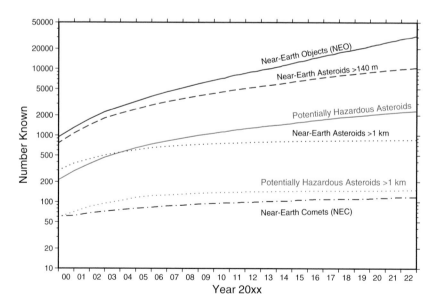

**FIGURE 5.18** Discovery statistics of Near-Earth Objects in the twenty-first century. The total number of objects is rising rapidly, but essentially all Near-Earth Asteroids larger than 1 km in diameter seem to have been found (author, data from NASA/JPL).

an assumption about reflectivity. Since about 2005, very few Near-Earth Asteroids (NEAs) more than 1 km in diameter have been found. Barring exceptional circumstances, for example, objects largely staying inside Earth's orbit and this being only in the daytime sky, the population of larger NEAs is thought to be well-surveyed and known. Smaller ones represent a very large population, which is not yet fully sampled.

Given that the population of NEAs is necessarily dynamic since they can hit planets, where do new ones come from? It turns out that simple collisions in the asteroid belt cannot give a sufficient rate of production of new asteroids, and more refined studies such as the time since last impact (determined by various means) indicate that the lifetime of such material would not match the ages inferred. Also mentioned above, the asteroid belt is a zone of deficiency of mass in the Solar System. Asteroid collisions are rare events. This, coupled with the low probability of sending material Earth-ward (in part related to low collision speeds among the asteroids, which are all traveling prograde, thus more or less in the same direction), means that a different mechanism must be found.

Already in 1868, Kirkwood wrote that the orbit of "A planetary particle at the distance 2.5 … would become more and more eccentric". Seeing this very clear gap at 2.5 AU in Figure 5.16, we know that indeed the orbits in that range did become so eccentric that they crossed planetary paths and were removed by interaction with the inner planets (or possibly went into the Sun). Modern calculations, such as those the results of which are shown later in this chapter, indicate that the increase in eccentricity for an exactly resonant object takes place on a timescale of thousands of years, very rapid indeed by astronomical standards. So, once an asteroid is in a resonance, it can get "pumped up" to enter the inner planet region. The problem becomes one of how to get asteroids into such resonances, which is also discussed below.

Apart from very close encounters that will be discussed in Chapter 8, gravitational interactions with planets do not work well to change the semimajor axis. If it is changed by a small amount in one "encounter" then usually the next encounter will have the opposite effect. We use "encounter" in quotes since the resonant interaction may take place at a fairly large distance from a planet, often Jupiter. In the final chapter, we will show that on close passage by a planet, a small body's orbit may be completely changed, including its semimajor axis, which directly reflects energy. Changing the semimajor axis results in a net change of energy of the small body. On the other hand, eccentricities of elliptical orbits are not associated with the overall energy and thus are easy to change by resonant effects. Indeed, the reason that the Kirkwood gaps exist is subtle and a little chilling for inhabitants of an inner planet: eccentricities get large enough for asteroid orbits, normally found in the zone or belt between Mars and Jupiter, to cross the orbits of inner planets like Earth. There, they are removed by collision! This would not be so chilling if we could regard the Kirkwood gaps as artifacts from the early Solar System, with the asteroids located at the various resonances as long gone. The modern view attributes the small effects of Yarkovsky drift to the constant resupply of projectiles to hit the inner planets. That small drift does not make much difference to the bulk distribution of asteroids, but for objects already near one of the resonances, the drift can eventually bring them into the zone where the resonant effect rapidly changes their eccentricity. We now know of several asteroids that are in Kirkwood gaps: they are not completely empty. Although there is a bit of a selection effect in that we tend to detect the asteroids that come nearest to Earth, some of the most hazardous known Earth-approaching asteroids are in resonance with Jupiter and have had their eccentricity pumped up so that they now cross Earth's orbit.

## 5.9 HAZARDOUS ASTEROIDS

The most famous hazardous impactor on Earth was responsible for the extinction of many species, including those dinosaurs which had not secured a favorable ecological niche by becoming birds, about 66 million years ago. It is not known with certainty what the impactor was, but evidence of the global influence of the impact is a widely distributed layer of the extraterrestrial tracer element iridium, at a time in the geological record when many species went extinct. This marked the end of the Cretaceous Period and the beginning of the Paleogene Period (previously called the Tertiary Period). The impact crater is found in the northern Yucatan peninsula of Mexico (Figure 5.19), largely buried and detected only by geophysical surveys, although having an imprint on local patterns of sinkhole topography. The crater is large, approximately 180 km in diameter, but, due to extensive deposition of flat limestone, has hardly any topographical expression. An impactor about 10 km in diameter is suspected of having caused this crater, although there is very little left of it. The vaporization of the impactor and parts of the impact zone, and injection of dust and reactive chemicals into the atmosphere, rapidly changed the environment, making for a major extinction event. There is some indirect evidence that a collision

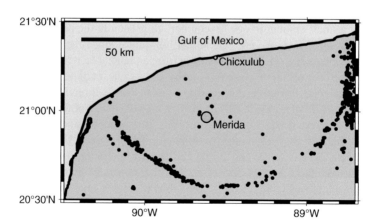

**FIGURE 5.19**  A ring of sinkholes, locally called *cenotes*, marks the outline of the 180 km diameter buried Chicxulub crater in Mexico's Yucatan peninsula. The impact happened 66 million years ago, famously marking the extinction of non-avian dinosaurs (author, data from Mexican topographic maps).

in the asteroid belt set the stage by creating a group of asteroids susceptible to drifting into resonance and being pumped up to an Earth-crossing orbit.

Such large impact events are rare as evidenced by geological evidence, or lack thereof, in the form of many craters. It is argued whether specific circumstances of the Chicxulub impact (named after a fishing village near the crater center) made it especially effective at causing extinction, but it appears that major impacts like this do not occur more frequently than about every 100 million years. As in any game of chance, those low odds do not prevent an event happening tomorrow; it simply would be unexpected. Furthermore, of course, the average period of large impacts being 100 million years, and one having happened 66 million years ago, does not lead us to "expect" one in 34 million years. Since even much lesser impacts could have devastating effects, including extreme local damage or more widespread damage through possible tsunamis, it has been important for mankind to assess the hazard. The large discovery rate of hazardous objects evidenced in Figure 5.18 has resulted from directed search programs with relatively small telescopes with large fields of view, assisted by object identification "pipelines" run on powerful computers.

Upon first detection of a new near-Earth object, its poorly-defined orbit usually includes a possible trajectory impacting Earth. In some rare cases, even the refined orbit continues to indicate a hit, and impacts have been observed and even in some cases meteorites gathered. It was mentioned above that Anaxagoras in about 500 BC could not predict a fireball, but with a bit of luck, we now can! These impact events are rare, and the most notable recent impact event, the Chelyabinsk impact in Russia on February 15, 2013, was not observed ahead of time since it came from the sunward side. Most detections, of which there are hundreds per night at times when the Moon's light does not interfere, result in trajectories not immediately leading to Earth. The paths of near-Earth objects are computed for hundreds of years into the future so that hazards may be assessed. For objects that do come close and appear dangerous, radar studies may be done. Radar is not well suited to discovering asteroids since the broadcast radar power goes down quickly with distance and only a very narrow field of view is possible. However, for asteroids which already have well-determined orbits allowing radar targeting, those orbits can be "refined" to be very precise.

The most recent asteroid to attract attention due to its initially determined hazard is 99942 Apophis, named after an Egyptian demon of chaos, and with a possible impact date of Friday the 13th, in April of 2029. Radar studies have now shown that although this 340-m equivalent diameter asteroid will pass close to Earth on that date, it will definitely not collide. As of late 2022, the asteroid known most likely to impact Earth is designated 2010 $RF_{12}$, with a 10% chance of impact in the year 2095. This asteroid is only about 7 m in diameter and would have an explosion in the high upper atmosphere with about half the energy of the Hiroshima nuclear weapon. Since mass scales with the third power of diameter, 140 m is regarded as a threshold for major damage, which would be regional. In Figure 5.18, it is seen that the discovery rate for NEAs greater than 140 m in diameter continues to rise. The red curve for potentially hazardous asteroids (PHAs) is also going up: these are not classed by diameter but most are likely small, like 2010 $RF_{12}$. Asteroids 1 km or greater

**FIGURE 5.20**   Fireball track of the Chelyabinsk meteor of 15 February, 2013 as shown by a dust trail (Plekhanov, 2013: CC BY-SA 3.0).

in diameter would cause a large disruption to life on Earth, although direct effects would be regional. The red curve showing the discovery of potentially hazardous asteroids in this size class is very flat: not many are being found. The only asteroid in this size range in the NASA "Sentry" list of hazardous asteroids is 29075 1950 DA, which has a 0.01% chance of impact in the year 2880.

The two largest impact events in recent history took place over Siberia in the morning: one on June 30, 1908, in a remote region known as Tunguska, and the other over the city of Chelyabinsk on February 15, 2013 (Figure 5.20). The Tunguska event caused extensive regional damage, with little debris left. The much smaller Chelyabinsk event did produce a large meteorite. Damage took place in both cases due to a shock wave from the airbursts. Combining our record of impact events with asteroid surveys indicates that the day-to-day risk to an individual from such events is small. However, on geological timescales, there is a definite connection between the heavens and the Earth. As indicated above, collisions in the asteroid belt are not likely the main way in which a connection to Earth via meteorites is made. We will first briefly review what such collisions produced, and then look at the seemingly unlikely mechanisms operating to bring us meteorites.

## 5.10 COLLISIONS IN THE ASTEROID BELT

To understand the role of collisions in the asteroid belt, it is essential first to understand the size distribution there. Our surveys are likely complete down to roughly stellar magnitude 18 as viewed from Earth. Unfortunately, the way to measure asteroid brightness is a bit complex and corresponds to a geometrically impossible situation. It is best to think that a low-albedo asteroid would be about 1 km in diameter if its "H" magnitude was 18, the right-hand limit of Figure 5.21; H = 15 corresponds to about 4 km in diameter. One can see that there is a power law fit (faint gray line) from H = 11 to H = 15, and this is an expected result for a population that has interacted. We expect that the fit continues beyond the range shown so that there remain a huge number of small asteroids yet to be discovered. Since there are such a huge number of small asteroids, and few large ones, collisions between asteroids are largely at small sizes, and any disruption produces yet smaller asteroids. These smaller asteroids also obey power laws, so that any given collision produces a lot of gravel and dust. As we have seen above, anything in this small size range is soon removed, either inward by the Poynting-Robertson effect, or outward by radiation pressure. Thus we can conclude that the vast majority of present-day collisions in the Solar System have little effect, with the collision products quickly removed. In recent years, evidence of such impacts has been observed, although some effects may be attributed to some asteroids having more than normal amounts of ice and other volatiles in them, making them into active bodies approaching the properties of comets. However, evidence collected over 100 years ago from asteroid statistics, the field pioneered by Kirkwood, showed that large collisions had at least at one time been important in the asteroid belt.

Kiyotsugu Hirayama of Tokyo first noted similarities in *proper elements* among groups of asteroids in 1918. Initially, in the dangerous territory of working with a small sample size (as with Kirkwood), he deduced that some groupings of asteroids were due to past collisions. He was right, and such a group of asteroids is now called

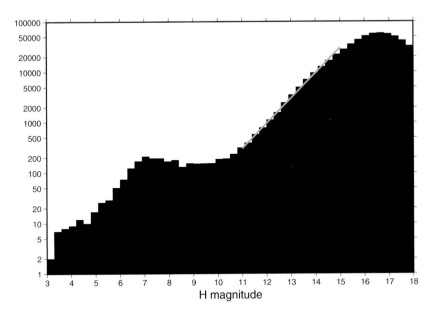

**FIGURE 5.21** Distribution of asteroid brightness H for about 620,000 asteroids, with bins 0.3 magnitude wide. H = 16 corresponds to about a 2-km-diameter body. Gray line is a power law (author, data from Minor Planet Center).

a Hirayama family. Proper elements are such things as eccentricity and inclination, corrected for current effects on the orbits, largely from Jupiter, allowing seeing a broader relationship. Figure 5.22 shows projections onto proper element axes with a large number of now-known asteroids, and the groupings are clear. In proper *a-e* space, as in the uncorrected *a-e* space of Figure 5.17, the Kirkwood gaps are clear. Often, family groups have similar observed colors as well as similar orbits, indicating the same parent material. A well-known family is associated with the bright asteroid Vesta, shown in red, included in 3-D hanging in front of the projections. Another very distinct family, very clearly bounded by Kirkwood gaps, is the Koronis family, shown in blue.

The case of Vesta is interesting since it was visited by the Dawn spacecraft (in orbit from July 2011 to September 2012). More than 50 years ago, the spectrum of Vesta was studied by Thomas McCord at MIT. It was found to be unique among the

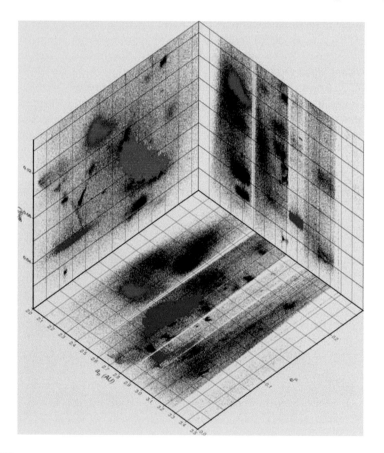

**FIGURE 5.22** Asteroid families shown using proper elements, semimajor axis at bottom left, eccentricity at bottom right, and sin(inclination) vertically. Families are seen on each set of axis projections, and the Vesta (red) and Koronis (blue) families are highlighted (author, data from Minor Planet Center).

then-known asteroids in resembling that of the volcanic rock basalt. Although we think of basalt as dark, as for example is the case in the maria of the Moon, on Vesta it is quite reflective. It has distinctive absorption bands, one of which at a wavelength of 1 μm is visible in Figure 5.23. More amazingly, the spectrum resembled that of an unusual type of meteorite. One of the usual identifying characteristics of meteorites is the presence of "chondrules", small spheres whose origin is not clear but which are never found in Earth rocks. The very rare Howardite-Eucrite-Diogenite (HED) family of meteorites is distinguished by being basalt-like and not containing chondrules, and thus are also called achondrites (the a- prefix in Greek meaning "without"). Forty years before the first small sample of an asteroid was brought to Earth in 2010 by the Japanese Hayabusa mission, it was realized that we had direct samples of the asteroid Vesta in our meteorite collections. Although the Hubble space telescope had suggested its presence, the Dawn spacecraft allowed detailed studies (Figure 5.24) of the very crater from which some of these meteorites originated. Known as Rheasilvia,

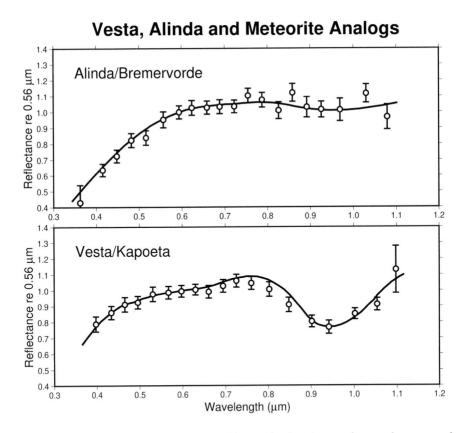

**FIGURE 5.23** Spectra of asteroids (dots with error bars) and meteorite samples measured in the laboratory (lines), for ordinary chondrite analog Alinda and a chondrite, and for Vesta and an achondrite (author: from digitized published data).

this crater, over 500 km in diameter, is almost as large as Vesta and dominates its Southern Hemisphere.

Between McCord's detective work and the visit by Dawn, the dynamical family of Vesta (Figure 5.22), consisting of much smaller asteroids than Vesta itself, had been discovered, with spectra taken showing the same unusual basaltic composition. These family members are commonly referred to as "vestoids". There is so much volume of Vesta blown away from Rheasilvia alone that the volume of all of the vestoids can be accounted for and little doubt that that is where they came from. In Figure 5.22, it may be seen that the outermost boundary of the Vesta family is the strong 1/3 period resonance at $a = 2.5$ AU. Close inspection of Figure 5.17 shows that there are known asteroids in that Kirkwood gap, mostly with high eccentricities that have been pumped up as originally suggested by Kirkwood. None of them appear to be vestoids, but most are small and have not yet had their spectra determined.

The large collisions that disrupt entire bodies, or at least excavate large proportions of them as in the case of Vesta (and Mimas), are rare since large bodies as impactors are rare. Various ages have been given for Rheasilvia, but none are younger than one billion years. The Meteoritical Society lists 70 observed "falls" of HED meteorites from 1803 until the present, so they are coming in steadily now, not a billion years ago. What is the mechanism getting them to Earth? Lacking known Vestoids/HEDs presently in the 1/3 resonance to study, we will turn to an asteroid of more common composition but with interesting dynamics. This is 887 Alinda, shown in Figure 5.23 to be an analog of a very common "ordinary chondrite", whose orbital evolution is shown in Figure 5.25. Because there are so many of this type of meteorite, a linkage to a specific asteroid is not possible for any of them, but Alinda very nicely shows what happens to an asteroid when in a resonance.

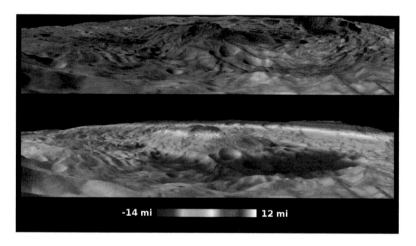

**FIGURE 5.24** View of the south polar region of Vesta, dominated by the large crater Rheasilvia (500 km diameter). False color indicates height. 14 miles = 22.5 km; 12 miles = 19.3 km (NASA).

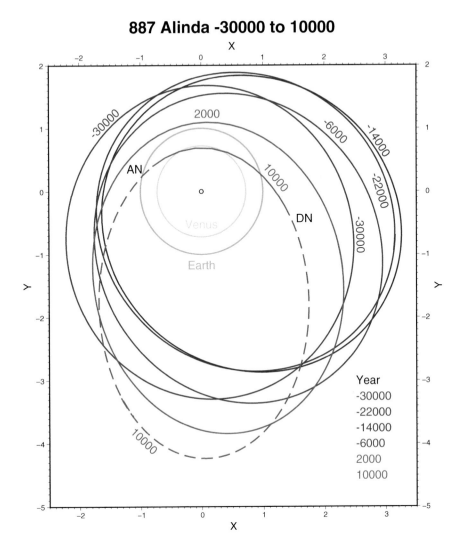

**FIGURE 5.25**    Evolution of the orbit of 1/3 resonant asteorid Alinda over 40,000 years. The initial near-circular orbit becomes very elliptical and crosses that of Earth a bit after the year 2000, and of Venus before the year 10000. Dashed portions of that orbit are below the ecliptic plane (author).

## 5.11 SPECIAL DELIVERY

Kirkwood's sparse writings do not indicate *why* he thought that eccentricities might get pumped up, and he appeared to believe that the main results of this would be collisions within the asteroid belt, leading to larger bodies being formed from initially numerous small ones. In fact, the intellectual background to this concept can be found in the work of Laplace around the turn of the nineteenth century, which focused on the stability of the Solar System. The work was taken further by Le Verrier, mentioned in Chapter 2 as being the French champion in the international competition to find the planet Neptune. In 1856, Le Verrier specifically mentioned that, although the larger bodies in the Solar System mutually limited each other in eccentricity that could be attained, the same did not apply to bodies of small mass. The work of Lagrange, the other important French *mécanicien* from the turn of the nineteenth century, had established planetary equations for the changes in orbital parameters. They broke down in cases of resonance, but showed that changes in eccentricity and inclination were far easier to attain than changes in semimajor axis. Whatever Kirkwood's inspiration, we now know that eccentricities do increase in certain Kirkwood gaps, leading to collisions with inner planets or perhaps the Sun in extreme cases. Figure 5.25 illustrates this for the asteroid Alinda, based on numerical integration from the present state. The orbits are shown at 8,000-year intervals, each marked and color coded. At time −30000 years (32,000 years before the present), the orbit had an eccentricity of around 0.3, which is fairly typical for asteroids. By the year −6000, the orbit had become noticeably more elliptical (eccentric), whereas in the present epoch (year 2000) its perihelion is just outside Earth's orbit (shown in green). By the year 10000, this drops down to be inside the orbit of Venus. Calculations for Alinda forward in time are chaotic, thus a bit uncertain, since its planetary interactions can be quite close, and the orbit can change a lot in any given one of them.

Indeed, the most dramatic of interactions, which is planetary impact, could take place. Since Figure 5.25 looks down on the orbit, its inclination is not clear, but in fact the inclination of Alinda varies between about 4° and 10° over the period shown. For clarity, some indication of this is given only on the orbit for the year 10000, where the portion of the orbit which is below the ecliptic is shown with a dashed line. The positions where an orbit passes from being below to being above the ecliptic, or vice versa, are known as *nodes*. The former of these is called the ascending node and marked in the figure by "AN", while the latter is the descending node, marked by "DN". A very important point is that for impact with Earth to occur, the node must be on Earth's orbit as well as that of the asteroid, *and* Earth must be at that point in the orbit when the asteroid also is. These requirements are quite restrictive. In the case shown, only the ascending node is even near Earth's orbit. The period of an asteroid in the 1/3 resonance is about 3.95 years (1/3 that of Jupiter's 11.86 years), not an exact multiple of Earth's period (1 year). As a result, the position of the asteroid relative to Earth will vary and eventually, the two may be in the same place at the same time. A meaningful statement cannot be made about this in the distant future for Alinda, but it would definitely be on the list of "potentially hazardous asteroids" 8,000 years from now. The converse is also true. Although Alinda's orbit goes inside that of Venus, it cannot possibly hit the planet, since its orbit is above the plane of Venus at all times when near it. A similar consideration applies to most present PHAs. Only with some orbital evolution could they hit Earth even if their present orbits cross ours.

These considerations still leave the major, recently solved problem of how to get asteroids into resonance. Being in resonance involves having a period that is a small integer fraction of that of Jupiter, like the 1/3 ratio of Alinda. There are other resonances possible, but those of Jupiter dominate due to its large mass and location near the asteroid belt. Perturbations due to planets do not have a net effect of changing the semimajor axis of asteroids, so if they do not have the right semimajor axis to be in resonance, how do they get into resonance, since that involves changing the semimajor axis? By reference back to Figure 5.17 and Figure 5.22, it may be seen that there are huge numbers of asteroids *near* in semimajor axis to the Kirkwood gaps, but not in them. The Vesta family is bounded at its outer boundary by the very 1/3 resonance we have most discussed. We saw with the example of Alinda that the 1/3 resonance is a "conveyor belt" which rapidly increases eccentricity, if and when an asteroid gets into it. The answer to asteroid and meteoroid delivery to Earth is that, whereas gravitational perturbations cannot bring objects into resonance, very subtle mechanisms involving radiation pressure can. Direct radiation pressure acts mainly on dust-sized particles, not relevant to the impact problem. The Poynting-Robertson drag acts, whether or not aided by a resonance, to bring gravel to small boulder class material inward, but again this size range easily burns up in Earth's atmosphere. Having many asteroids and billions of years to wait, an even less effective mechanism is fine for getting asteroid-sized objects into resonance, and this is the Yarkovsky effect mentioned above. This very subtle effect has now been directly measured with radar for close approaches to Earth of several asteroids. It depends on many parameters but basically is a reaction to radiation pressure from infrared emission from heated parts of a rotating asteroid. This can slightly accelerate or decelerate an asteroid along its orbit, and speed changes correspond to changes in semimajor axis. It appears that the fate of asteroids near the border of a strong resonance is to wait for possibly billions of years with small changes in semimajor axis until they are in resonance, whereupon their eccentricity can increase rapidly to bring them into the inner Solar System.

After a chaotic stage of bombardment following the initial formation of the Solar System, which left its mark on ancient and undisturbed surfaces like those of our Moon, the Solar System came into the state it now has. The asteroid belt is largely a fossil relic of the early times, now without major collisions, allowing only small ones able to produce gravel and dust which are rapidly cleared, and slow processes operating to inject kilometer-scale objects into resonances for a fast ride to the inner Solar System. Astronomy has become geology even for the Invisible Solar System, largely dominated by very slow processes but with the occasional reminder that catastrophic large events are indeed possible.

## 5.12 COMETS

Comets have been long known as celestial phenomena, although usually not as "invisible" denizens of the Solar System. Indeed, until Tycho demonstrated that they did not have appreciable parallax, and thus must be very far away, they were often regarded as atmospheric phenomena. Until the time of Edmond Halley, comets were only known to become visible at unpredictable times. Of course, there was usually some event in progress to which to tie them, more often than not a disaster. Even the word disaster comes from dis- (a negative prefix as in dishonest) and -aster for star.

**FIGURE 5.26**  Monochrome drawing of what is likely Halley's Comet in the Bayeux Tapestry (actually an embroidery), taken to foretell the doom of King Harold (out of frame at right), who was slain at the Battle of Hastings in 1066 (Public Domain, Wikipedia).

Perhaps the most famous depiction of a comet in history is on the Bayeux Tapestry (which is actually an embroidery), as shown in Figure 5.26, depicting a group of men pointing at what is likely Halley's Comet in 1066. For their English king, Harold, shown nearby on his throne, this comet was definitely a bad omen. For William the Conqueror, who sponsored the embroidery after conquering England and seeing Harold killed, it was a good one.

To some extent, the reputation of comets was blackened by the English language's most famous playwright, who featured a conversation with Julius Caesar in the play of the same name, in which his wife warned him that "When beggars die, there are no comets seen; The heavens themselves blaze forth the death of princes". Caesar was assassinated on the ides of March (March 15), 44 BCE, but it is likely that the associated comet was that of May–June of that year, also seen in the Far East. This was taken instead as a good sign that Caesar had become a god, and was commemorated on coinage (Figure 5.27) by his successor Octavian/Augustus. The name of comets used in the West derives from their tails. They (Figure 5.28) resembled flowing hair as seen by the unaided eye, and "comet" derives from the Greek *comes* for hair. In oriental languages, they are often called "broom stars", also a fitting allusion. Aristotle thought they were in the atmosphere and the idea prevailed in the West for about 2,000 years.

A common feature of very bright comets is that they get to be that way by going closer to the Sun than Earth's orbit. Usually, this means that they disappear into the

FIGURE 5.27   Denarius of Augustus showing the comet of 44 BCE as "Divine Julius", minted years after the fact (Classical Numismatic Group, 2006: CC BY-SA 2.5).

FIGURE 5.28   Comet Hale-Bopp. One of the brightest comets of the twentieth century, it was visible in 1996–1997 for several months. The ion (blue) and dust (white) tails are clear (Shutterstock).

daylight sky and then are observed a second time when they appear to move into the night sky. It is not always clear in such cases that the same comet is involved, and the apparent path in the sky can be complicated. Although a couple of earlier workers had suggested that comets move on parabolic orbits, Isaac Newton was slow to adopt this model. When he did, he incorporated the results into the *Principia* with a full explanation based on the same laws as he used for other motions. Although Newton was a contemporary of Halley, it was not Halley's comet but rather the exceptionally bright comet of 1680 that he studied in detail (Figure 5.29). He was challenged by this problem, noting that "This being a Problem of very great difficulty, I tried many methods of resolving it". A little-known fact is that he made his own observations using a telescope seven feet long and equipped with a micrometer, allowing positions to be taken with respect to nearby stars. In the figure, the comet, moving counter-clockwise, has a smaller tail before perihelion (top) than after. The very great length of the tail, about 1 AU after perihelion and the extreme solar heating there, is clear in the figure and likely quite accurate. The tail generally points away from the Sun, and Newton gave a description of heating of the body of the comet in considerable detail. He also notes that the length of the tail was 70° as observed on December 12 (first point after perihelion). Once this understanding of the true nature of comet orbits was attained, it was quite clear why comets might be seen in the evening or morning and then on the other side of the Sun.

The Great Comet of 1680 had a parabolic orbit, essentially falling in from a very large distance and returning with about the same energy. We now know that there is a large reservoir of comets at very large distances from the Sun, named the Oort cloud after the Dutch originator of the concept, Jan Oort. Painstaking analysis of comet orbits to see if the "parabolas" of irregularly appearing comets actually were very extended ellipses suggested that this was the case, but, as Oort published in 1950, with semimajor axes of the order of 150,000 astronomical units. This is

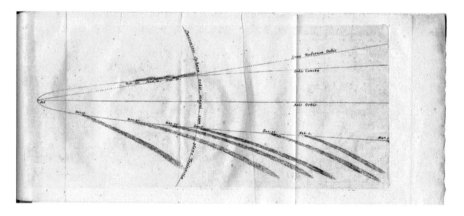

**FIGURE 5.29** Parabolic orbit of the Comet of 1680 from Newton's *Principia*. Dates in November at the top, outbound in December–March (1681) at the bottom. Circle is Earth's orbit (Newton, 1687: public domain).

approximately half the distance to the nearest star! From this reservoir, the long-period (thousands of years or longer) comets appear to fall in irregularly, and, as often as not, with retrograde orbits. As Newton already noticed, comets activate when near the Sun. The exceptional comet Hale-Bopp of Figure 5.28 activated outside the orbit of Jupiter, heating even there being sufficient to drive off some volatile gas. However, Kepler's law of areas dictates that comets spend only an infinitesimal fraction of their period in the inner Solar System, languishing in the cold depths of space with no gas emission. Then, they are solid bodies (Figure 5.30), comparable in size to small asteroids. Their surfaces can have impact craters and pits from which volatiles are emitted, scars of visits to the active inner Solar System. Since Oort cloud comets are not observable, their properties are obtained statistically, but there are likely billions of them, with a total mass perhaps approaching hundreds of Earth masses.

Newton's colleague Edmond Halley applied the fruits of his labors as published in the *Principia* (1687) to compile a catalog of 24 comets in 1705. Using historical observations, he determined orbits using Newton's parabola method. He noted that comets observed in 1531, 1607, and 1682 had very similar orbits, although with

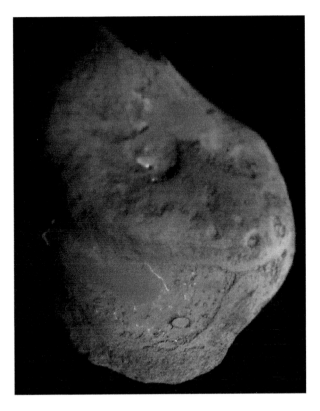

**FIGURE 5.30**   Nucleus of Comet Tempel-I, about 5 km by 7 km, imaged by the Deep Impact spacecraft on July 4, 2005 (NASA/JPL/UMD).

slightly different periods, a change which he attributed to interaction with Jupiter. He could thus conclude that the observations were of the same comet, which would have been on an elliptical orbit with a period of roughly 75 years. He also noted that, within the period of historical observations used, there should have been a comet seen in 1456, but could calculate that its path would have mostly been unobservable due to it being near the Sun in the sky. Subsequently, Halley modified his methods to use a retrograde elliptical orbit, and taking account of Jupiter perturbations, predicted a return in 1758. Most people will see Halley's comet only once in their lifetime, and Halley presumably did see the rather unspectacular comet of 1682 (likely a disappointment after the Great Comet of 1680). He passed away in 1742, with the periodic comet indeed later re-observed (as part of a complicated story of secret observations and amateur astronomer farmers) in 1758 and then in 1759 on its outward leg. After this, the comet became widely known as Halley's comet. It is the most famous comet not because of its brightness but due to its historical significance. Those who saw its most recent return, in 1986, may have shared the disappointment Halley likely felt in 1682, although at that time he did not know the comet would bear his name. Halley's was only the first of many comets now known to be periodic, likely due to having had their original Oort cloud orbit modified by Solar System bodies so that they no longer return that far out into space. The 2061 visit of Halley's Comet is expected to be more impressive than in 1986, but the unpredicted fresh interlopers like Comet Hale-Bopp are likely in the meanwhile to give an even better show.

## 5.13 COMETS AND METEORS

One class of material that regularly impacts Earth is relatively benign, at least at the surface of Earth. Almost no object observed as a meteor hits the surface of our planet. Regular meteor "showers" occur throughout the year, the most noted being the Perseids observable in summer. It typically brings more than 60 meteors per hour, and they appear to come out of a position in the sky in the constellation Perseus. Sometimes, the showers are much more intense, and a very strong showing of the Leonids in 1833 (Figure 5.31) understandably stimulated a lot of work in meteor astronomy. The fact that each meteor shower seems to come from one point (in Perseus or Leo for the examples mentioned) is due to the meteoroids being on essentially parallel paths before entering the atmosphere and is an effect of perspective. A similar effect occurs if one has a clear view of a highway or railroad line going off into the distance: the two edges seem to converge on a point. In the case of meteors, that point is called the "radiant".

The Italian astronomer Giovanni Schiaparelli is best known as the inventor of the concept of "canals" on Mars. As director of the Brera Observatory in Milan, he had access to refracting telescopes of 22 and 49 cm aperture, and, when Mars was at opposition and closest to Earth, made detailed maps. These showed linear features which we now regard as illusory. We probably do not wish to extend our study of the Invisible Solar System to the "imaginary" Solar System. Suffice it to say that, despite an American review of Schiaparelli's Mars work carefully noting that "The correct translation of the Italian word *canale*, used in reference to the streaks

FIGURE 5.31   Leonid meteor shower of 1833, woodcut by Adolf Vollmy, 1888, based on indirect information (Vollmy, 1888: public domain).

on Mars, is channel or strait, and not canal", American observers, including that reviewer, took the translation as canals totally literally, including its implications that the canals were made by intelligent life. This mistranslation is often used to exculpate Schiaparelli, but one must note that he also wrote a book in 1893 entitled "*La vita sul pianete Marte*", i.e., "Life on Mars". Schiaparelli's contributions to science are mostly now thought to have been through stimulating interest in Mars, but one notable exception stands out.

By careful studies of meteors, using the fact that their motion relative to Earth makes them more likely to be seen in the morning, and the motion of the radiant, Schiaparelli was able to deduce their orbits. When the comet Swift-Tuttle was discovered in 1862, he was able to compare its orbit with those of Perseid meteors and conclude that the comet was the source of the meteors. Urbain Le Verrier of Neptune fame, by then the Director of the Paris Observatory, in 1867 showed that the Leonids were subject to outbursts such as that shown in Figure 5.31, due to clumping of material along the comet's orbit. Also in 1867, the orbit of comet Tempel-Tuttle, the first comet of 1866, was published and noted by several researchers to be similar to that of the Leonids. Kirkwood's 1867 book attributed this to Schiaparelli, and graphically depicted the meteor orbits (Figure 5.32, left). From this diagram, one can make

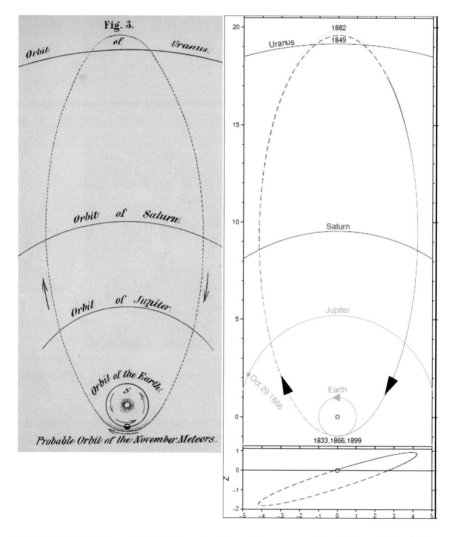

FIGURE 5.32    Kirkwood's (left) November (Leonid) meteors orbit and the calculated 1833–1899 path of Comet Tempel-Tuttle (right; side view at bottom 1833–1866, showing inclination of the orbit). Dashed portions are below the plane of the ecliptic. Arrows show directions and relative sizes of velocities. Sun size exaggerated in both panels (left: Kirkwood, 1867; right, author based on data from JPL Horizons).

a quantitative inference about the speed of comet-derived meteoroids, most of very small size. Earth progresses counterclockwise in its orbit at about 30 km/s (arrows in figure). The meteoroid has a very elongated retrograde orbit, so that the speed of objects at Earth's distance is basically that of free fall from a large distance, which is $\sqrt{2}$ times greater than the circular orbit speed. Thus the meteoroid speed relative to the Sun is about $1.4 \times 30 = 42$ km/s. Since these come almost directly at Earth, their speed relative to us is about 72 km/s. For this reason, meteors from cometary sources appear to observers to flash by very quickly. In contrast, asteroidal material which does manage to hit Earth is moving in the same direction, and from sources much closer than the extremes of a comet orbit. Some such material comes in at only a few km/s, making a much slower meteor and increasing the chances of material not being burned up and able to fall as meteorites.

Kirkwood's part (left) of Figure 5.32 shows Earth near a clump of meteoric particles which would have been released by the comet when near the Sun in 1866. That year did not have an unusually strong display like that of 1833: the particles were clumped irregularly. Nevertheless, it is true that when the comet makes perihelion passes every approximately 33 years, particles are released, and the density along the path is then much higher. As Le Verrier explained, this explains meteor showers which have regular outbursts, like the Leonids. In contrast, the stream producing the Perseids appears to be much older, and likely the comet itself is old and no longer as active when near the Sun. This gives a much more regular stream of particles and a more repeatable meteor shower. Particles spread along cometary orbits due to small changes in their velocities and varying responses to planetary gravitational influence, so that, without fresh injections, they spread along the whole orbit.

Earth passes through streams associated with different comets at different times of year as it moves around its orbit. For the Leonids, Earth passes through the stream in mid-November. For the Perseids, a different stream is passed through in August. The right side of the figure shows the orbit of comet Tempel-Tuttle from 1833 to 1866 (bottom) and 1833 to 1899 (top). The side view shows the inclination of the comet orbit. As with asteroids that may impact, Earth must be near a node of the comet's orbit for its particles to strike the planet. In the figure, the direction of motion of the meteoroids is toward the left and down, so that they appear to come out of a point in the northern sky, which explains why there is a radiant. After the perihelion passage of 1866, the comet passed about 2 AU from Jupiter, with closest approach on October 29 of that year. Although on less extreme orbits, planetary influences usually average out, it is easy to envisage the effect of Jupiter in this case. It constantly attracted the comet, but its relative direction of motion took it away from the comet's orbit after the closest point. As a result, there was an overall increase in the comet's speed, allowing it to climb higher out of the Sun's gravity well at its extreme point, as indicated by the aphelion of 1882 being further out than that of 1849. Such influences explain the changing period of such comets as already noted by Halley in the eighteenth century.

Of course, very few comets have paths that intersect Earth's orbit. Meteor showers typically last a few days, allowing us to infer that the width of the streams of comet debris are comparable to the distance traveled by Earth in that time. Some very old and spread-out showers have low activity, so are hard to distinguish from the background "sporadic" level of meteors, but can last a month or so. Ironically, there is another class of body strongly associated with planetary orbits, but in such a way that the planet is "dynamically protected" from being hit. These asteroids bear the exotic name, "Trojans".

## 5.14 TROJANS AND TRAPPED ASTEROIDS

Nominally, the two-body problem of Kepler is the only exactly soluble general problem in celestial mechanics. Two bodies orbit their mutual center of mass on conic section orbits. For bound bodies these orbits are ellipses or circles. In the case of the planets, the Sun is so massive that it is very close to the center of mass and it is not a great error to say that the planets orbit it. Earth happens to have a large companion moon, which, although much smaller in mass than the Sun, is 400 times closer, so that its gravitational force (and tidal stretching force) cannot be neglected in any refined calculations. Thus, the "three-body" problem is relevant to situations near Earth's surface, such as tides, and also in determining the apparent path of the Moon accurately. Indeed, during the period of Western expansion into the world by sea, commercial and military interest in position determination led to great interest in the Moon's orbit, since for at least part of each month it was an easily observed celestial object. Accurate tables and observations gave the promise of a valuable navigational aid if a precise time could be determined from the Moon's exact position. Already in the late eighteenth century, the Sun-Earth-Moon three-body problem drew a lot of interest because of its practical implications as well as due to its theoretical interest. As a result, the Paris Academy offered a prize three times in a row for "perfecting the methods on which the lunar theory is founded".

Lagrange won the prize in 1772 with his *Essai sur le problème des trois corps* (Essay on the three-body problem), which showed that there were five special points in the planar problem of circular motion involving three bodies. Three of these are along a line joining the two main bodies, two of which, near the secondary body, have modern practical applications. The third of these is in the "anti-planet" position on the opposite side of the Sun, and apart from speculative science fiction, has had little application. The positions of those "Lagrange points" which are on the same side of the Sun as Jupiter are shown in Figure 5.33. The $L_1$ point is between the Sun and Jupiter, and the $L_2$ about an equal distance beyond it. In the case of Earth, similar points exist. Although these points are unstable, only small amounts of properly directed thrust enable spacecraft to stay near them. The $L_1$ point is a very good place to put spacecraft observing the Sun continuously (which may also look out toward an always-full Earth disk). The $L_2$ point is a good place for spacecraft looking outward into space, such as the recently launched JWST infrared observatory. Of interest in Figure 5.33 are the thousands of "Trojan" asteroids which have gathered at the $L_4$ and $L_5$ points naturally. There is kind of an "anti-Kirkwood-gap" effect when in 1/1 resonance, as can be seen by revisiting Figure 5.17. It may be noted there that the 2/3 resonance also tends to trap asteroids: these are known as "Hildas". At the Lagrange points 60° from the planet, it is possible for an asteroid to orbit the

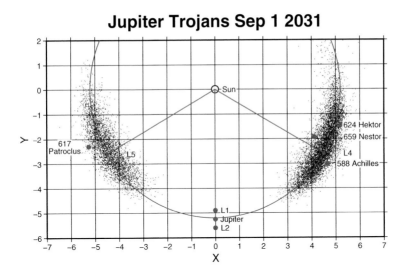

FIGURE 5.33   Jupiter Trojan asteroids as they will be on September 1, 2031. The first four discovered Trojans are shown and labeled, along with the four Lagrange points on the same side of the Sun as Jupiter (author, computed from initial positions from Minor Planet Center).

Lagrange points (the technical term is "librate") stably. Lagrange believed that his discovery of stable points in the three-body problem had no significance in the real universe. Asteroids had not been discovered in 1772, and comets were clearly too dynamic to have much to do with Lagrange points. However, in 1906, Max Wolf at the Heidelberg Observatory in Germany, at the time the world's leading asteroid discoverer due to using photographic techniques, found a very distant asteroid. Later in the year, a second was found and it was eventually realized that they resided near the two Lagrange triangular points as they and Jupiter orbited the Sun. Two more were soon found, as shown on Figure 5.33. This first group of seemingly associated asteroids brought about the name "Trojan" since they, much as the armies in the Greek epic poem of the Trojan War, the Iliad, were grouped.

One significance of Trojan asteroids may be that the ability to stay at Lagrange points might preserve ancient material that has not participated in the dynamics described earlier in this chapter. As such, Trojan asteroids may hold "fossil" material. An ambitious mission including visits to five Trojan asteroids was launched in late 2021. It is called "Lucy" after the homonin (early human) fossil of that name, based on the idea that it may be exploring the fossil remnants of an earlier phase of the Solar System. It will visit four $L_4$ Trojans before swinging over to end its mission visiting 617 Patroclus, now known to be a binary asteroid.

## 5.15  SUMMARY

The small bodies of the Solar System range from those like Ceres, which are almost planets (in fact now bearing the designation "dwarf planet"), down to very small dust particles. The Solar System was a dusty place in its early history, but most of that dust agglomerated into planetary bodies with some admixture of gas. Remaining solid materials up to boulder size are subject to removal due to interaction with the light and radiant heat from the Sun. Even larger bodies can interact with light to slowly drift into resonances, which can very effectively change the orbital properties so that they come into the inner Solar System and can interact with planets. Semimajor axis is not readily changed by resonances, but, by driving material to high eccentricity where it is removed by collisions with planets, gaps appear in the distribution of asteroid semimajor axis. Impacts once played a large role in changing planet-like surfaces, although geological activity on Earth has removed most traces of this. Impacts are now rare, but do pose the threat of a "very bad day" for life on Earth. The hazard has been largely assessed, although the space inward of Earth has not been well surveyed since the Sun interferes with ground-based searching in that region. Comets form part of the Invisible Solar System since their elongated orbits dictate that they spend most of their lives as small bodies in its outer reaches. When they do come into the inner Solar System, they can be spectacular, but, unless on a direct course toward Earth, do not pose much hazard since they release mostly small particles and gas which our atmosphere protects us from. We can now target spacecraft missions to small bodies, which may hold clues about the early history of the Solar System.

# 6 Wind from the Sun

In trying to figure out why the Solar System between the planets looks empty, it was mentioned that there are usually about ten million particles in each cubic meter near the Earth, mostly protons and electrons, and that they are of solar origin. What was not emphasized is that these particles are hurtling into deep space at supersonic speed. The Sun is ejecting its atmosphere into space!

As discussed in Chapter 3, the earliest spectroscopic observations of light emission from the chromosphere of the Sun, and the lower corona above it, in 1868 and 1869, suggested the presence of two new elements, one of which was found to be helium within thirty years of its detection. The identity of the other was more resistant to solution, having been given the name "coronium" since that seemed to be the only place that it was seen. It was only in the 1940s that the green line at 530.3 nm (green in Figure 6.1), and some others subsequently detected as the deepening mystery drew more observations, were finally identified. The Swedish spectroscopist Bengt Edlén, working through analogies with other atomic structures, concluded that some of the lines came from iron. The surprising thing was that it was iron in very high ionization

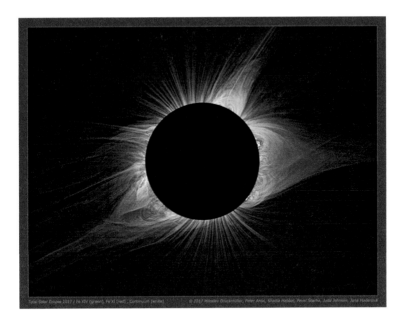

**FIGURE 6.1** The solar corona imaged with emission lines of iron and in white light, processed to allow magnetic field control to be seen and the million K temperatures to be determined (Habbal et al., 2021: CC BY 4.0).

DOI: 10.1201/9781003451433-6                                                                    **183**

states, the green line coming from Fe XIV, where the Roman numeral XIV (14) indicates that 13 of the 26 electrons of this heavy element have been removed. This takes a lot of energy, so Edlén concluded that the corona must have a temperature in the millions of degrees. This outrageous finding seemed unphysical since the Sun's effective surface temperature is only about 5,800 K, so how could the corona above it be so hot? In familiar physical systems, heat flows from hot to cold, so one should not be able to heat the coronal gas from below where there is a much lower temperature. More time has now elapsed since Edlén's time than from the discovery of the coronal lines to their identification, yet we still do not have a firm idea of how this heating can happen. The skepticism at Edlén's time was met with the secondary observation that the coronal lines were very broad, an indication that the ions were moving very fast, which happens at high temperatures. Eventually, all the coronal lines were explained in a consistent manner involving very high temperatures, so the explanation was accepted: the corona is very hot, with temperatures in the millions of degrees. It is only recently that laboratory techniques have allowed the creation of such highly ionized atoms for study.

Figure 6.1 shows the use of the coronal lines (530.3 nm and a red line at 789.2 nm due to 10-times ionized iron) to take the corona's temperature during an eclipse, with special techniques also applied to white light imaging to reveal how plasma follows the magnetic field. The green, being due to a higher ionization state, indicates a hotter temperature of about 2 million K, while the red shows about 1 million K. The green is further out, showing that the corona gets hotter as one goes up. In the polar regions (near the top and bottom), the structure seems very regular and resembles that of a magnet as seen in Figure 6.2. However, in the equatorial region, the structure seems much more complex. This is due to the interaction of hot and flowing plasma with the magnetic field. Figure 6.1, although more colorful than what one sees during an eclipse, also makes it clear why the term "*corona*" is appropriate, since that is simply Latin for "crown".

The observed fact of the high temperature of the Sun's corona leads inevitably to its evaporation into space, although it turns out that "evaporation" is too gentle a word for what really happens. In this upper part of the Sun's atmosphere, too much material is present, and at too high a temperature, to form a static structure. We will discuss more about ionization in the next chapter since it is an important topic in the vicinity of planets, but in the corona the temperature is so high that atoms cannot remain neutral; they have many electrons knocked off to form ions. As they flow out, they move into a lower-density region where they cannot get back together with the electrons again, so the gas remains ionized, forming the "plasma" that characterizes interplanetary space. Heavy elements, like iron, will retain a fair number of electrons but are minor constituents. The more common hydrogen and helium, without such a highly charged nucleus to tightly retain electrons, are fully ripped apart into electrons and nuclei, with the latter referred to as ions (a word originally introduced in electrolysis as pointed out earlier).

In the Sun's interior, below the photosphere, which is blocked by the Moon in the eclipse photo of Figure 6.1, the temperature rises with depth, and a normal sort of heat flow prevails, from the hotter interior toward the cooler exterior. This characterizes

**FIGURE 6.2**   Iron filings align along the "magnetic field lines" of a magnet, allowing direct comparison to the Sun. Magnets always have two poles but the field lines radiate out from one to the other (Shutterstock).

the entire solar interior, although the heat flow can be either by radiation or convection, depending on where one is. In either the radiative inner zone or the convective outer layers, the normal rule, that heat flows from hotter to cooler, applies. Once past the photosphere, *light* can flow outward almost unimpeded and has little interaction with matter. As discussed in Chapter 3, the temperatures of bodies able to intercept light are determined by that major source of energy input, while their energy output is by re-radiation in the infrared. The balance of these two determines the temperature of solid bodies in the Solar System. The corona is not dominantly heated by light, since it is a thin gas. As mentioned above, the light passing through typifies a surface at a much lower temperature than it has, so it cannot heat the corona anyway. We still do not know all the details of the heating of the corona, but in this part of the solar atmosphere, radiation dominates the energy transport outward but not the heating of the gas. In the larger scheme of things, the plasma heating and outflow that originate above the photosphere are secondary effects, bearing about one percent of the energy that outflowing light does. Although the corona is very hot, it is not very dense. Various processes involving energy exchange between magnetic fields and plasma, with energy transport by waves, are felt to be responsible for heating it, but a mechanism is not yet well established. The current Parker Solar Probe mission, named after physicist Eugene Parker, who passed away in 2022, is getting closer to the Sun to probe some of these questions with *in-situ* observations, although even it may not get down to where most of the heating takes place.

The outflow of hot gas is called the *solar wind*. In addition to being essentially invisible, it carries a magnetic field, which optically can be inferred only through the structures it imposes by interacting with gas (as for example in Figure 6.1 and Figure 3.12). Far from the Sun, even those structures cannot be detected optically, but, at very few favored spots, we have spacecraft to detect the magnetic field directly. The complex solar wind plasma outflow is of great interest, and in recent years, of practical consequence. The solar wind, always a gale, can become turbulent and disruptive in ways that affect modern civilization.

## 6.1 HISTORY OF THE SOLAR WIND

Spectroscopic observations and certain other aspects, such as the fact that prominences were seen to move with the Sun, not the Moon, during eclipses, showed by the late nineteenth century that there was a solar atmosphere. The explanation of the unusual spectral lines in the 1940s did not change the concept that this atmosphere, like that of Earth, may have had local flows but otherwise was static. Progress in understanding the magnetic effects of the Sun on Earth was possible without knowing that there was a solar wind. Various considerations, some described below, had led to the conclusion that there was a Sun-Earth link, but most of the evidence was only empirical. For example, activity on the Sun, such as solar flares, seemed to be followed within days by magnetic disturbance. An earlier concept, that the Sun emitted beams of electrons which could cause an electric current near Earth and, in turn, magnetic effects, had been discarded by the 1930s. This notion basically saw the Sun-Earth space as a large Crookes tube. However, over the vast distances of space, electron beams would disperse rapidly, in what was assumed otherwise to be a vacuum as in a Crookes tube, due to self-repulsion. In 1930 and 1931, Sydney Chapman, a noted expert on geomagnetism, and Vincent Ferraro, still regarding space as a complete vacuum, proposed that during magnetic storms a fast neutral gas blob impacted Earth but was held off by its magnetic field. A particle feels a force perpendicular to its velocity and to the magnetic field when in a magnetic field, as particles from the Sun would be if moving near Earth. This would cause positive particles to swing around the Earth in such a way (Figure 6.3) as to generate a magnetic field, explaining magnetic storms. Many details of this model do not hold up today and our fuller picture is presented in the next chapter; however, it did include a boundary outside of which the (temporary) solar outflow would be, and inside which a current similar to that now known as the "ring current" would flow. Because there was no magnetic field outside the boundary, but Earth's field inside it, the boundary had to carry a current, still known now as a Chapman-Ferraro current.

Because there is always a solar wind, the Chapman-Ferraro current is always active, but its strength varies with solar wind conditions, as will be described in the next chapter. It is, perhaps surprisingly, in the behavior of comet tails that the first hint of the continuous solar wind was deduced.

It had long been known that comet tails point generally away from the Sun. Since most comets are bright near perihelion, they are often seen in the dawn or dusk sky and it is apparent that the tails rise upward from the horizon and get broader away from the brightest part of the comet, the nucleus and surrounding coma (Figure 6.4). As the Sun is below the horizon at these times, the tail points away from the Sun.

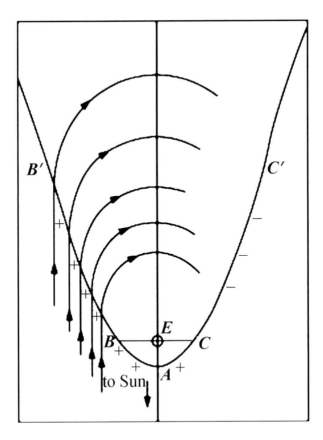

**FIGURE 6.3** Chapman-Ferraro view of material from the Sun (bottom) interacting with Earth's magnetic field (1931: redrawn by NASA, 1965 and P. Song, 2010).

Taking an overall view of the larger system of Earth, Sun, and Comet, Newton (Figure 5.29) shows this in his depiction of the orbit of the comet of 1680. As a question of detail, there are at least two types of tail, as can be seen in Figure 5.27 since Comet Hale-Bopp very distinctly showed both. In that figure, the blue tail is the ion tail, also known as a Type I tail. The whitish tail (Type II) has the color of the Sun, which illuminates it, and is composed of dust. The ion tail points almost directly away from the Sun, whereas the dust trail lags behind the comet a bit. In early spectroscopic work, in about the same epoch as the solar eclipse observations, comets were seen to show the solar spectrum, readily interpreted as that of the Sun reflected from dust, with emission lines or bands that were hard to identify. Since comets are cold bodies for much of their existence, they boil off "vapors" when near the Sun at perihelion, as already dramatically described by Newton in the *Principia*. These typically have molecular composition. For example, the blue in Comet Hale-Bopp and others is commonly attributed to carbon monoxide. In the strong solar light, including its unfiltered UV component, such molecules often become ionized. Early observations were thus puzzling, but now cometary spectra are well understood and can be used to study the early Solar System's composition. In 1951, Ludwig Biermann, working

**FIGURE 6.4**  Comet NEOWISE in a pre-dawn sky over Utah in July, 2020 (NASA/Bill Dunford, cropped).

at the recently founded Max-Planck Institute at Göttingen, Germany, suggested that the radiation pressure from the Sun could push dust back only rather slowly, forming the dust tail behind the comet's motion, but that solar radiation would not affect the motion of gas very much. The ion tail had to be pushed back instead by some particle emission from the Sun. A follow-on publication, in 1952, shows a charming 3-D metal model of the motion of Comet Halley in 1910, with a little tail that could be turned to show how it pointed with respect to the Sun. After a visually disappointing showing in 1986, Halley's Comet will return in 2061 in a manner that should dramatically show the orientation of its tails to northerly observers (Figure 6.5).

This figure should be used carefully: the comet's orbit for the 2061 return will not be known well until it is recovered telescopically as it returns toward the inner Solar System. Nevertheless, its perihelion pass should be in summer 2061, with viewing conditions like those in the figure. The figure is presented mainly to show the ion tail (blue) and the dust tail (white). Note that the ion tail points toward the Sun, while the dust tail lags behind the comet.

From subtle considerations of the angle observed for the Type I or ion tail, Biermann could conclude not only that particles must be coming from the Sun and pushing it back, but that they must have an incredible speed of about 400 km/s. Recalling that

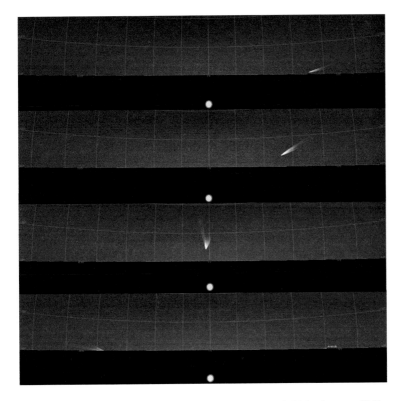

FIGURE 6.5   Halley's Comet on (top to bottom) June 4, 10, 16, 22 in the year 2061, as seen at 1:50 am local time from 66.5 degrees north. The Sun's (invisible) position below the horizon is shown by a yellow dot. Exact dates are not possible to specify now, but this will be the general behavior (Stellarium, montage and overlay by author).

Earth moves at 30 km/s, this is a very fast speed relative to those of orbiting bodies. Biermann's work stimulated a breakthrough paper in 1958 on the "Dynamics of the Interplanetary Gas and Magnetic Fields" by Eugene Parker, then a young scientist at the University of Chicago. The article was not well received, with the two main points being that there must be a supersonic flow of gas radially outward from the Sun, and that there was a magnetic field in the gas in the form of a spiral, each point being revolutionary. The noted astrophysicist Subramanyan Chandrasekhar, editor of the "Astrophysical Journal", overruled two negative reviews, while verifying that the paper was mathematically correct, and published it. One may speculate that he recalled his fundamental work on white dwarf stars having been rejected by Sir Arthur Eddington, the noted astrophysicist mentioned in earlier chapters. In a follow-up paper about the fact that such a fast and magnetized gas would have to interact with Earth's magnetic field, Parker coined the term "solar wind". Figure 6.6 shows how Parker calculated the solar wind speed to vary with radial distance and with the temperature of the corona. In a region near the Sun (left in Figure 6.6) corresponding to the corona, the acceleration is very rapid, but for all temperatures, by the time it passes Earth at 1 AU, the solar wind is not accelerating much.

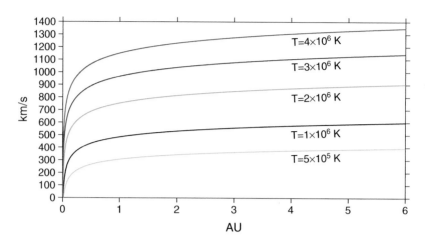

**FIGURE 6.6**   Solar wind radial speed as a function of distance in Parker theory, for various coronal temperatures (author).

This is a good place to note that Biermann had viewed a flow of particles out from the Sun, whereas Parker used hydrodynamical equations involving fluids. Many of the perhaps unintuitive properties of space plasmas arise from effects due to the interaction of particles with magnetic fields. It has been pointed out that, in interplanetary space, ionized particles cannot neutralize themselves by colliding with particles of the opposite sign. This is for the two reasons that density is so low that there are few collisions of any sort, yet it is high enough that, from a distance, the plasma seems neutral. So, interplanetary gas is made up of ions and electrons and is an ionized yet "quasi-neutral" plasma. How can a gas so thin behave as a fluid at all? Normal fluids interact because their particles are so close together! The answer is that fields cause interactions among the particles, and, on a large scale, they do interact in bulk. As a result, it is correct to use fluid equations as long as the scales are large enough. Also, one frequently distinguishes and separately studies particle effects. For example, the rare high-energy particles are often studied with particle equations in fields arising from the vast bulk of lower-energy particles also present.

Parker's theory made some simplifications, but the general idea that there was rapid acceleration to supersonic speeds in the corona was inescapable. As we shall soon see, Earth is shielded from the solar wind, largely through the influence of its magnetic field. As a result, early investigations had to be done from lunar or interplanetary spacecraft. The concept of the solar wind came at the dawn of the space age. Sputnik 1 had shocked the Western world upon its launch on October 4, 1957. The radio experiment, which simply emitted a "beep", was developed by Soviet physicist Konstantin Gringauz. The first spacecraft to attain interplanetary space (although inadvertently by missing its target, the Moon) was Luna 1, launched on January 2, 1959. It carried an ion detector instrument designed by Gringauz which was primitive by today's standards but is credited with the first detection of the solar wind. Similar instruments were on other Soviet spacecraft in 1959 and 1960, but the first systematic study of the solar wind was done by the US spacecraft Mariner II on its way to Venus from August to December 1962. This fully vindicated Parker's model, with additional measurements of the magnetic field, yet to be discussed. Another accomplishment of Mariner II was to bolster US confidence during the "space race". It is also likely that the solar wind data had a more lasting impact than that from the target planet, Venus. One of the Soviet spacecraft which detected the solar wind in 1960 was Venera I, which failed to reach Venus. This was the first of 14 unsuccessful Soviet Venus missions, the losing streak only being broken in 1967. Mariner II, however, had established the US ability to successfully navigate to other planets and to sample the interplanetary medium on the way there. By spending a long time in space, it established not only the presence of the solar wind, but also its high variability, responding to active regions on the Sun and the 27-day rotation period.

The Parker model in its initial form made interesting predictions about the solar wind magnetic field but did not incorporate it into the dynamics of the gas. Such a purely hydrodynamic model gives the results shown in Figure 6.7. The transition from subsonic flow near the Sun to supersonic flow takes place at a "critical radius" indicated by dots in the figure, which is an enlargement of the leftmost portion of Figure 6.6. The rightmost boundary of Figure 6.7 is 7% of one AU, so the solar wind is supersonic in all regions of the Solar System not very close to the Sun. As observed at 1 AU, typical solar wind flow speed is about 350 km/s under quiet conditions, corresponding to a coronal temperature of about 750,000 K. In the more detailed picture, this would put the critical radius at about 8.5 solar radii. The Parker Solar Probe is steadily using Venus encounters (see Chapter 9 for how such encounters work) to reduce its perihelion, with the aim to get within ten solar radii by 2025.

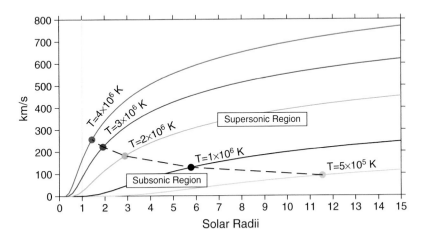

**FIGURE 6.7**   Details of the Parker model. From the solar surface (yellow vertical line) outward, the solar wind becomes supersonic very rapidly, especially at high coronal temperatures. The dashed line separates the subsonic and supersonic regions for various temperatures. Since 1 AU is about 210 solar radii, this figure shows the extreme left side of Figure 6.6 (author).

As may be seen from Figure 6.6, this may not be sufficient to enter the subsonic zone. Ironically, it is not easy to lose energy and fall into the Sun, much as we would like to have this spacecraft do so!

Before moving on to the important topic of the solar wind magnetic field, as initially laid out in Parker's paper, we now look at deviations from the assumption of radial smooth motion implicit in it. Figure 6.8 shows the view of the Sun from the Solar and Heliospheric Observatory (SOHO) which sits at the $L_1$ Lagrange point of Earth, the analog of that of Jupiter shown in Figure 5.33. From this vantage point, the Sun may be continuously imaged without the nuisance of Earth getting in the way. Furthermore, under average conditions, the radially outflowing solar wind may be sampled about an hour before it gets to Earth. SOHO's and other space-based observations of the Sun are possible due to "coronography". If one merely places an obscuring disc in front of the Sun as seen from the surface of the Earth, atmospheric scattering will usually prevent seeing the faint corona. (The reader is strongly discouraged from looking in the direction of the Sun under any circumstances.) In 1930, the French astronomer Bernard Lyot developed a method to use the polarization of the scattered sunlight in the corona, while that in the atmosphere is not polarized, to be able to view the corona if such a disk obscured the photosphere. In what must count as one of the great ironies in astronomical writing, Lyot wrote that an advantage of such instruments would be to replace eclipse observations, which require "long and expensive expeditions...in desert countries, which are very difficult";

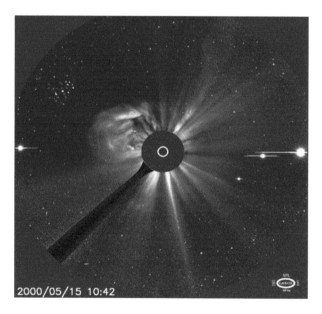

**FIGURE 6.8**   Solar and Heliospheric Observatory (SOHO) image of the solar corona incidentally showing the Pleiades star cluster at upper left, the brightest individual objects being Mercury (left), Saturn, Jupiter and Venus (in order toward right). A coronal mass ejection is seen at upper left (NASA/ESA).

having invented a way to avoid such trips, Lyot passed away prematurely while on one in Sudan in 1952. Coronographic observations from space rely on an occulting disk in front of a telescope, in principle much easier than on the ground. Instead of the short views of the corona possible during an eclipse, one could get observing sessions of about 45 minutes (half an orbital period) from near-Earth orbit, as first tried on the US manned orbiting station Skylab in 1973. From the $L_1$ point, continuous monitoring is possible. In Figure 6.8, the boom used to support the occulting disk is visible at the lower left. The disk covers much more than the photosphere: the apparent diameter of the Sun is marked as a small white circle. The transition radius where the flow goes supersonic would be perhaps one-third to halfway toward the side of such an image and nothing particular is seen there: the solar wind ramps up steadily without, for example, a shock wave. The flow in most parts of this image appears to be essentially radial, although it is hard to tell if this is simply due to flow or the configuration of the magnetic field. However, in the upper left quadrant, the corona is more distorted than in the simple loops shown in Figure 6.1. With continuous monitoring, such structures may be seen to move outward from the Sun rapidly and are known as "coronal mass ejections". They will be revisited in Chapter 7, since they have an important effect near planets they may hit. Much of the background of this image shows stars: for example, the "Little Dipper" of the Pleiades cluster can be seen at the upper left. At times, the detectors respond to "radioactive" particles in space, which are discussed in Chapter 8.

## 6.2 THE SUN'S MAGNETIC PERSONALITY

Sunspots, mentioned already in the first few pages as having been observed by Galileo soon after the invention of the telescope in about 1610, were closely followed for about 300 years before their physical nature could be understood. Until about 1910, there was no particular reason to associate them with magnetism. That association largely arose through the work of one man, George Ellery Hale. There can be little doubt that he was one of the most significant personalities in American astronomy in the twentieth century. Born in Chicago, his first major accomplishment was the installation of the 40-inch (ca. 1 m aperture) refracting telescope, still the world's largest, at nearby Yerkes Observatory in 1897.

Moving to southern California, he oversaw the installation of the 100-inch (2.5 m) Hooker telescope on Mt. Wilson, overlooking Los Angeles. He was also a prime mover behind the 200-inch (5.1 m) telescope on Palomar Mountain near San Diego, built in 1948. It was the world's largest until 1993 and was posthumously named after him. The many discoveries in the night sky that these telescopes fostered underpinned much of our present understanding of the universe, but Hale was originally a solar astronomer, and made pioneering discoveries in the field of solar magnetism, which it would not be incorrect to say that he invented.

While modern instrumental techniques allow us to look at Figure 6.1 and immediately conclude that magnetic fields are important on the Sun, the fleeting glances of the corona that had been seen up until the turn of the twentieth century did not give rise to much discussion about possible magnetic fields there. Some statistical studies had linked magnetic disturbances on Earth to solar sources, as will soon be discussed, but these remained controversial. Hale was an expert on instrumentation and had invented (simultaneously but independently of Frenchman Henri-Alexandre Deslandres) the *spectroheliograph* in 1892. In Figure 3.10, the dark lines of the solar spectrum are shown. The lines indicate less light due to absorption of the "white" light of the photosphere at specific wavelengths by constituents of the atmosphere. The lines are dark, but not black. What light is present in them carries a lot of information about the solar atmosphere they pass through. In some cases (prominence spectrum at top in Figure 3.10), the lines are not dark but in emission. In the most general case, both absorption and emission contribute to the details of a line. By scanning the spread-out spectrum, made by an initial slit, over the detector (with usually a second slit), one can make an image in a specific line. The solar chromosphere features red emission at the hydrogen beta red wavelength of 656 nm, for example making the prominence of Figure 3.12 red. By working at this wavelength, features high in the solar atmosphere can be imaged as seen in Figure 6.9. As early as 1908, Hale, observing on Mt. Wilson with a large solar telescope he had constructed there, used the increased sensitivity of photographic emulsions to red light, to do pioneering studies of upper solar atmosphere forms which made him think of "solar vortices". He noted in a paper of that name that, near sunspots, there was a "Decided definiteness of structure, indicated by radial or curving lines, … as iron filings present in a magnetic field". In the same year, he used a magnetic effect on light known as the Zeeman splitting (Figure 6.10) to definitively conclude that sunspots harbored magnetic fields of up to about 0.5 tesla. This is about 10,000 times the field in Earth's polar regions.

The details of the Zeeman effect will not be gone into here, but it is not surprising that atomic energy levels, which, in a classical view, involve electrons going around a nucleus, constituting an electric current, should change in energy if that current is in an external magnetic field. The result is that, if the fine structure of lines is looked at with very high resolution, the lines shift in response to a magnetic field. Figure 6.10 shows details of an iron line at 617.3 nm, which can be picked out in the color part of Figure 3.10. In high resolution, the line is seen to be shifted (i,j strips) or split (k,l strips). In regions near the top and bottom, the line is neither shifted nor split and these correspond to the normal photosphere with low magnetic fields. If the line is shifted, it indicates one sign of a magnetic field present, with the shift being positive or negative depending on the line-of-sight magnetic field. If the line is split, it indicates that the field of view had both directions of magnetic field present. Polarization is also involved but is ignored at this level of description.

These techniques can be applied from Earth's surface or from space to look at the whole surface of the Sun to detect the magnetic field strength and whether it is up or down. They cannot, however, detect all components of the magnetic field. Figure 6.11 shows the resulting "magnetogram" with a comparative white light image for the same day. The large sunspots have a complex pattern of magnetic field, while some small ones seem to have more effect on the magnetic field than their size would suggest. It is also possible to note that regions of positive and negative (outward and

FIG. 3.—Hydrogen ($H\alpha$) vortices surrounding the bipolar sunspot of September 1924, photographed by Lewis Humason at Mount Wilson with the 13-foot spectroheliograph.

FIGURE 6.9    Hale's Hydrogen-alpha imaging showing vortical structure above a sunspot (Hale, 1927).

inward) magnetic fields are paired near sunspots, an effect discovered by Hale. In this sense, sunspots really are like magnets, as he envisaged.

Sunspots appear dark alongside the adjoining, less magnetic photosphere since the strong magnetic fields in plasma contribute to the pressure of the plasma. Gases respond quickly to pressure differences, and, as a result, the surface of the Sun is all at near the same pressure. A similar situation is observed on Earth, where weather is mostly due to gas moving around to try to equalize pressure. Steady pressure changes are, with height, perpendicular to the surface. Since sunspots have a magnetic field and thus an extra, non-thermal source of pressure, their temperature can be lower than that of adjoining gas, but their pressure is still the same. Viewed the other way around, without this extra source of pressure, the adjoining gas must have a higher temperature to give the pressure, since pressure in a normal gas arises from its thermal motions and thus temperature. The emission from photospheric gas, following blackbody rules, goes up as the fourth power of temperature. As a result, the sunspots, with lower temperatures, appear dark, compared with adjoining hotter regions of the photosphere. They nevertheless have a temperature of about 5,000 K, and, if viewed against a dark background, would be brilliant.

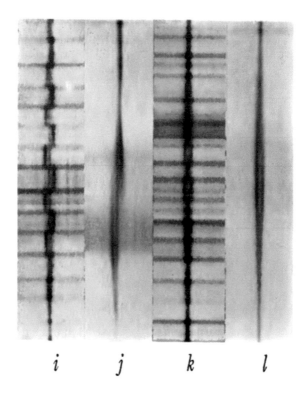

$$i \qquad j \qquad k \qquad l$$

**FIGURE 6.10** Zeeman effect in sunspot spectra. Strip j shows reversal of polarity (thus, magnetic field) in top vs. bottom sunspot (wider parts of trace) (Hale et al., 1919).

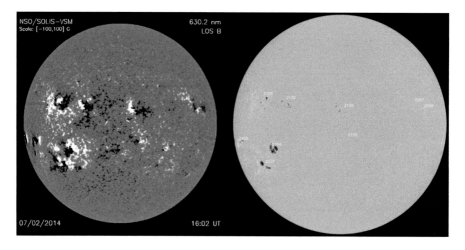

**FIGURE 6.11** Magnetogram (left) in 630.2 nm iron line, and sunspots (right) on July 2, 2014. Scale differs slightly (National Solar Observatory/SOHO-NASA, author montage).

Now, about one hundred years after his pioneering work, Hale's original insights into magnetic vortices, or, more precisely, loops, are made clear by ongoing monitoring of the Sun at X-ray wavelengths. Loops both against the background of the Sun and off its edge (corresponding to a prominence, likely also giving off H-alpha light as in Figure 3.12) are clear in Figure 6.12, taken by SOHO on September 6, 2017. Hard to miss in the image is the X-ray flare which is marked also by instrumental artifacts in the form of an "X" and a vertical bar. Not all flares lead to a coronal mass ejection (which can arise from above the level of flares, which are near the surface) but, in this case, one did. The mechanism for flares is believed to be the conversion of magnetic energy into particle energy by "magnetic reconnection". In this sense, flares have a relation to processes in Earth's magnetosphere which will be discussed in the next chapter. Figure 4.11 showed the detection of X-rays from flares by the GOES (weather) spacecraft in Earth orbit. GOES does not make images but simply detects the amount of incoming power, much like an exposure meter in a camera does. Nevertheless, the sharp rises in a short time, which, bearing in mind the logarithmic scale, can be by a factor of 1,000 in only a few minutes, show that the term "flare" is appropriate for these events. Flares are close to the photosphere and do not necessarily eject much material into space, but they are often associated with magnetic structures higher up which do. Such a coronal mass ejection, as seen in Figure 6.8, often has a curved structure, revealing its origin in magnetic field loops.

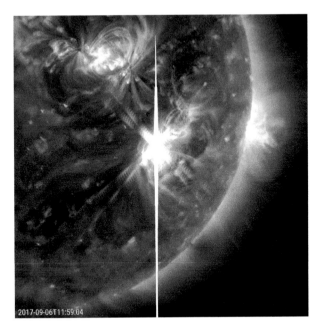

2017-09-06T11:59:04

**FIGURE 6.12**   X-ray image highlighting features in the low corona. The reader's attention is drawn to the loops at the upper left and off the edge of the Sun to the middle right but will inevitably turn to the major flare near the center (SOHO/NASA).

The corona in general features loop-like structures, and loops are often a sign of magnetic fields. Particles flow along field lines by orbiting them, and field lines usually appear to begin and end at the solar surface. The concentration of particles makes the field line visible because where there are more particles, there is more electromagnetic energy emission. The presence of loops can trace the magnetic fields that are present in the corona. Loop-like structures are more common near the equator of the Sun than near the poles. Where the outflowing solar wind gets entangled with loops, it is often slowed down. Near the poles, the emission seems darker and there are fewer loops. Here, the solar wind flows out more rapidly. It has been teased out with difficulty, because it is difficult to launch spacecraft to regions of space above the ecliptic plane, that there are two types of solar wind: slow and fast. Although the slow wind characterizes equatorial regions, and the fast wind polar regions, there can be pole-like regions at low latitudes, too. In this case, they are called *coronal holes*, since they appear dark. A small coronal hole is visible near the bottom left of Figure 6.12. Rather than being holes into which material is falling, coronal holes emit fast solar wind, not slowed down by being tangled up in magnetic loops.

Before visiting flare and coronal mass ejection effects in space, we will return to the more placid configuration of the Sun's magnetic field as seen in the polar regions in Figure 6.1. Those striations that look similar to the field lines near the poles of a magnet are due to average fields of under $10^{-4}$ T, a bit stronger than that of Earth. Locally, small (about Earth-sized) magnetic concentrations near the poles may have fields up to about 0.1 T, about as much as in a weak sunspot. Despite the weaker field than in sunspots and the active regions surrounding them, the polar fields have greater correlation with what goes on in the invisible Solar System gas dominated by the Sun's magnetic field, known as the *heliosphere*. Generally, the sunspot and active region fields close back on themselves, just as the field around a magnet does, and do not extend far into space. It is the larger-scale magnetic field that can get dragged out into space by the solar wind. One piece of evidence for this is the 11-year period for reversal of the Sun's large-scale magnetic field, also observed in the magnetic fields out in the solar wind. Parker's pioneering paper predicted and explained how this happens, although we do not actually know well how the Sun's overall field reverses.

Parker used considerations involving the special theory of relativity in explaining how the Sun's magnetic field spreads into a spiral form in the solar wind. Here, we will use a more intuitive approach. Recall that Parker's theory of the solar wind involved material moving radially outward, through the critical radius, where its motion became supersonic, and then continuing outward at high speed. However, the Sun is rotating, as Galileo's drawings of Figure 1.3 showed. The rotation is slow: relative to the stars, the (sidereal) rotation takes 24.5 days at the equator. Sunspots are usually at higher latitudes than this and in fact their average latitude changes throughout the solar cycle. The fluid body of the Sun has differential rotation, and higher latitudes rotate more slowly. So 25.4 days is usually used as the sidereal rotation period, and the rotation is in the same direction as Earth orbits. It takes 27.3 days for a typical sunspot to catch up with the Earth, and this period shows up in the solar wind. As with planets, it is referred to as the synodic period. A rotating body emitting material that goes radially outward may be looked at as a rotating water sprinkler as shown in Figure 6.13. A spiral *pattern* is found in the drops, but the individual drops are moving radially, not in a spiral. The solar wind flow is essentially radial.

Parker derived the magnetic field in his 1958 paper without referring to the current that causes it. In fact, there is a large-scale current sheet in the plane of the Solar System, made as part of a circuit caused by the Sun dragging its magnetic field through the plasma near it. These strong inner magnetic fields control the plasma flow and it essentially rotates along with the Sun. Although this rotation is slow it generates an electric field. In turn, this generates a current, although the details are complicated in the very high conductivity plasma of the solar wind. Figure 6.14 shows how this spiral current sheet generates a magnetic field *above* it. This diagram is appropriate for a stage of the solar cycle where the magnetic dipole field points to the north pole of the Sun. Not shown but below the current sheet, the field would be in the opposite direction but still in a spiral. We may think of the Sun as a large magnet with a field leaving the south pole, when this figure is appropriate, and entering the north pole. What happens in between is not simple like it is for a magnet with nothing surrounding it. There is a complex interaction with the radially flowing solar wind plasma distorting the field in the equatorial region. Mariner II's first long interplanetary voyage saw it equipped with a magnetometer with the initial goal of seeing

**FIGURE 6.13**   Water sprinkler. By "connecting the dots" one can follow a spiral pattern of drops back to the nozzle. The drops themselves, however, are moving radially outward, not on a spiral path (Shutterstock).

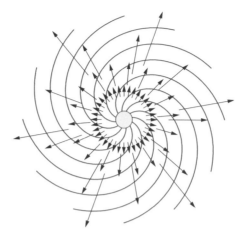

**FIGURE 6.14**  Magnetic field lines (red) in the solar wind and near-radial outflow of plasma (blue vectors). The arrow spacing decreases outward, indicating decreasing density, although the vectors become longer to show increasing flow speed. The magnetic field direction depends on position with respect to the current sheet as discussed in the text. The flow speed increases outward, although the magnetic field strength decreases. This is not indicated, only the spiral shape of the field lines, with the magnetic field being tangential to them (author).

whether Venus had a magnetic field like that of the similarly sized Earth. It does not! However, by operating during the flight to Venus, the magnetometer showed the spiral nature of the solar wind magnetic field for the first time.

Although Mariner II in 1962 established the spiral nature of the heliospheric magnetic field, an initially puzzling result came from the "Interplanetary Monitoring Platform" or IMP-1 spacecraft launched in late 1963. Its highly elliptical Earth orbit allowed it to "dwell" in the solar wind for long periods of time, monitoring its properties near our planet. The direction of the magnetic field, although about 45° from the Earth–Sun line, alternatively pointed either inward or outward. This was quickly realized to be due to the presence of a *heliospheric current sheet* nearly but not quite in the ecliptic plane. When Earth is above the current sheet, the field points in one direction, but if it goes below the current sheet, the direction reverses. The basic geometry of magnetic fields near a current sheet is shown in Figure 6.15.

Since the Sun's dipole magnetic field changes sign every 11 years, the magnetic field shown in Figure 6.14 would point outward above the heliospheric current sheet for about that period of time, and, for the next 11 years, it would point inward. In between these reversed periods, the dipole field is weak. We see that the solar cycle is actually 22 years long rather than the 11 usually discussed. The *number* of sunspots goes up and down with an 11-year cycle.

The solar wind current sheet is pushed up and down by plasma coming from active regions on the Sun. This is sometimes referred to as the "ballerina skirt", for reasons that may be obvious in looking at Figure 6.16. This view is a bit misleading since a skirt's material twirls with the ballerina, whereas, in the solar wind, the

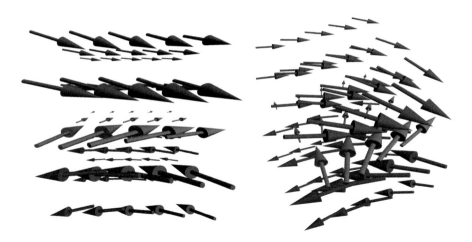

**FIGURE 6.15** Magnetic field (blue vectors) above and below a current sheet (yellow vectors) going out from the Sun in a partial perspective view. The view is from very slightly above the current sheet at left and from higher above at right. The magnetic vectors are at right angles to the nearby current, and both the field and current decrease in strength with distance from the Sun. Magnetic field direction is reversed above and below the current sheet (author).

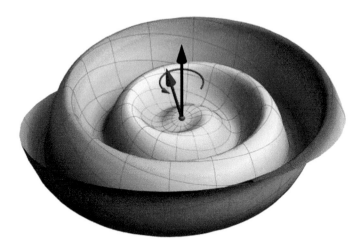

**FIGURE 6.16** Depiction of the heliospheric current sheet in three dimensions, also showing the Sun's rotational (black) and magnetic dipole (red) axes as vectors. Although plasma flow is radial, this pattern is established by conditions near the surface and thus rotates with a solar rotation period of 27.5 days. A planet or spacecraft near the ecliptic plane (between blue and red) will alternately have the current sheet above and below it, making the local magnetic field reverse (Orcinha et al., 2019: CC BY 3.0).

plasma does not rotate in this manner; rather, it flows directly outward. If the current sheet is blown upward by an active region, the field will reverse sign compared with when it may have been below the observer. Since those regions rotate with the Sun, there can be several current sheet crossings every solar rotation period (about a month). Now that we have spacecraft, including at the $L_1$ point, continuously monitoring, it is easy to see the current sheet move above and below us, as IMP-1 originally detected, as shown in Figure 6.17.

In the Parker model, for average solar wind speeds of about 400 km/s, the magnetic field at 1 AU should be either at 45° from the Earth–Sun line, or opposite this, which is 135°. Figure 6.17 shows this angle using real one-hour averaged solar wind data from near Earth in 2018, a year near the solar minimum, little disturbed by solar activity, and thus a good time to observe regular behavior. It can readily be seen that the angle is either −45° or 135°, with few values in between. This corresponds to the expected "Parker spiral" angle. Multiples of 654 hours plotted on the horizontal axis correspond to the synodic solar rotation period of 27.25 days. Apart from in the first interval (the plot starts January 1), when the pattern is a bit disturbed, there is a repeating up-down pattern in the angle. This corresponds to the current sheet being alternately above and below the Earth, with that pattern repeating as the same long-lived active region pushes the current sheet up or down.

When the Sun is active, this regular pattern is disrupted. Such periods are typified by the expulsion of coronal mass ejections, such as the one seen in Figure 6.8. Sometimes, these are in the same region as a flare lower down, but not always. The modern age of space science began on September 1, 1859, with a combination of flare and coronal mass ejection, of which the first was observed, and the other felt through magnetic and auroral effects at Earth about a day later. The link between these two, however, took one hundred years to firmly establish.

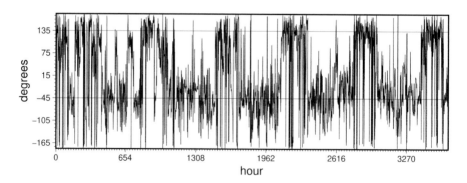

**FIGURE 6.17** Hourly average values of the solar wind angle from the direction toward the Sun for the first 152 days of 2018 (author, from data supplied by Papatashvili and King, 2020).

## 6.3 CARRINGTON SPEED

Having profited from the Industrial Revolution and the building of an Empire, the United Kingdom had many wealthy people in the mid-nineteenth century. While many of them led lives of self-indulgence, others spent their abundant free time in more serious pursuits. Lord Rosse's personal telescope in Armagh, Northern Ireland, was the largest in the world when built in 1845. Most famously, it had revealed spiral structure in "nebulae" later found to be galaxies far outside our star system. The wealthy amateur observer Lord Carrington studied the much closer Sun by day. The Sun's surface (which must *never* be observed without adequately protected instruments) presents many details to study. Carrington began solar studies in 1853, at the age of only 27, using a 10-cm aperture refracting telescope to project a large solar image safely onto a screen. This allowed a large amount of detail to be seen, and Earth's rotation caused the image to cross the screen to allow precise determination (through timing) of positions of features on the Sun's photosphere. Carrington was fascinated by large and complex sunspots, which had decorated the Sun in the summer of 1859, repeating their crossings of the disk every month or so. In this epoch, when photography was still young, Carrington made detailed sketches of the dark spots against the bright photosphere. Suddenly, on September 1, in the region between two large spots, a very bright patch appeared: a rare *white light solar flare* (Figure 6.18). While flares were known before this time, this was the brightest Carrington had ever seen. It was also very transient; having gone to find a witness to this remarkable

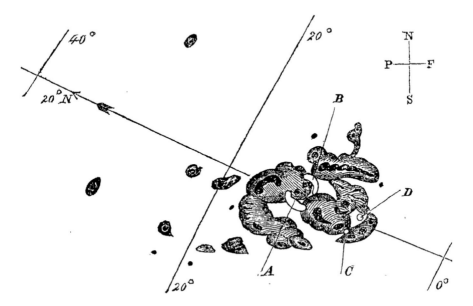

FIGURE 6.18    The Carrington white light flare of September 1, 1859. Regions A and B lit up brilliantly but briefly (Carrington, 1859, courtesy: Royal Astronomical Society).

event, Carrington found that it was no longer visible when he came back to the telescope a few minutes later. He was nevertheless able to record the event in a detailed sketch. Two associated phenomena could not be seen by Carrington: *X-rays*, very short wavelength electromagnetic waves emitted by the hot gas of the flare (as in Figure 6.12), as well as by electrons spiraling in its magnetic field, and a *coronal mass ejection*. The latter arose far above the surface, and, as the name implies, a large ball of gas was ejected from the corona into interplanetary space. Unbeknownst to Carrington, the nearby magnetic observatory at Kew, near London, recorded a small magnetic spike at the time of the flare. Such a so-called *crochet* is due to increased ionization in Earth's *ionosphere*, leading to electric current patterns changing. This ionization is due to X-ray electromagnetic radiation, traveling at the speed of light and arriving at the same time as the visible light photons, which allow the flare to be seen. The light travel time from the Sun to Earth is seven minutes. The coronal mass ejection (henceforth, CME) consisted of ordinary matter, entangled with magnetic fields, and thus had to travel much slower, arriving about 18 hours later. When it did arrive, the largest *magnetic storm* ever recorded took place. No longer showing only a small crochet, the magnetic recordings at Kew went off the charts.

Taking about 18 hours to traverse one AU means that the gas of the CME traveled at about 2,000 km/s. This remains the fastest one ever clocked, adding to the superlatives about what is now known as the "Carrington event". Its effects on nascent technologies of the time were notable and would have wrought havoc on modern technologies had they existed. Telegraph communication was disrupted, with sparks flying in telegraph offices. Magnetic compasses, then still essential for navigation, were greatly disturbed. A worldwide auroral light show took place, with auroras seen as far south as Havana.

Strange to say, the Carrington Event's evident connection between events on the Sun and on Earth did not seal a debate that had been going on at a low level for decades. Although Newton's invisible gravity force had forever linked the heavens and Earth, a Greek-like separation of these domains prevailed in other regards. The void between the planets was believed to be exactly that: a void or perfect vacuum. Nothing could be transmitted through it. The Sun was a mysterious place: the Industrial Revolution was powered by coal, so that seemed a natural source of its energy. The sunspots so meticulously sketched by Carrington could not be explained by any contemporary science; in this respect, not much had changed since Galileo almost exactly 250 years earlier had shown that such blemishes existed on the Sun's otherwise perfect surface. Magnetism was a mysterious and invisible force. Its link to electricity was well known, since Faraday had practical views and promoted devices in which his eponymous law saw rotating magnets produce electricity in generators, and motors which caused rotation by a related effect. The full theory of electromagnetism remained in obscure publications of the Scot Maxwell. The ionosphere was unknown, so even the magnetic crochet directly associated with the Carrington flare had no obvious mechanism. A flash of light on the Sun and a magnetic needle on Earth deviating could more plausibly be explained by chance than by any theory. Similarly, if a magnetic storm and great auroras followed such a flare, there was no clear evidence of a link since CMEs are invisible.

Magnetic storms were already well known at the time of Carrington. Earth's magnetic field had been proposed by William Gilbert, court physician to Elizabeth I of England and Ireland, in the sixteenth century. Lodestones, or natural magnets, had been known to the Chinese at least 500 years previously, and their transformation for practical uses in compasses was perfected by Gilbert's time. Long-ranging voyages established not only that there was a global magnetic field, making compass needles always point more or less northward, but that there were some complexities to this. It was known that the north magnetic pole to which they in fact pointed was not at the north geographic pole, giving rise to *declination*, the angle between true and magnetic north. With the age of global navigation beginning, the determination of longitude was a major commercial and military problem, whose partial solution lay in determining charts of declination. In other words, although magnetism was an invisible force, as the influencer of compass needles, the characteristics of the Earth's magnetism were well established. Magnetic instruments, although mechanical in nature, were relatively sophisticated and accurate. The compass holds a magnetized needle horizontally, and it responds to only the horizontal part of magnetic force on it. With some small amount of damping (supplied for example by having a fluid-filled compass), the needle will settle down pointing to magnetic north. Gilbert's theory of Earth as a giant lodestone also indicated that magnetic field lines would point into the Earth in the northern hemisphere (ironically, in modern magnetic parlance, this means that the north magnetic pole is, magnetically, a south pole).

The angle of field lines from the vertical is known as the *magnetic inclination* or *dip*. A "dip needle" is a device like a compass, with the needle suspended in such a way as to show the angle of magnetic field lines from the vertical. Figure 6.19 shows Gilbert's diagram overplotted with a dipole field. He drew small bars to show what a dip needle's angle would be, and these line up very well with the modern calculated dipole. On the closest field lines to the pole, a color scheme has been used to indicate field strength. Gilbert thought that the field was felt only within a sphere of influence marked "*Orbis Virtutis*". This was likely due to observations that magnets brought close to each other seem to very suddenly start to attract. A magnetized needle suspended so as to be able to rotate only upward and downward will align with the magnetic field in such a way as to indicate the inclination, and dip needles were carried worldwide for global surveys. The magnetic field also varies in intensity, being stronger near the poles than near the equator. This was also measured early in the age of navigation. A needle will overshoot and oscillate about the direction of the field, a tendency mentioned above that is damped out in a compass. If, instead, the needle swings freely, its frequency of oscillation is higher in a larger field. By this method, a map of intensity of the global field could also be made. Not only were spatial changes in the field studied, but also their changes with time. The declination changes, especially in northerly regions, since the north magnetic pole moves. We now know that this is due to changes in the generating regions of the main geomagnetic field deep in the Earth.

The rate of change of declination is important in navigation and is indicated on charts so that one can find true north on a certain date from a measurement of magnetic north. This change is very slow, but the deep study of Earth's magnetic field

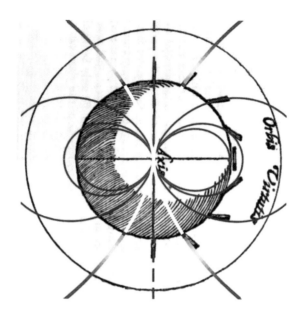

FIGURE 6.19  Gilbert's portrayal of Earth as a magnet from 1600 overplotted with dipole field lines. His dip needles line up well with the inclination. North at top (public domain, with overlay by author).

also showed that there are times when the magnetic field changes over much shorter periods. Usually, the changes were so small that microscopes were used to observe the motion of compass needles. Such observations showed that there could be periods of up to several days in which the needle deviated from its usual position. By the time of Carrington, some correlations between times of magnetic storm prevalence and the *sunspot cycle* had been noted. However, such circumstantial evidence did not prevail against the common-sense notion that Earth was connected to the Sun through gravitational attraction and nothing else. That Earth had a second invisible force field, magnetism, was firmly established, but it took the realization, due essentially to Hale in the early twentieth century, that the Sun was also magnetic to move a small step toward understanding the origin of magnetic storms.

Prior to Hale's advance in the physics of sunspots, they had received attention from astronomers since their discovery in 1610 by Galileo. Small telescopes were adequate to count them, the abundant light of the Sun making it possible to project images onto a screen. Strangely, one of the first things the newly counted sunspots did was disappear. For several decades, a period around 1700, now known as the *Maunder minimum*, no sunspots were observed despite adequate observational equipment being widespread. However, apart from this, an overall pattern emerged, which was a *sunspot cycle* or, more generically, a *solar cycle*, of 11 years duration, at the minimum of which few or no sunspots were visible, and at the maximum a number up to several hundred, as shown in Figure 6.20. The depths and peaks of this cycle varied, and the length also could vary, although usually not as much.

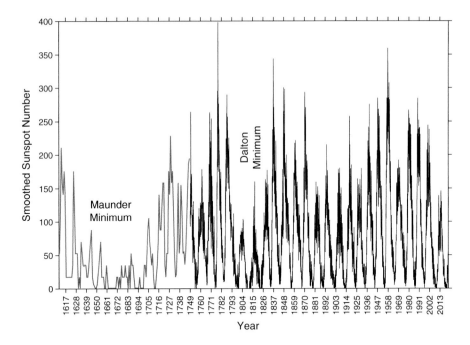

FIGURE 6.20 Sunspot number by year (red) and by month (black). Date labels are at 11-year spacing (author, data source: WDC-SILSO, Royal Observatory of Belgium, Brussels).

The distribution of sunspots in latitude was found to vary systematically through the cycle. After the discovery of the magnetic nature of sunspots by Hale, and the ability to investigate this in detail, it was found that the overall magnetic nature of the Sun reversed every second cycle, so that the solar cycle is really 22 years long. Much like the declination drift of the Earth's magnetic pole, the solar cycle appears to be related to the generation mechanism of the overall solar magnetic field within its body. In both cases, the exact mechanisms appear to be complex, but as mysteries of the hidden Solar System, rather than the invisible Solar System, they are not discussed further here.

As of the early twentieth century, magnetism had been extended to the Sun, explaining sunspots. Circumstantial evidence linked the sunspot cycle to magnetic storms. But the evidence was not yet in place to make a connection across the perceived perfect vacuum of the invisible Solar System. Earth was magnetic, and the Sun was magnetic, but how were the two linked? We have one more visit to make, to the outer reaches of the Solar System, before returning to this question in the next chapter.

## 6.4 BEYOND THE BARRIER

One might say that we have a privileged view of the solar wind since we are near the heliospheric current sheet. We have little choice since the orbital mechanics of the

Solar System place Earth in the same plane as the other planets, and the solar equatorial regions, and it takes a lot of energy to get out of that plane. Similarly, of course, Earth is in the inner Solar System, so finding out what is at its outer boundary is difficult and requires a huge amount of energy to climb out of the solar gravity well. The techniques to overcome these barriers to explore both the out-of-plane and far-out limits of the Solar System are triumphs of space navigation, and are described in the final chapter. However, we will start our voyage to these extreme limits in a surprising place: the kitchen sink!

The flow of water in Figure 6.21 is very like the solar wind. Rather than from the Sun, it flows out radially from where the faucet stream impacts the flat pan. Although recent studies have shown the situation to be a bit more complex, the flow may be regarded as supersonic; any small ripples starting near the edge are not going to be able to move in toward the center since they would be moving more slowly than the outflowing water. A striking feature is that the outer water does not appear to be flowing very fast, and it is much deeper than in the central fast-flowing region. More subtly, but also like the solar wind, near the center there is no discrete feature marking the flow becoming supersonic. We can only infer such a change at the other boundary where the flow slows down, and there is a step in depth. Such a feature is called a "hydraulic jump". In a steady flow, this jump, or shock wave, is stationary. It marks a transition from supersonic to subsonic flow. Clearly, there is also some complex interaction leading to the formation of bubbles. A lot of physics in the kitchen sink!

FIGURE 6.21   Water flow from a faucet onto a flat pan in a kitchen sink. The radial flow is like the supersonic solar wind, and the sudden change in depth is called a "hydraulic jump" (author photo).

Such an analogy must stimulate thought about what happens at the outer edge of the Solar System. The solar wind is supersonic. We know through means beyond description here (except to note that one of them involves Faraday, indeed related to the device he is holding in Figure 3.2) that interstellar space, like our own Solar System in defiance of Newton's concept, is not a vacuum. Even if that interstellar gas is flowing, there must be at least one place where the supersonic solar wind hits it, much like the supersonic water hits the deeper, slower water at a hydraulic jump. In other words, although the solar wind has a quiet (albeit hot) beginning, it should end in a shock wave.

Through gravitational boosts of the sort described in detail in the final chapter, the twin Voyager spacecraft have both crossed out of the heliosphere (Figure 6.22) at about 120 AU from the Sun. The "termination" shock which marks the end of the supersonic solar wind was crossed at about 94 AU out by Voyager 1 and 84 AU out by Voyager 2. Boundaries are often indicated by the suffix "-pause", and the outer boundary marking leaving the heliosphere is the *heliopause*. Outward of the shock, one finds compressed solar wind, much as in the kitchen sink, where one finds slower-moving but deeper water. This region is called the *heliosheath*. Analogous to the turbulence-creating bubbles in the water flow, the shock is marked by plasma waves. However, we will focus on bulk properties. Some analogies can also be drawn to the interaction of the solar wind with planets as discussed in the next chapter. It is also worth noting that the heliopause is a frontier of research as well as of the Solar System, so some aspects like its overall shape are controversial.

The plasma instrument of Voyager 2 measured out to the heliopause as shown in Figure 6.23. The Parker theory calls for the solar wind speed to level off, as may be seen in Figure 6.6, which only goes to 6 AU. With variations on the time scale

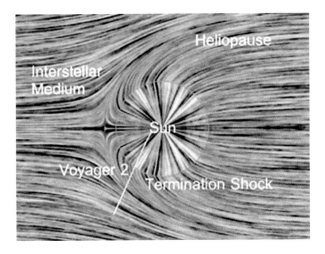

**FIGURE 6.22**   One view of the heliosphere and interstellar medium, showing the solar wind radial flow (yellow), the heliosheath (orange) and the path of Voyager 2 (white line) (modified from Richardson et al., 2022).

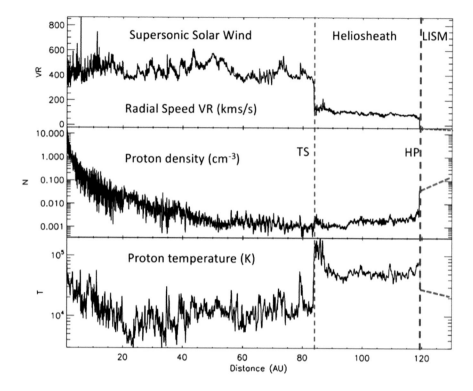

**FIGURE 6.23** Solar wind radial speed, proton density, and temperature. Black lines are from Voyager 2. Transition shock (TS) and heliopause (HP) are marked. Lower limit is 1 AU (Earth) (modifed from Richardson et al., 2022).

of years typifying Voyager 2's flight out to the transition shock (launch 1977; shock crossing 2007), the solar wind speed as seen in the top panel of Figure 6.23 was about 400 km/s. The density in the Parker theory falls off as the inverse square of distance; in the semilog plot of the middle panel, this rate seemed to vary a bit but certainly trended downward as the density of an expanding flow must. Although on average higher near Earth, proton temperatures (bottom panel) got down to about $10^4$ K in the outer Solar System. High coronal temperatures (of the order of $10^6$ K) force the solar wind to become supersonic, which is a directed high-speed flow. On the other hand, the gas temperature itself is reduced considerably. Until its heliopause crossing, which took place in 2018, Voyager 2 was in the heliosheath. Here the flow speed was reduced to being subsonic, about 100 km/s, with a slight increase in density, but with an increase in temperature due to the gas being heated going through the shock. Not shown here, the magnetic field increased in the heliosheath, due to compression with the magnetic field "frozen in" to the gas. The magnetic field as measured in the local interstellar medium is slightly higher than in the heliosheath, possibly because it is also compressed through motion. It is expected that the termination shock is basically stationary, much like the hydraulic jump in a kitchen sink.

## 6.5 FAR OUT (OF THE PLANE)

Already in Figure 6.1 it was clear that the corona in the polar regions of the Sun was different from that in the equatorial regions. Most of the discussion above had to do with the equatorial plane of the Sun, which is not tilted very much, so that is also approximately the ecliptic plane in which planetary orbits lie. Since the beginning of the Space Age, but especially since the Mariner II flight of 1962, we have monitored the solar wind near the Earth almost continuously. We now regularly have spacecraft at the Earth $L_1$ point continuously monitoring both the Sun and the solar wind. Many interplanetary spacecrafts carry solar-wind-measuring instruments. These measurements are, however, restricted to the equatorial plane.

Due to the conservation of energy, it is just as difficult to get spacecraft in near the Sun as it is to get them to go outward. The Parker Solar Probe uses multiple encounters with Venus to get closer to the Sun. Since Earth goes around the Sun at 30 km/s in the prograde (counterclockwise as seen from the north) sense, spacecraft that escape Earth's own gravity find themselves in space on an Earth-like orbit, which is in the ecliptic plane and traveling at about this speed. Getting out of the plane in order to sample the solar wind, expected to be different there, is difficult. The Ulysses mission of the European Space Agency and NASA, launched in 1990, used a gravity assist from Jupiter (see final chapter) to be boosted out of the ecliptic plane to directly sample the high-latitude solar wind. The basic results are shown in Figure 6.24. The 300 to 400 km/s flows in the ecliptic give way to much faster flows once one is at higher latitude (above about 25°) than where sunspots and active regions are typically found. The poles of the Sun are like coronal holes (a small one was visible in the UV in Figure 6.12). Lacking magnetic loops (recall Figure 6.1), which constrain plasma outflow, the polar regions have high-speed flow, about 1,000 km/s. This is referred to as "fast solar wind".

The effects of the solar wind on Earth are due to the solar wind in the ecliptic plane. Even in equatorial regions, there can be large coronal holes. They also emit fast solar wind. As the Sun rotates, the passage of such regions can be regular. To some degree, if one sees activity at one time, similar activity may be expected at one solar synodic period of about 27 ½ days later. Much of the repetitive character seen in Figure 6.15 is due to patterns involving coronal holes near the Sun's equator.

## 6.6 ANOTHER WAY OF SEEING

Although the story of how we have done so is rather amazing, involving space missions lasting over 45 years, caroming off giant planet gravity fields, and listening to signals as weak as those from a cell phone across the whole Solar System, there have been only two probes to date, the Voyagers, that have crossed into interstellar space. The New Horizons mission, launched in 2006, had as its main target the dwarf planet 134340 Pluto. Pluto is part of the outer Solar System's Kuiper belt of small objects, which have not been discussed in detail but have some relation to asteroids and short-period comets. Famously, Pluto was "demoted" from being a planet in 2006 as it was realized that it was more related to these small bodies than to "real" planets.

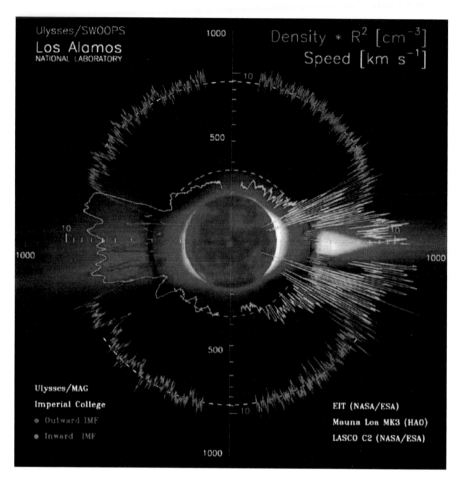

**FIGURE 6.24** Solar wind speed as a function of angle around the Sun through the poles, also color-coded by direction of magnetic field (McComas et al., 2000). Data is from the Ulysses spacecraft, placed in polar orbit around the Sun as described in Chapter 9. The speed, shown with axis labels reaching 1,000 km/s, is high and remarkably constant at high latitudes. The solar magnetic field direction, which reverses every 11 years approximately is mostly outward (red) in the northern hemisphere, and inward (blue) in the southern. The density (scaled to 1 AU) is shown in green as being low at high latitudes, and higher but variable at low latitudes. The general appearance of the Sun and corona near solar minimum is shown as a background image.

Having explored Pluto and its satellites, and Kuiper Belt object 485968 Arrokoth, New Horizons will cross the termination shock in the mid-2020s. It carries plasma instruments but not a magnetometer. Since *in-situ* measurements are so few, a way to sample the extreme reaches of the Solar System remotely is needed. In common with many things discussed in this book, the region is characterized by thin gas which is basically invisible. We have not yet discussed this effect from magnetic fields much, but charged particles cannot travel long distances in a straight line in magnetic fields since the fields deviate them. The best answer for remote sensing using particles is to use neutral particles. Fortunately, these do exist, although in very small numbers.

The topic of charged particles being tied to the magnetic field is treated in more detail in the next chapter. Here, we simply note that particles move in a spiral and cannot move very far perpendicular to the magnetic field direction. Although the gas in deep space is not dense and collisions are rare, they do occur. A common constituent is ionized hydrogen, shown spiraling around the vertical magnetic field in the left panel of Figure 6.25 as a small red sphere. It hits a neutral atom at rest, shown by a larger blue sphere, and, in this collision, a small kick is given to the neutral, but more importantly, an electron is taken from it to make the hydrogen neutral. Now, the atom will spiral around the magnetic field, but the hydrogen is now a neutral atom (thus shown as blue) and can move off totally free from being bound by the magnetic field. It still retains a lot of energy and so is called an "energetic neutral atom" or "ENA". Since there are not many collisions, the neutral atom may move large distances, even traversing the entire Solar System. Of course, since there are not many collisions in the source region (at least in the outer reaches), not very many ENAs are generated. Nevertheless, it is possible to build ENA detectors to operate in space and make images of the totality of ENAs coming in from various directions.

ENAs have been observed from Earth's magnetosphere, and an ENA imager was on the Cassini mission to Saturn. The Interstellar Boundary Explorer (IBEX) satellite, launched in 2008, made all-sky maps of the emissions from the heliosheath. Almost immediately, a concentration of emission in a ring around the sky, which became known as "the ribbon", was apparent in the data (Figure 6.26). Subsequently, it has been possible to observe changes in the ribbon over a solar cycle. Inferences have been made of a possible link to the local galactic magnetic field (which as

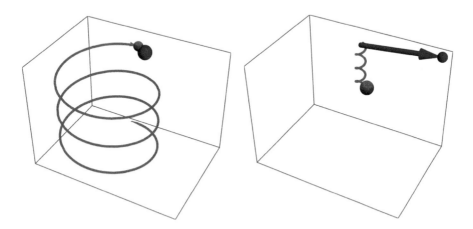

**FIGURE 6.25** Liberation of an ion through charge exchange. Positive charges are red and neutrals are blue. An upward-spiraling proton (left) hits a neutral atom at rest. Charge exchange has an electron move to the hydrogen, which then becomes neutral and can move far (author).

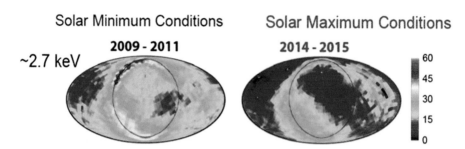

**FIGURE 6.26** IBEX observations of 2.7 keV energetic neutral atoms under varying solar wind conditions. The striking feature of the "ribbon" is not fully explained (Galli et al., 2022).

mentioned above can now be measured directly by the Voyager spacecraft), but the explanation of the concentration of ENA emission in the ribbon is still not completely settled.

## 6.7 SUMMARY

Although the energy in the solar wind is only a small fraction of a percent of that emitted by the Sun as light, it has more complex interactions with the Invisible Solar System. The combined particle and magnetic fields in the solar wind lead to interesting phenomena which may include shocks and anomalous temperature structures. Near the Sun, the corona is too hot to be static yet we do not have a full explanation for its heating. The solar wind is supersonic from near the Sun until it comes to the outer boundary shock that in many ways marks the edge of the Solar System and, by definition, is the edge of the heliosphere. Plasma is not able to cross magnetic fields very well, and outflow from the Sun is mainly from the poles or from "coronal holes" where the magnetic field structure is not looped. Energy can be released impulsively from areas that have strong magnetic fields, likely through magnetic reconnection, giving rise to flares, or higher in the corona, to coronal mass ejections. Similar to how plasma cannot readily cross magnetic fields, individual charged particles cannot either. However, there are ways in which they can convert to energetic neutral atoms and convey information across long distances. The Carrington event suggests that interactions become more complex, and may have an impact on our technological civilization, when a magnetic planet, like the Earth, is involved. This is part of the subject of the next chapter.

# 7 Blowback

In the previous chapter, we made the surprising observation that all the planets are in the outer atmosphere of the Sun. Unlike the neutral gas that characterizes at least the lower parts of those atmospheres clinging to planets, that of the Sun is electrically conducting and carries a magnetic field. Because it also is moving very fast, it is natural to expect that it would form downstream disturbances in the form of eddies or "tails" (Figure 7.1) when it encounters an obstacle like a planet. Since it also has a magnetic character, this interaction is more complex when it involves a planetary magnetic field such as that of the Earth. An important aspect of solar wind flow is that it is supersonic.

Although they may be hard to detect on a slowly flowing river, in principle, small waves generated where the water splashes on hitting an obstructing rock can travel upstream since they move faster than the river flow. In supersonic flow, this cannot happen since the flow is faster than the wave speed. In air or plasma, like the solar wind, one type of wave is the sound wave. Strictly speaking, in a magnetized plasma, the analogous wave is called a *magnetosonic* wave. In sound, the fluid is compressed and rarified alternately, with this disturbance moving out from the source. In a magnetosonic wave, both the gas and magnetic field compress together and then become less compressed, much as in a sound wave. A loudspeaker operates by having a surface move in and out at the rate of the sound being generated, alternately compressing and expanding the air near it. This goes out at the speed of sound, which, in air, is conveniently about 333 m/s or 1,234 km/h. For an emitter of sound traveling faster than this, the sound cannot get ahead of its source, and "piles up" into a shock wave.

FIGURE 7.1 In a smooth subsonic flow, like that of a slow river, water gently flows around obstacles, making a wake analogous to that seen behind planets in the solar wind (author photo).

DOI: 10.1201/9781003451433-7

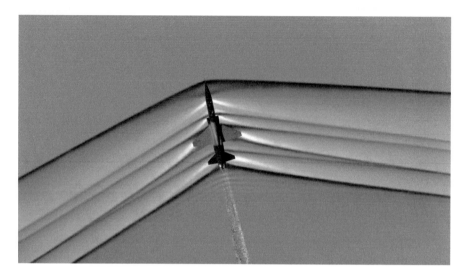

**FIGURE 7.2**    The shock waves associated with a sonic boom from a jet aircraft, made visible through modern Schlieren photography techniques (NASA).

Ahead of this "sonic boom" (Figure 7.2), no sound is heard. In a plasma, things are a bit more complex but similar phenomena occur, including most notably the formation of shock waves.

Although there are some common features with water flow around obstacles, the interaction of the solar wind with solid bodies in the Solar System is more complex due to it being supersonic and magnetized. The degree of complications of the interaction due to magnetism largely depends on whether the body with which the solar wind interacts is itself magnetized. Internally generated magnetism is a subject whose details are beyond the scope of this book. Suffice it to say that small bodies cool quickly, and we observe that the planet-sized bodies with internally generated magnetic fields are large. Mars is smaller than Earth and no longer has a "main" magnetic field, although there is evidence that it once did. Some remanent magnetism at the surface leads to interesting magnetic phenomena at low altitudes, but these are a detail we will not consider here.

Earth, of course, does have a magnetic field, as evidenced by the use of compasses or their modern electronic equivalent. The fact that Venus, Earth's "twin" in being almost exactly the same size, does not have a sizable field, as shown by Mariner II, is likely related to its very slow rotation, which does not lead to the "stirring" of its interior. Continuing down to bodies the size of Earth's Moon, cooling happened long ago, so that the internal dynamos needed to make a magnetic field are no longer possible. Some planetary moons are in this size range, but, not being alone, have not cooled. For example, Jupiter's innermost large moon Io (about the size of Earth's Moon) has active volcanoes. This is certainly evidence that Io's interior has not cooled, which is an effect due to tidal flexing by Jupiter. At the other extreme, all of the giant planets have magnetic fields, having not cooled over the lifetime of the Solar System, and having internal properties favorable to active dynamo field generators. Obviously, very small bodies like asteroids and comets do not generate their own fields. This does not preclude some interesting interactions with the solar wind, especially for comets, which release gas when near the Sun.

## 7.1 UNPROTECTED BODIES

The solar wind near Earth contains, on average, 10 particles per cc, i.e., $10^7$ particles per cubic meter, so at the nominal speed of 400 km/s or $4 \times 10^5$ m/s, we get $4 \times 10^{12}$ particles flowing through each square meter of a surface every second. Most of these particles are protons, each with a mass of about $2 \times 10^{-27}$ kg, so that the solar wind mass flux is about $8 \times 10^{-15}$ kg/(s m$^2$). One cubic meter of a rocky surface has a mass of 3,000 kg/m$^3$ and a further bit of math (with $3 \times 10^7$ s in a year) shows that the solar wind would need to flow onto a surface for about 100 million years before bringing in a mass even vaguely comparable to that near the surface. We can thus simply imagine that the solar wind is absorbed into rocky surfaces, perhaps some protons recoiling a bit, others just going into the material. The upshot of this is that a cavity is created *behind* the body; if the solar wind is already very close to a vacuum, then behind a rocky body this is even more true. In the direction away from the Sun, the cavity will eventually fill from adjacent solar wind. How fast it fills is determined by the magnetic field direction since particles or plasma do not readily travel across magnetic fields, but readily flow along them.

A good example of solar wind particle loss to the Moon is clear in Figure 7.3, showing data from two Acceleration, Reconnection, Turbulence and Electrodynamics of the Moon's Interaction with the Sun (ARTEMIS: an unmanned project, not to be confused with the mid-2020s Artemis manned mission) probes, one near the Moon and one passing behind it. The lunar "shadow" in both solar wind ions and electrons is very clear. If one examines solar wind magnetic data (not shown), there are signs of an interaction, which, to some extent, is caused by ions ejected from the surface, usually due to interaction with solar wind electrons.

While we have emphasized above that the solar wind brings negligible mass to the lunar surface, its cumulative effect over the billions of years available for it to act can be detected. This effect, which takes place on all airless bodies in the Solar System, is an example of *space weathering*. We must be careful to distinguish this term from *space weather*, which is an effect on technological systems. Space weathering is more analogous to weathering in geology, with the effect of meteorological phenomena on rock mainly being to break it down through the action of water and frost. Similarly, the solar wind is not the only agent of space weathering; such things as micrometeorite impact are also important on unprotected surfaces in space.

Figure 7.4 shows part of the Chaplygin crater, as photographed by Apollo 11 in 1969. This crater is about 100 km in diameter, larger than many that can be seen on the lunar nearside with binoculars. It is, however, found in the lunar highlands of the far side of the Moon. The far side, of course, is not a "dark" side and faces the Sun and the solar wind for about two weeks of the month. Notably, the action of more recent impacts than the ancient one which excavated the large crater can be seen

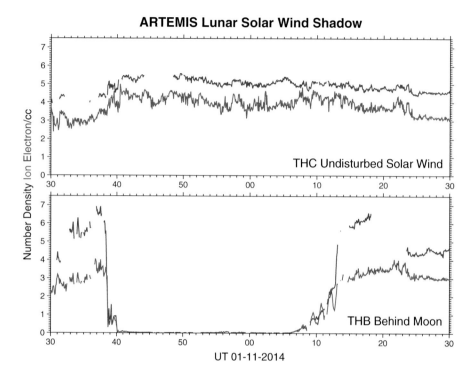

**FIGURE 7.3**  Solar wind-charged particle density from THC probe in solar wind and THB which passed behind the Moon (author, from NASA/ARTEMIS data).

through the presence of the much smaller (about 10 km diameter) Chaplygin B crater on its rim. The material excavated by this more recent impact is much brighter than the older material surrounding it. This is evidence of space weathering; exposed surfaces tend to get darker when exposed to the Solar System environment, including the solar wind. On the near side, this effect may be directly seen with binoculars since the crater Tycho, smaller than Chaplygin, has bright "rays" extending from it. These are most noticeable at times of the full Moon (Figure 7.5). The dark areas making up the "man in the Moon" are ancient basalt lava flows in basins made by ancient impacts (Late Heavy Bombardment of Chapter 4), whereas the areas that are brighter overall are highland terrain.

We have far more information about the interaction of the solar wind with the Moon than with any other airless body, in part due to the long presence of the ARTEMIS probes. Nominally, the small planet Mercury is in the same size class as Earth and several other large moons of the Solar System. For reasons not fully explained, it has a magnetic field large enough that its interaction with the solar wind is more like that of Earth than of our Moon. Several of the other large moons are inside the magnetospheres that magnetized planets produce, leading to a further degree of complication. Indeed, Earth's Moon sometimes enters our own magnetosphere, but for only a brief period of its month-long orbit. Our basic conclusion

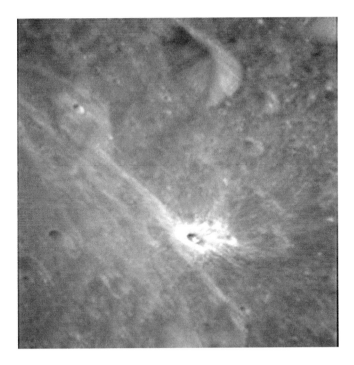

FIGURE 7.4    Large Chaplygin crater (lower left) with a small, fresh crater on its rim (NASA).

FIGURE 7.5    Contrast-enhanced full Moon photo to show the rays of crater Tycho (bottom right). The Moon is usually in the solar wind, so surfaces darken over time. Young impact crater ejecta, including rays, is still bright (author photo).

is that airless bodies directly interacting with the solar wind will basically absorb it, creating a cavity behind them that gradually fills in with distance downstream. They may further have ions ejected from their surfaces and may slightly modify the magnetic field around them. Long-term space weathering occurs on such surfaces and tends to darken them. These are the simplest interactions possible, and bodies with atmospheres and magnetic fields have more complex interactions. When close to the Sun, comets, although having a nucleus about the size of an asteroid, emit gas, and thus will be considered among bodies with atmospheres as we look at how that affects interactions with the solar wind.

## 7.2 BODIES WITH ATMOSPHERES

Atmospheres are fragile shells around bodies with sufficient gravity. Our Moon, for example, cannot retain gas of any significant density near its surface. We are now all too aware of the fragility of Earth's atmosphere, with small changes in composition able to affect the heat balance at the surface where we live. Other atmospheres may have changed with time also. Many changes on Earth are for intrinsic reasons. Humans are not the first living beings able to affect our atmosphere; for one thing, the oxygen we all enjoy likely arose due to the geologically ancient development of plants that liberate this gas through photosynthesis. In other cases, such as that of Mars, the solar wind may have played a large role in stripping away the atmosphere. On a shorter timescale, and forming part of the invisible Solar System, atmospheres can deviate the solar wind if this is not already done by a magnetic field.

Atmospheres on planets or moons with significant gravity are held up by gas pressure. Near the surface, as in our own atmosphere in regions that have "weather", atmospheric gas is well regarded as a fluid. As we have used the habitual measure of ions per cubic centimeter (cc) in discussing the solar wind, we note that air at sea level mostly holds molecules (nitrogen, $N_2$, and oxygen, $O_2$) and there are about $2.5 \times 10^{19}$ of them in one cc, as compared with about 10 ions/cc in the solar wind. Although there is a distribution of speeds among molecules in air near sea level, a typical speed would be about 400 m/s rather than the (outward-directed) 400 km/s typical of the solar wind. We consider 100 km/h (60 mph), which is about 28 m/s, as a reasonable speed for driving a car. However, we would consider this to be a rather strong wind: in fact, it is the threshold of hurricane strength. Wind is a bulk movement in one direction in some geographical region, but the molecular velocities in air are in all directions, and the typical path of a molecule before colliding with another is only about 60 nm (about ten times less than the wavelength of light). As a result of the collisions, air has pressure, directed in all directions, and it is characterized by a temperature related to the average speed. The pressure near the surface supports the weight of the atmosphere above it (more correctly, the gradient or pressure changes with height does so). And so it goes, with pressure gradients from below supporting less and less overlying air as one goes upward. These concepts are from gas theory, but eventually one gets to a level where gas theory breaks down, largely through the path length of the fastest molecules getting large. This gives essentially the definition for the "edge of space", which is also known as the "exobase".

For molecules such as oxygen and nitrogen significantly present in the atmosphere, the exobase is at about 500 km altitude, although this varies with solar activity. Since many satellites orbit at altitudes less than this, technically they are in the upper atmosphere. These low-Earth orbit satellites will either have their orbits decay through interaction with the rarified upper atmosphere or must occasionally "boost" themselves using small rocket engines. An example of solar activity affecting satellites was seen in the case of a SpaceX Corporation launch of 49 "Starlink" satellites in February, 2022. Such satellites are initially launched into a low orbit, at about 300 km altitude, for an initial check. Those that pass inspection on-orbit are boosted into a higher orbit (above 500 km) where atmospheric drag is minimal. Those not passing inspection can either be de-orbited, using thrusters, or, in the case of more severe failure, will come down due to the drag present in the low orbit. In this instance, a solar storm heated and expanded the upper atmosphere so that 40 of the 49 satellites experienced unexpected drag and were taken out of orbit before being able to be boosted to safety.

Having molecules of much lower mass, hydrogen, which is only marginally present (0.00005%) in Earth's atmosphere at sea level, moves much faster than other molecules and, due to that low mass, rises in the atmosphere. In the uppermost parts, Earth's atmosphere is almost all hydrogen atoms (H), not molecules ($H_2$), as at sea level. Hydrogen atoms interact with the ultraviolet part of the solar spectrum strongly, absorbing light at a wavelength of 121.6 nm. This is then re-emitted, allowing detection of the thin hydrogen upper atmosphere of the Earth. Figure 7.6 shows an early detection of the hydrogen atmosphere of the Earth, known as the geocorona. This image was taken by astronauts on the Moon, where the lack of atmosphere allows ultraviolet light to be detected. In addition to the geocorona around the Earth, the reflection/reemission of solar UV is seen on the dayside of Earth and UV auroras near the poles.

The geocorona has recently been detected even beyond lunar orbit! The Apollo astronauts who took Figure 7.6 from the lunar surface likely did not realize that they were still in Earth's atmosphere (Figure 7.7) since it is not very dense there, about 0.2 H atoms/cc. The solar wind itself is usually denser, with about 5 to 10 ions and electrons per cc (with most of the ions, $H^+$). However, as an object lesson that the thin plasmas of space do not have many collisions, Earth's hydrogen cloud is not pushed back much by the solar wind and is roughly spherical. The neutral hydrogen atoms of the geocorona do not interact with the overall neutral solar wind flow which has about equal numbers of positive (ion) charges and surrounding electrons.

At this point, it is useful to consider how atoms work and why there is a Lyman $\alpha$ geocorona in the first place. Since there is basically *no* interaction of the solar wind with Earth's hydrogen outer atmosphere, this is an even simpler interaction than even the absorption of the solar wind by exposed rocky surfaces. So, it is also fitting to study the atomic level interaction, which is also about the simplest possible, to prepare for more complex cases. Tantalizingly, we can state here that understanding how auroras can be so bright first requires understanding why simple arguments show that they cannot even happen! The hydrogen atoms that give rise to the geocorona are the simplest atoms that exist, and they interact with single photons of

**FIGURE 7.6** Apollo 16 1972 UV view of Earth showing the inner geocorona and UV aurora near the poles. Sun is to lower left (NASA).

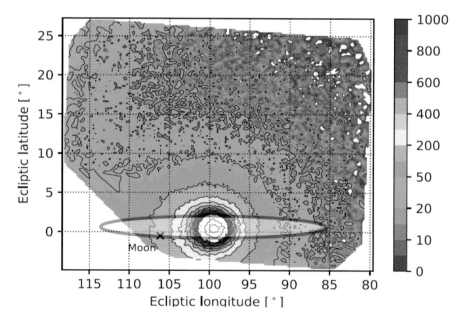

**FIGURE 7.7** Lyman α geocorona as imaged by SOHO in 2019. Gray ring is lunar orbit in projection. Color scale is Lyman α flux (Baliukin et al., 2019, modified).

light, but a lot of fundamental physics is revealed by that interaction. We will take a modern quantum mechanical approach, but link this to classical physics where appropriate. What we do *not* want to do is think of atoms as little solar systems with electrons whirling around!

Hydrogen atoms are made up of one proton and one electron. The proton is 1,836 times more massive than the electron, so it is not a bad approximation to regard it as stationary in the center of the atom. However, it is harder to "nail down" the position of the electron. Heisenberg's uncertainty principle tells us that we cannot simultaneously know both its accurate position and momentum. The level of uncertainty is (not coincidentally) about the "size times momentum" of an atom. This fundamental principle is related to the dual nature that all particles also have wave-like properties. The best we can do about the electron is solve quantum mechanical equations that dictate how it behaves in the electric field of the proton, which, like gravity, falls off following an inverse square law. Those equations give a *wave function* of the electron, whose value squared gives the probability of finding it in some small volume near the proton. Other quantities, such as the energy of the electron, may be found by taking appropriate averages with the probability. Despite what might seem like vagueness in the definition of where the electron is, the solutions of the quantum mechanical equations give very clearly defined states in which the electron may exist. Much like in classical mechanics, these are characterized by a well-defined energy and angular momentum. In fact, the angular momentum requires two numbers to define in quantum mechanics, and the energy has levels also defined by a number. For the purposes of this discussion, we will minimize the fact that the individual particles have an intrinsic property called *spin*, which acts like angular momentum as well. This detail is important in lower-energy interactions, such as those involving radio waves. For example, the famous 21-cm line, relevant mostly to discussions about galaxies, is generated by spin interactions. Going beyond hydrogen, we find that rules about particle spin basically create all of chemistry but let us stick to understanding hydrogen for the moment. All we need to know about spin to go on is that it must be added onto the angular momentum, and photons carry one unit of it. This corresponds to polarization, specifically "circular" polarization, which was briefly alluded to in Chapter 6 when discussing how magnetism on the Sun is determined using the Zeeman effect.

If a hydrogen atom in its undisturbed "ground" or 100 state (Figure 7.8, top) absorbs a photon, it is only possible for it to transition to a state in which it has gained the energy of the photon and *also* its angular momentum. Cold hydrogen, which typifies Earth's exosphere, is usually in this ground state, a symmetric electron cloud around the single proton nucleus. A photon being absorbed adds angular momentum to the system as well as the precisely defined amount of energy corresponding to an energy level jump. On being absorbed, a photon would be taken from the light coming toward a possible observer, contributing to a dark "line" in a spectrum such as those seen in Figure 3.10 for visible light. At the atomic level, to satisfy both the need to conserve angular momentum and energy, a transition to a *symmetric* electron cloud at a higher energy is *not* possible: the target state must have angular momentum since the photon brings some (as spin). If the transition is Lyman alpha, the energy level is

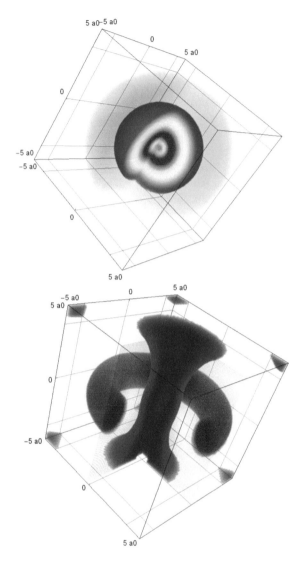

**FIGURE 7.8**  Hydrogen 100 (top) and 211 states. Reddish color indicates higher probability of the electron being in that position. The 211 state has angular momentum (red ring) and the electron can transfer to the 100 state by emitting a photon, which carries energy and angular momentum away. The electron then being nearer the nucleus, excess energy is carried away by the photon, and the final state has zero angular momentum (author).

the next one up, so the energy label changes from one to two. This particular energy change corresponds to a wavelength of about 121.5 nm, in the deep ultraviolet, as shown in Figure 7.9. The second energy level has two states, one with and one without angular momentum, without it making any difference to the energy if the atom is isolated. However, by absorbing the photon's angular momentum, the atom's angular momentum number, originally zero, must change to one (there is also a second number associated with angular momentum, which is not discussed here). The reverse process happens very quickly: in about one hundred millionth of a second ($10^{-8}$ s) the electron will seek a lower level and the atom will "decay" from the excited state (note that this is not a nuclear decay: the nucleus is unaffected). Governed by similar rules on angular momentum, this decay from an asymmetric state can only be to a symmetric one (and, in the case of Lyman alpha, there is only one possibility anyway, the symmetric ground state). In terms of light going out from the Sun, this interaction makes it possible to "see" Earth's exosphere since the emitted light can be in any direction. The original light was only moving directly out from the Sun. In fact, seen against a dark background, the spectrum of the Lyman alpha geocorona would be a single emission line at about 121.5 nm wavelength. Figure 7.7 shows the exosphere while looking outward from the Sun from SOHO's vantage point between Earth and the Sun, so it shows photons that have been re-emitted almost in the opposite direction to that of the original Lyman alpha photons coming from the Sun. On the other hand, Figure 7.6 is roughly looking at right angles to the original light from the Sun.

## 7.3 IONIZATION

In the discussion about the solar wind, we noted that it is ionized quite close to the Sun, indeed it likely leaves the photosphere already ionized. That ionization may occur due to collisions. It is also possible for light to create ions. If photons of high-enough energy are available, hydrogen atoms can absorb them to totally remove the electron from the atom, leaving a positive proton as an "ion". In ionization, angular momentum considerations that are important in transitions between atomic energy states are not important since the electron can carry off angular momentum. Rather, energy considerations dominate. To speak of energy at the atomic level, a convenient unit is the "electron volt" already mentioned. This unit is connected to the concept of a volt (V) giving the electric potential difference between two points in space. For example, if you connect a DC voltage meter to the ends of a single-cell battery, you will see a voltage reading in the range of about 1 to 3 V, depending on the battery chemistry. If a single electron moves toward the positive side from the negative side (say, in a conducting wire), it will gain energy. The amount gained will be the battery voltage in units of electron volts (which we refer to as "eV", a non-SI but handy unit). The definition of the eV is that it is the energy gained by an electron in traversing one V of potential difference. The very fact that batteries have voltages measured in volts indicates that this is a natural unit for processes (in the case of batteries, chemical processes) involving atoms or molecules. Being associated with individual atoms, the eV is a very tiny amount, and expressed in SI units is $1.602 \times 10^{-19}$ joules (J). As a reminder, one joule is the energy of a force of 1 newton (N) acting through one meter. Newton is, of course, associated with apples in the minds of many people. A fairly small apple has a mass of about 100 g, and, in the approximately 10 m/s$^2$ gravitational acceleration field, would feel a downward force (weight) of about 1 N. Lifting such an apple 1 meter would need 1 J. Compared with common energies like those involved in a person climbing stairs, or modestly accelerating a car, the joule is a small unit. On the other hand, if you think of the number of atoms involved in these activities, which is of course huge, they involve energy changes of order one eV *per atom*.

Figure 7.9 shows the full set of energy levels for hydrogen in the $l = 1$ and 2 angular momentum states. If cool hydrogen is in a general radiation field with all wavelengths of far-UV present, the many hydrogen atoms in the $n = 1$ energy level (the ground state) will be able to absorb specific energies of light and have an electron move to a higher energy level, also with a change in angular momentum since the absorbed photon brings that in as well as energy. These states would have angular momentum $l = 1$ and specific energies. The lowest energy jump (of about 10 eV as shown in Figure 7.9) corresponds to a wavelength of about 121.5 nm, which is referred to as Lyman α following the convention mentioned in Chapter 3 for the lower-energy (visible light) Balmer series. Similar to that series, transitions to higher levels than that giving the α line are possible, involving shorter wavelengths. Once the photon is absorbed, the atom, if in free space, cannot remain in the higher energy state for long, and emits a similar photon to that absorbed, which is the emission process shown for Lyman α in the figure. The spacing of energy levels gets very small for higher energies within the atom. There are many transitions characterizing the Lyman series of transitions that leave the electron still in the atom, with the change in energy approaching 13.6 eV, which corresponds to that wavelength of 91.2 nm. The many lines become densely spaced near this wavelength to form the "Lyman limit".

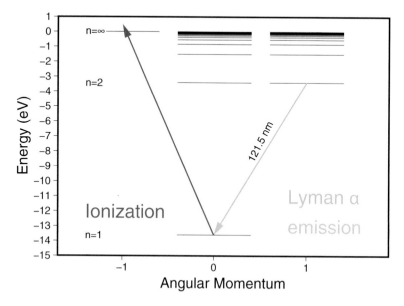

**FIGURE 7.9** Lyman emission (right) and ejection of an electron from the ground state to ionize a hydrogen atom (left). Angular momentum is the "l" quantum number.. (author).

The amount of energy required to completely remove an electron from the ground state of hydrogen is 13.6 eV or more. A far-ultraviolet photon (91.2 nm or less wavelength) can do this. The action of one such photon, of about 14.1 eV, is illustrated in red in the figure. Extra energy in that photon beyond the 13.6 eV shows up as (kinetic) energy, mostly of the electron. In this case, the kinetic energy of the freed electron would be about 0.5 eV. The atom left behind, now without an electron and thus bearing a positive charge, is referred to as a positive ion.

The solar wind itself is fully ionized partly since there is UV light to remove electrons, but also because the corona, heated by mechanisms that remain basically unknown, is very hot and has collisions, which are also effective at ionizing if the incoming particle brings enough energy. Farther from the Sun, the solar wind has very few collisions to allow electrons to rejoin protons (or the roughly 10%, by number, of helium nuclei also present) to form atoms. Thus, once ionized near the Sun in the corona, it remains ionized. Ionization due to photons is more important in forming *ionospheres* near planets, as we will discuss in the next section.

The ability to recombine depends on the density because, despite the attraction a positive ion has for electrons, under low-density conditions, they may not be able to link up quickly. However, even if the collision rate is too slow for electrons and positive ions to combine, the solar wind plasma remains *neutral* on large scales. In Figure 7.3, the undisturbed solar wind seems to have unequal numbers of ions and electrons: this is because 10% of the ions are helium, which contributes two electrons. The solar wind is, however, extremely conductive electrically since any small electric field present would cause electrons (mainly) to move around and remove that field. In the previous chapter, we saw that charged particles are forced to circle the magnetic field and, in that sense, are bound to it. So charged particles move in an anisotropic manner. Nevertheless, viewed moving along with the plasma, particles move around to neutralize any electric fields and remain neutral. Usually we do electrical measurements and see electrons move (as in a wire between battery terminals) from a stationary point of view. However, we will see below that a magnetized plasma measured as flowing by will be observed to also have an electric field in it. The same plasma, when measured in a frame traveling with it, will have no electric field.

## 7.4 IONOSPHERES

We have now seen the two simplest possible solar wind interactions: to pass right into a solid surface and be absorbed, as at Earth's Moon, or to pass right through a thin neutral atmosphere, as in the case of Earth's geocorona. We might imagine that the interaction with a dense neutral atmosphere might be intermediate between these cases, and that may well be the case. However, there are no such outer atmospheres in the Solar System! At least, not in the direction facing the Sun. Relatively dense atmospheres exposed to the impact of the solar wind are also exposed to solar UV light, and rapidly form "ionospheres". Referring back to the discussion of the solar light output in Chapter 3, we note that the photosphere does not emit much UV light due to its comparatively low temperature of 5,800 K. Anomalous heating of the corona and even the upper chromosphere does result in high enough temperatures to generate UV light, and nonthermal processes can also generate some. These hot gases are thin and do not in total generate much electromagnetic radiation. Where the photosphere generates about 1,400 W/m² (1.4 kW/m²) of visible radiant energy at Earth, that in the far-UV is measured in mW/m², roughly one million times less. However, some UV light is energetic enough to remove electrons (although usually just one) from common atmospheric constituents. This typically takes about as much energy as ionizing hydrogen. The remaining mix of positive ions and electrons is highly electrically conducting. Ionospheres are thus gravitationally bound to the planet but may be idealized as perfect conducting layers. Although Figure 7.10 gives the impression of not much variation in the production of $O_2^+$, $N_2^+$, and $O^+$ with height, the height scale is logarithmic so there is huge variation on the upper side as well as on the lower side, where below 90 km there is hardly any ionization. The general trend is that photons come in from above and are consumed in producing ionization, while there is more material to ionize in the denser layers below. The result is indeed a layer, very sharp on the bottom but rapidly thinning with height. Above 150 km height, the ionosphere is dominated by ionized atomic oxygen ($O^+$).

A slightly subtle aspect of ionospheres bears discussion. They are dense enough that chemical reactions can occur. The basic initial process sees atmospheric constituents ionized. It is possible to ionize molecules without disrupting them. In the

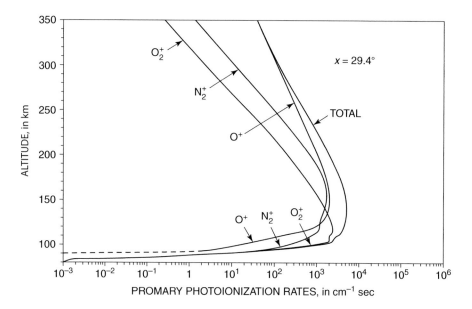

**FIGURE 7.10**   The production rates of the dominant ions in the ionosphere by dayside UV light. The production rate (horizontal) scale is logarithmic. At high altitudes, ionized atomic oxygen ($O^+$) dominates by a factor of about 100. At low altitudes, most ions are from normal atmospheric molecules ($O_2^+$ and $N_2^+$), but neutral molecules dominate the composition (NASA, 1965, Hess and Mead).

case of $CO_2$, once an electron is removed, we refer to a positive $CO_2$ ion, or $CO_2^+$. It takes 13.8 eV to ionize $CO_2$, comparable to what is needed for hydrogen. Because there is relatively little solar UV with photons of this energy, one should not think of ionospheres as having "only ions": there may remain many neutral atoms or molecules. Despite this, the electrons (and, in some cases, ions) are quite able to move in response to electric fields, so the gas is quite conductive. The electrons remain in the ionosphere, so that overall the plasma remains neutral, made up of positive ions, electrons, and neutral molecules or atoms. If an electron bearing enough energy recombines (re- in this case does not necessarily mean with the original molecule from which it was liberated) with a $CO_2^+$ ion, it may bring in enough energy to liberate an oxygen atom. Mechanisms such as this allow the conversion of $CO_2$ to oxygen in its atomic form (O rather than the $O_2$ molecule found lower in some atmospheres). Thus, the chemical composition of an atmosphere may change with height due to what are referred to as photochemical reactions, since at least one of their steps involves light. Another important mechanism in Earth's ionosphere involves nitric oxide (NO) in the lower layers. This molecule has a lower ionization potential than most, and once ionized can participate in chemistry that may result in other constituents becoming ionized. Overall, the photochemistry of ionospheres can be complex. Another complication is that, in polar regions, charged particles are injected into the ionosphere, as will be discussed later in this chapter. In Figure 7.6, there are circles around the north and south poles of Earth corresponding to UV aurora excited by input of charged particles. On the night side of Earth, where there is no solar input, and even on the dayside, where the winter hemisphere at high latitudes may be in darkness, charged particles may be the dominant factor in producing ionization.

The solar wind arriving in the vicinity of an ionosphere finds itself encountering an electrical conductor. Magnetic fields cannot penetrate a conductor, and, when already in one, remain bound to it. The solar wind usually carries a magnetic field. As a result, ionospheres exclude the magnetic solar wind to a large degree; the magnetic field cannot leave the solar wind plasma it is already in, nor can it penetrate the conducting ionosphere it hits.

The resulting diversion of the solar wind cannot take place without it transitioning from being supersonic to being subsonic, allowing normal gas processes like diversion through a pressure gradient to be possible. Referring back to the concept of a sonic boom generated by a moving aircraft (Figure 7.2), this transition takes place with a shock wave (Figure 7.11). However, unlike in that case, in which the cause of the shock wave moves while the medium (air) is at rest, in the case of the solar wind, the obstacle is stationary (or at least slowly moving, compared with the solar wind flow) while the medium flows rapidly. We have already discussed stationary shock waves at the outer boundary of the heliosphere, and analogously in the hydraulic jump in a kitchen sink. In the solar wind near a diverting planet, as in the case of a rock diverting the water of a river, a stationary wave forms. However, unlike in a river, small waves cannot propagate upstream due to the incoming supersonic flow. The waves carrying pressure information all pile up to form a "bow" shock, where the term bow arises from the somewhat analogous forward part of a ship generating a pressure wave. The solar wind plasma transitioning the shock is usually compressed,

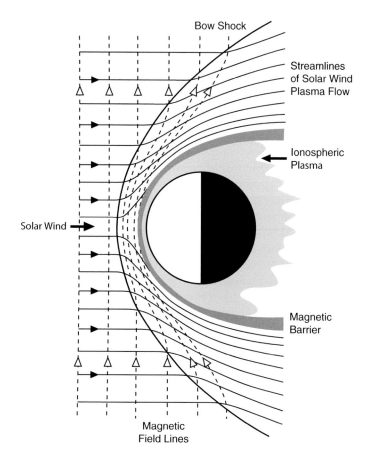

**FIGURE 7.11**  Diversion of solar wind by a nonmagnetic planet with an ionosphere, like Venus or Mars (C. T. Russell and M. Sowmendran, UCLA).

forming a layer above the ionosphere but separated from the incoming solar wind by the shock. The planet finds itself sheathed in this plasma, referred to as an ionosheath, which is denser and also slower than the solar wind. Since the ionosheath is above the conducting ionosphere, it can have a magnetic field in it, which is usually enhanced (by compression of the plasma into which it is frozen) over solar wind values. This is sometimes referred to as a magnetic barrier.

Figure 7.12 dramatically illustrates the exclusion of the solar wind magnetic field from the vicinity of Venus by its ionosphere. Above about 420 km, the Pioneer Venus orbiter was in the solar wind and detected a magnetic field (bottom scale, solid line) of about 60 nT, whereas below this height, the field was close to zero. In contrast, the number density of electrons was very low (top scale, dots) in the solar wind and high in the ionosphere. The pressure of the atmosphere/ionosphere could exclude the solar wind, although the mechanism for this was a current layer at the thin ionopause. This

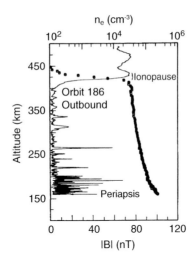

**FIGURE 7.12**  Pioneer Venus Orbiter data for the ionosphere of Venus from June, 1979. Top scale $n_e$ ($cc^{-3}$), shown as dots, bottom |B| (nT), shown as a line (C. T. Russell and M. Sowmendran, UCLA).

can be inferred from the change in magnetic field, since there is a jump in magnetic field at a current layer.

To summarize, denser atmospheres directly exposed to the solar wind interact strongly with it, but this is mostly due to them being ionized themselves. The interaction is complicated by the supersonic nature of the solar wind flow, resulting in a bow shock with a compressed layer beneath it. We now move on to the most complex possible solar wind interaction with a planet, that involving a planetary magnetic field. Most planets of the Solar System do have a strong enough magnetic field that the interaction of the solar wind with it is important. Among them, only Mercury does not have a significant atmosphere (and thus no ionosphere). All these planets have "magnetospheres" formed mainly by solar wind interaction with the planetary magnetic fields, but each has peculiarities due to features of the planets and their environments. Earth is, of course, one of these planets, and its magnetosphere is the best explored. It also has important implications for technological society in the form of "space weather" arising mostly through the solar wind–magnetosphere interaction, which can be very complex.

## 7.5 MAGNETOSPHERES

We examined the simplest possible interaction of gas bound to a planet interacting with the solar wind: no interaction, as in the case of the geocorona. We considered a non-conducting solid surface, which can absorb the solar wind impinging on it and create effectively a solar wind shadow behind the body. Bodies with atmospheres have the upper parts of them ionized, mainly by UV on the side hit by the solar wind.

Since such ionospheres are electrically conducting, they do not allow the magnetized solar wind to penetrate. The inability to enter the atmosphere means that it exerts a force on the solar wind, and the solar wind on it. One may regard this as causing a "standoff" of the solar wind. The most complex solar wind interaction is found when the body being struck is significantly magnetized. In this case, the standoff of the solar wind is outside the body (we consider only planets here, but a similar situation applies to some moons). The planets that have a significant magnetic field are Mercury, Earth, Jupiter, Saturn, Uranus, and Neptune, in order out from the Sun. The standoff distance is determined by the pressure of the solar wind at the location of the planet and the strength of the magnetic field of the planet. The higher the pressure, the more the solar wind can push back the magnetic field of the planet. Conversely, if the planet has a strong magnetic field, the standoff distance will be larger. There is a tendency to think of magnets as attracting each other, but they can also repel, effectively giving evidence of magnetic field pressure.

Figure 7.13 shows a stack of strong magnets above an even stronger spherical magnet (a tiny "Earth" in Gilbert's terms from 1600). These magnets are about 2 cm above the other one despite supporting the weight of a broom (not visible). The bottom magnet is held in place by duct tape, and one can see that the magnet stack is turned and held in place by the walls of a clear tube. Magnet experiments are difficult since magnets exert a turning force on each other in addition to an attractive or repulsive direct force. If not restrained in this manner, the magnets would flip over so that opposite poles would be near each other, and then attract each other very strongly. A disk magnet can be shown to be equivalent to a small coil of wire, making a magnetic field that is dipolar a large distance away. This may be taken even further to compare a single particle spiraling around the magnetic field, which is equivalent to a current in a wire and in turn to a disk magnet. So, the suspension of disk magnets shown in the figure shows exactly the force that causes particles in a dipole field to move in toward the poles and then "bounce" back. We will return to this point in the next chapter, but since there is a force upward and we can regard much of it as spread over the area of the lowest disk magnet, that area, which is, by definition, pressure. Thus, a magnetic field is demonstrated by this little experiment to exert a pressure, at least on other magnets. Clearly, a magnetic field does not exert pressure on all substances: one can easily imagine that in pouring water over a magnet, there would be no "pressure" opposing it. However, a moving conductor in a magnetic field will have electric currents induced in it. The solar wind is a conductor due to being almost 100% ionized and thus responds to magnetic pressure. This allows us to take a first look at how magnetized planets interact with the solar wind.

The best indicator of a planet's influence in the space around it is the "magnetic moment". This allows calculating the magnetic field far away from it, as a function of distance from the center and of the angle between where the pole points and where the observer is located. When one is in a plane at 90° from the pole, one is in the "magnetic equator" of the body. Near the surface of the body, as we already saw in the discussion about historic magnetic declination charts, the field can be complex, but these "higher order" effects die off with distance faster than the dipole

FIGURE 7.13  Magnet stack suspended in midair above a black "mini-Earth" spherical magnet. Not seen is that the stack is supporting the full weight of a broom inserted into the glass tube (author photo).

field, which falls off as the third power. At a large distance, any magnetic body or current distribution (except in very special, one might say contrived, cases) looks like a dipole. Despite the experimentation of Coulomb with "magnetic poles", there is no such thing as a single magnetic pole: they always come in pairs. His long, thin needles gave an approximation to magnetic poles, but in fact the field lines looped around to close at the other end of the rods. Break a magnet in two and you do not have two "monopoles", but instead two complete magnets with two poles each. This property of magnetism contrasts with electric fields, in which field lines begin and end on charges. By convention, they go out from positive charges. This being said, how are planetary magnetic fields generated?

Charming as Gilbert's idea of 1600 may be, planet-sized objects cannot be magnets. The special solid-state properties of magnetism disappear with heating, a fact shown by Pierre Curie in 1895 just before changing, with his new wife Marie, his scientific field to radioactivity. At temperatures higher than the "Curie temperature", typically a few hundred to a thousand kelvins, "permanent" magnetism ceases. Since most of the interiors of the Earth and planets are above this temperature, they cannot be permanent magnets. A magnetic field can move through an imperfect conductor but not through a very conductive body, as was mentioned above in regard to ionospheres. This important concept will be revisited, as will its violation which can happen under special circumstances. Magnetic fields in materials change through the presence of electric currents, and in an imperfect conductor, these dissipate energy, making the field go away. Much like a drop of ink in a cup of water, this can be regarded as a diffusion process. The timescales for diffusion of magnetic fields in bodies of planetary size and conductivity are long, but much less than the age of the Solar System. Thus, possible initial fields cannot persist to any great degree. As mentioned above, small bodies have cooled, which is a process related to magnetic field diffusion. Large bodies can retain heat and remain hot enough to be liquid inside. Recall that, as proved by its reserves of helium, Earth is also heated by radioactive materials. Larger planets may also generate internal heat through continued gravitational separation of substances inside them. Since the interiors of even "solid" planets are liquid, it is felt that self-sustaining generators, known as "dynamos", cause the magnetic fields of all planets which have them, of which a selection is shown in Figure 7.14.

The presence of a dynamo in the Sun is inferred from its 22-year cycle of magnetism. The number of sunspots rises and falls on an 11-year cycle, but the overall magnetic polarity change takes place over two of these periods. This is the true physical period of the solar magnetism cycle. Earth is also known to have magnetic pole reversals, on a timescale of hundreds of thousands of years. We have not yet observed such cycles on other planets. Because in what follows it will be clear that Earth's magnetosphere largely fills a protective function, a field reversal could be a cause for concern since there may be a brief period with essentially no field. Although research on this tantalizing question continues, it can be stated that no major mass extinction in the geological record coincides with a magnetic reversal of Earth's field as recorded in the rocks. This is in contrast to the overwhelming evidence of impact-related extinctions discussed in Chapter 3.

Although Mercury and Earth have iron cores, those of Jupiter and Saturn are thought to be made of liquid metallic hydrogen, a form in which it conducts electricity. The extreme pressure inside these planets causes what is effectively ionization at high density, with electrons free to move past protons. This mechanism does not appear to operate in Neptune and Uranus. These planets, which have been visited only once on flybys by Voyager 2, do have magnetic moments, respectively 27 and 49 times that of the Earth, and magnetospheres were detected. They will not, however, be further discussed here.

The intrinsic factor of magnetic moment alone leads to a large variation in the size of the magnetospheres in the Solar System shown in Figure 7.14. Measured by magnetic moment relative to Earth, Mercury has a paltry 0.07% field, while that of Saturn is 580 times as high, with Jupiter totally the giant planet at 20,000 times as much. The Sun's magnetic moment at maximum is about 30 million times that of Earth, dominating the structure of the heliosphere. The Sun's weak polar field should not mislead in this regard: the magnetic moment is a better indicator of the overall

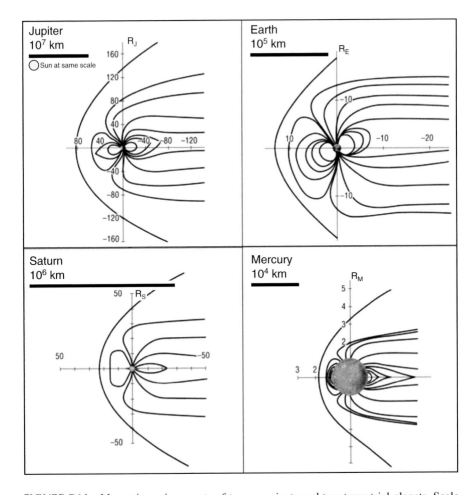

**FIGURE 7.14** Magnetic environments of two gas giants and two terrestrial planets. Scale bars are marked by length in km, with a white inset showing the scale bar length of the next smallest planet. The bow shocks are the leftmost curve in each panel and other curves are magnetic field lines. Between the bow shock and outermost magnetic field is the magnetosheath of compressed solar wind. Planet bodies are shown approximately to scale. Note the huge variation in size of magnetospheres, with Mercury being shown at about 1,000 times the scale of Jupiter (modified and corrected from Fujimoto et al., 2007).

effect. Among magnetospheres, Jupiter's is a giant with 104 million km (the scale bar size in the figure) a good scale to measure it on.

To a first approximation the supersonic solar wind is in balance with the magnetic pressure at the front, or "nose" of planetary magnetospheres, Figure 7.14 showing the magnetospheres resulting from the standoff of the solar wind by planetary magnetic fields. The solar wind is coming from the left in the plots. As was the case with ionosphere interaction, there is a bow shock. The supersonic solar wind, apart from some details arising from particle behavior, does not "know" that it is going to hit a magnetic obstacle, and, when it does hit, a bow shock is formed, much as a shock wave is formed in near-stationary air when a supersonic aircraft flies rapidly through it. In the case of magnetospheres, however, it is the medium outside which is moving supersonically. Compressed plasma is found between this shock and the second interface, known as the magnetopause, where the planetary magnetic field becomes dominant. Similar to how solar wind could not penetrate the conductive region of an ionosphere even if not protected by a planetary magnetic field, it cannot directly penetrate into the magnetospheres of magnetic planets.

The details of the frontside interaction with the solar wind are shown in Figure 7.15, giving data during the passage of the THEMIS "B" spacecraft from the magnetosphere outward into the solar wind on July 17, 2008. The data are shown by time: in terms of motion, this corresponds to right-to-left motion on the sunward side of Earth in Figure 7.14. At the beginning of the passage, the spacecraft was in a region with hot plasma not well shown by the instruments used. It was also not very dense; the flat portions of the top two curves may indicate that there was not enough plasma for a good detection. The magnetic field, however, was dominated by the northward or positive $B_z$ component (bottom panel, blue line), which was around +50 nT until about 20 UT. We will examine shortly why this field was northward, but what matters here is to note that it was fairly strong while the plasma density was low. Between 20:00 and 20:15 UT, this situation changed abruptly; the magnetic field declined to near-zero values, while the density increased about 100-fold (top panel; note that the scale is logarithmic) from about 0.1 to 10 ions per cubic centimeter. The temperature as measured in eV also declined considerably (second panel from top) although with a large spread (color spectrum in third panel). The magnetic field (bottom panel), although small, also had considerable fluctuations, so that this region seemed to have a lot of turbulence. Very abruptly, at 22:40 UT, the situation calmed down considerably: near-zero magnetic field, a density of about 1 particle per cc, and particle energy around 1 keV. These values correspond to being in the solar wind.

If we look at this more in terms of solar wind flow, rather than what direction the spacecraft happened to be going in this nice example, we reverse the above sequence in which the spacecraft was moving away from Earth. Coming in with the impinging solar wind, we first find undisturbed flow in the solar wind, a rapid increase in density at the bow shock, then dense, turbulent plasma in the magnetosheath, and finally strong magnetic fields dominating the region nearest Earth. The magnetosheath flow diverts plasma, which does not enter the magnetic region. The plasma that is in the magnetic regions is very hot and thin and will be discussed in the next chapter. Although this example, chosen for its clear features, does not show much magnetic field compression in the magnetosheath since the solar wind magnetic field was small to begin with, compression is a common feature. Since the plasma cannot diffuse out of the solar wind, if the wind gets compressed, so does the magnetic field. Neither can it usually penetrate the magnetosphere, which is dominated by Earth's own magnetic field.

Having now examined the sunward side in some detail, we revisit Figure 7.14 to note that, on the nightside, away from the solar direction, the planetary magnetic field is highly deformed by interaction with the solar wind and drawn back into a

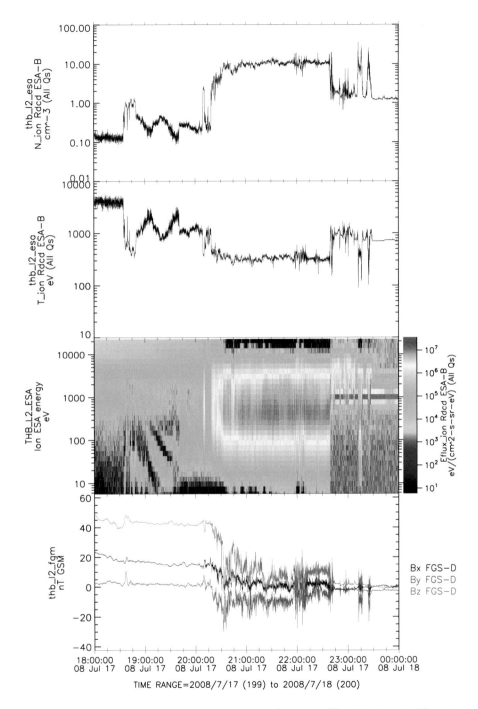

**FIGURE 7.15** THEMIS-B spacecraft data when passing outward into the solar wind through Earth's magnetic nose. See text for description (NASA, THEMIS project, and CDAWeb).

"magnetotail", the beginnings of which are shown in the figure. Magnetotails extend much further behind the planet than the standoff distance of the magnetopause on the front. We will make one important observation about the planetary magnetic field, using Earth as an example. Referring to the panel showing Earth, recall that magnetic field lines enter the planet in the Northern Hemisphere (in other words, the north region houses a south magnetic pole). The *subsolar* region is between Earth and the Sun: on the surface of Earth, the Sun would be overhead which is why this is called sub- (i.e., "under") solar. This is on the dayside of the Earth, and here the magnetic field lines are like those of a dipole but compressed by the pressure of the solar wind. Since the field lines curve around to go into the Earth at the pole, on the direct Earth-Sun line they point generally northward. We noted this in the THEMIS-B spacecraft passage just discussed, and it is always the case. Note that Earth is tilted so that its equator is not usually on the direct Earth-Sun line. We must now contrast the situation on the nightside: here, the field lines coming into the pole originate in deep space, off the right side of the diagram. In the Southern Hemisphere, the field lines go out from the pole and off into deep space. The important consequence of this is that in the upper part of this panel, field lines are coming in from deep space toward the Earth, and in the lower part they are going out into deep space from the planet. Thus, in the middle must be a region where field lines are oppositely directed, and, in a small region extending out in the equatorial region far from Earth, there is near-zero magnetic field. For the other planets, the magnetic fields may have reversed overall direction, but the same considerations apply: there is always a dipole-like inner region and a stretched magnetic tail.

## 7.6 RECONNECTION

In all discussions of magnetic-field-dominated regions so far, and this is most of the invisible Solar System, we have not dealt with regions of zero magnetic field. It turns out that such regions have special properties and allow a new physical process not previously discussed, but of extreme importance in space physics. This process is known as *magnetic reconnection*. The magnetotail always has oppositely directed magnetic fields so that situations of near-zero field are easy to imagine. However, on the sunward side, the solar wind may come in with magnetic fields either in the same basic direction (northward) as Earth's, or oppositely directed, by being southward. Since reconnection needs oppositely directed fields to operate, and Earth's field in the subsolar region is always northward, this process will occur on the dayside of Earth only if the solar wind magnetic field is southward. In the early 1970s, studies with spacecraft in the solar wind and magnetic detectors on Earth to indicate geomagnetic activity associated with currents in the ionosphere and near-Earth space empirically showed a strong connection between southward solar wind magnetic field and such activity. This was strong empirical proof for a process, which basically had to be reconnection, that allowed energy to flow into the Earth's magnetosphere when the solar wind magnetic field points in a southerly direction.

As a result of reconnection, the actual behavior of the magnetosphere deviates from what would happen if plasma-carrying magnetic fields were totally excluded from mixing with magnetospheric plasma. In the ideal case, such regions would be completely isolated from each other. This is referred to as a "closed magnetosphere", having a region dominated by Earth's magnetic field able to deflect the solar wind and form a magnetic bubble in space. During conditions of northward magnetic field in the solar wind, this is indeed what happens. There are some small regions that may be subject to reconnection under such conditions, but, for the large part, there is little interaction of the solar wind with Earth. The best that could be hoped for would be "viscous interaction", much like the interaction of fluids, but in collisionless plasma. Being very inefficient at transferring energy, this interaction would affect the details of the shape of the magnetosphere, but not much else. Even before the spacecraft observations of the 1970s, plasma theorists, notably England's James Dungey, had proposed that reconnection, both where the solar wind hits the Earth and in the magnetotail, could cause the magnetosphere to interact more strongly with the solar wind than viscous interaction would.

With the caveat that reconnection is still poorly understood, Figure 7.16 shows the basic configuration of a simple type of it. This is a "symmetric" reconnection as would operate in the magnetic tail, but a modified version of it could also operate in dayside regions. One should also be aware that other, important reconnection schemes rely on the presence of shock waves, which are not considered in this simple view. Without them, the observed rate of reconnection cannot be attained, but we can get an idea of how the reconnection process works.

In Figure 7.16, a current line is shown by circled x symbols. This symbol is often used to indicate a current flowing into the plane of the paper (by showing the "feathers" on the tail of an arrow going inward). The magnetic field is consistent with this by the "right-hand rule" but this means that Earth must be to the right in the figure, consistent with Figure 7.15 but opposite to how things are depicted in Figure 7.14. In a magnetotail with an oppositely directed magnetic field above and below the central plane, not only must the magnetic field in the middle have no horizontal parts (and, in the very middle, be zero), but it will have minimal or no magnetic pressure. It is

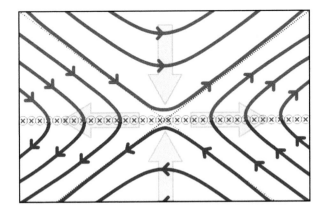

**FIGURE 7.16**   Plasma flow (yellow arrows) allowing and rising from reconnection, which takes place at a current sheet marked by circled "x"s. Red and blue lines are the magnetic field (frame from ChamouJacoN, 2009: public domain).

thus no surprise that combined gas and magnetic pressure above and below would force plasma to flow into the central region, as shown by the vertical yellow arrows. In the zero magnetic field region, in a fluid view, magnetic diffusion can occur since there is no magnetic field to prevent this. In a particle view, particle orbits become undefined. The net result is conversion of inflowing magnetic energy to fluid heating and acceleration and/or equivalent particle motion to jet out along the line of symmetry, as shown by horizontal yellow arrows. Although this is an ongoing process, under steady conditions it can be regarded as continuous, with the overall geometry not changing. In a magnetic tail, this would create hot jets of plasma directed toward Earth (here, toward the right) and in the antisolar or tailward direction (toward the left). Reconnection is observed to create such jets, which are referred to as "bursty bulk flows" when approaching Earth, and create bubbles of plasma known as "plasmoids" which move away from Earth in the magnetotail.

## 7.7 SUBSTORMS

On the dayside, with southward incoming solar wind magnetic field, the asymmetric reconnection can allow energy transfer into the magnetosphere that normally would not be possible. The injected energy and magnetic flux that cannot normally enter influence the whole structure of the magnetosphere, although its shape remains basically the same as in Figure 7.14. Cross-tail electric current, which must separate the oppositely directed magnetic fields in the tail, strengthens. In turn, of course, this enhanced current results in stronger magnetic fields in the tail above and below it. The extra pressure causes "thinning" and due to a more flattened, less dipole-like magnetic field, "stretching" of magnetic field lines (Figure 7.17). In principle, the energy input from the dayside reconnection could make changes overall, with a stable situation resulting, but, in practice, the exact balance of energy and magnetic flux moving around is rarely stable. Usually a "magnetospheric substorm" takes place, with particles sent toward Earth which cause auroras to light up as in Figure 7.18, and a plasmoid is sent down the tail. With this energy dissipated and magnetic tension (current) reduced, the magnetic field in the near-Earth tail returns to a more dipole-like configuration.

Part of the phenomenology of substorms is that the inner part of the electric current that crossed the magnetotail is instead diverted to the Earth's polar ionosphere, as shown in Figure 7.19. These currents can total 1 MA or more and cause local magnetic fields near the auroras that are also driven by substorms. The signals at the surface of the Earth may exceed 1,000 nT, or 2% of the main field. This gives rise to the ironic situation that "substorms", which were discovered after "magnetic storms", can have larger magnetic field changes than the storms can, despite the "sub-" prefix. Induction, as found by Faraday, depends on both the amplitude and rate of magnetic field change. The main field is essentially static but the dynamic fields of substorms, although of lesser amplitude, may induce electric fields in the Earth which can affect technological infrastructure. The impulsive changes that happen during substorms are thus a major component of "space weather" with potential to affect technological infrastructure. The electric fields induced are measured in mV/km normally, but since systems, such as power grids, are large, the total applied voltage can be high.

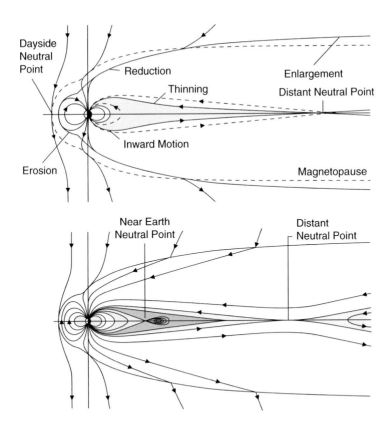

**FIGURE 7.17**  Changes due to reconnection forming a substorm event in Earth's magneto-tail, during the "growth" phase and near "onset". The dark grey pinched-off region in the bottom diagram later moves down the magnetotail as a "plasmoid" (C. T. Russell and M. Sowmendran, UCLA).

Furthermore, it is at a low frequency, whereas the networks are designed to operate at higher frequencies of typically 50 or 60 Hz. This alone can cause damage since what is effectively a short circuit can be created. Electric currents, usually not wanted, caused by the induced electric fields, are referred to as "geomagnetically induced currents", or "GIC".

We will close by examining a large space weather event that took place on April 5, 2010. This resulted in damage to the "Galaxy-15" communication satellite that led to the removal from service of the $270 million satellite. In space, that damage was due to effects discussed in the next chapter. Here, we will present data supporting the view that reconnection powered this space weather event, and look at effects on the ground, which do not appear to have caused damage.

The diversion of electric current from space into the near-Earth region during a substorm happens in a systematic manner. The temporary formation of a system of currents referred to as the "substorm current wedge", as depicted in Figure 7.19, can be regarded as an electrical discharge. The original sketches of this current system

**FIGURE 7.18** Auroras during a magnetic substorm on February 27, 2023, viewing toward the north from near Edmonton, Canada. The fluorescence of auroras is mostly due to electrons impacting atomic oxygen, which results in "forbidden" green emission in higher-density regions lower down, and red emission higher up. The purple color is an unusual emission that happens if electrons are very energetic (Dr. Brian Martin, photo).

viewed it from the top and resembled a wedge of pie, hence the name. The real structure is more complex than even the figure suggests, but the basic structure is one of electric current flowing down from space east of the region of midnight, crossing the midnight sector in the ionosphere, and returning to space west of the region of midnight. Upward electric currents are mostly carried by electrons, but, due to their negative charge, the particles themselves flow downward. The most energetic of them causes auroras. Figure 7.19 corresponds to much of the situation that would have given the auroras of Figure 7.18, with the upward current over western North America. Since a large current, of the order of MA, can flow in the conductive ionosphere, which is only about 100 km above the ground, the largest magnetic changes detectable on the ground come from substorm events. The substorm current wedge superposed on other current systems which are normally present, but it generally

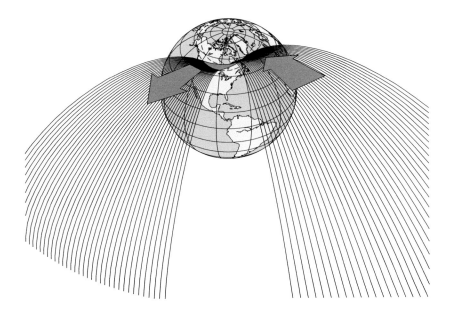

**FIGURE 7.19** The "substorm current wedge" diverts electric current that normally flows across the inner magnetotail into the ionosphere, where it flows briefly before returning to space (author).

carries stronger currents than they do. Both systems are associated with auroras, which are brightest in the midnight region straddled by the current wedge, and in the upward current region. Reconnection in the magnetotail powers such displays: they are *not* a direct result of the input of solar energy on the dayside. Although we can detect UV auroras even on the dayside (Figure 7.6 gives some indication of this), auroras are mainly a nightside phenomenon not only because they can be seen then, but also since the energy making them comes from the magnetotail, on the opposite side of the sky than where the Sun is.

Dayside reconnection, which brings energy into the magnetosphere, is enabled by a southward-pointing solar wind magnetic field since that can give rise to the opposed field geometry necessary for it to happen. Figure 7.20 shows observations from the Japanese Geotail satellite in the bottom panel when a density enhancement took place in the solar wind. In this event, it happened that the pressure from the solar wind and the density, shown in the bottom panel, had the same approximate numerical value. Their changes are more important: a large increase in density caused the pressure to increase fivefold. This increase in pressure compressed the magnetosphere overall, and also pushed back the magnetopause closer to the Earth than its usual ca. 10 Earth radius standoff position. Likely more important in this case, the magnetic field "northward" component (blue, in the same panel) turned southward, to a large negative value of −15 nT, sustained for about an hour from 8:20 to 9:20 UT. Almost immediately, the electric current in the magnetotail as detected by the THEMIS A satellite strongly increased, as indicated by a sudden increase of the X

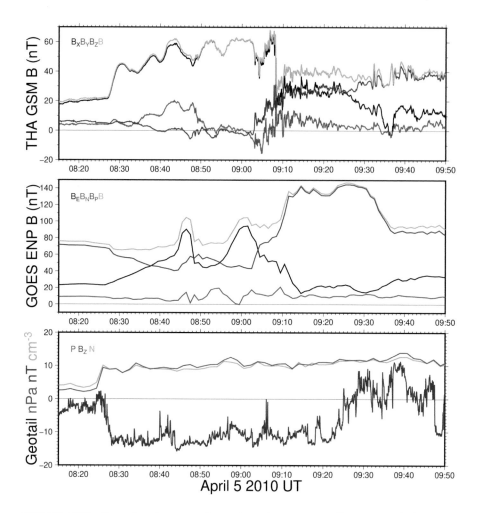

**FIGURE 7.20** Space-based causes of a major substorm event. Bottom panel shows the northward component of solar wind magnetic field, the pressure, and number density. Middle is the magnetic field at synchronous orbit, and top is the field in the magnetotail (author, based on CDAWeb data from NASA for THEMIS and OMNI, JAXA for Geotail).

magnetic component (black trace in top panel, coinciding with green trace since the X component totally dominated before 9:08 UT), The X component of the magnetic field is that part of the three-dimensional field vector that points toward the Sun. An increase in this component indicates that the current in the central plane of the magnetotail became stronger. The response at geosynchronous orbit as detected by a GOES weather satellite in the middle panel of Figure 7.19 was slower. Under continued forcing by reconnection, even there the X component became stronger than the normally dominant Z (here called P for poleward) component of the dipole-like field normally found there. For reasons that are hard to determine for an individual event, but which are like a stretched rubber band finally breaking, the strong magnetic field in the tail started to collapse a bit after 9 UT, and definitively collapsed about 9:08 UT. In part associated with strong plasma flows carrying magnetic field, but powered by tail reconnection at a near-Earth neutral point (see Figure 7.17, bottom panel), the magnetic field in the inner magnetosphere reverted to being dominated by the Z component, which approached being all of the magnetic field (making the blue and green traces nearly coincide). In this strong event, the field there became even stronger than the 100 nT typically found at synchronous orbit, although it soon relaxed to about that value. Although larger than a typical substorm, this one shows the cycle driven by dayside reconnection in response to solar wind southward magnetic field that typifies substorms. Tail reconnection allows a relaxation of the stresses that the dayside reconnection imposes on the magnetosphere. Usually, the solar wind's direction changes, and amplitude of variation, allow only small substorms, which tend to happen every few hours. This one was unusually large.

For this large event, a larger than typical magnetic signal of a substorm was detected by stations in the midnight region of Earth, one of which was Athabasca, Canada. In the map of Figure 7.19, this observing station would have been under the left (western) part of the westward ionospheric current flow shown by a grey region. A westward current overhead produces a southward magnetic field. Since, by convention, the northward field on the ground is positive, a southward field is a negative deviation on a graph such as Figure 7.21. The reduction of nearly 2,000 nT seen there is about 15% of the normal northward field at this location on Earth, about 13,200 nT

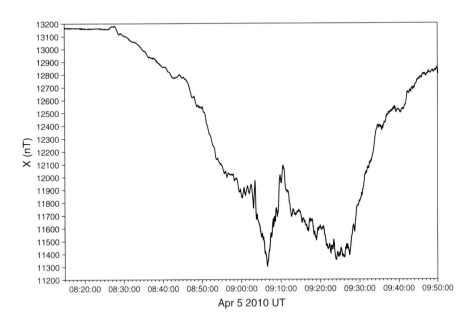

**FIGURE 7.21**  Northward component of magnetic fields at Athabasca, Canada, during the substorm period on April 5, 2010. Time shown is the same as that in the previous figure (author).

as seen near the beginning of the time period shown. At times, this change was rapid, and, at such periods, significant electric fields would have been induced according to the laws found by Faraday. While this is one of the largest events detected at this site, other substorms may feature more rapid changes, even if smaller. The combination of rapidity and amplitude of change of the magnetic field determines how much electric field would be generated that could affect ground infrastructure. A well-known space weather event on March 13, 1989 featured that combination to cause a large blackout in eastern North America. While electric grids have been hardened in response to lessons learned, the potential of space weather affecting our vital electrical networks remains a hazard.

## 7.8  SUMMARY

This chapter has mentioned several ways that the outflowing solar wind can interact with moon or planet-sized bodies. The most complex interactions are possible with magnetized planets, and Earth is the best studied and has been put forward as an example. The rapid rotation and complex arrays of moons of the giant planets bring about other complications in the solar wind–planet interaction, which were not discussed in detail. Analogs of substorm cycles have been observed at several magnetic planets. Although bulk plasma effects, which are of low energy, give rise to interesting and even threatening phenomena, particle effects can also be pronounced. This is the topic of the next chapter.

# 8 Radiation in Space!

The word "radiation" likely bears generally negative meanings for non-scientific people, whereas, for physicists in particular, it is a generic label. The roots of the word are in the concept of radiating out from somewhere: radiation has a source. Something as innocuous as ripples in shallow water, made by vibrating a finger as in Figure 8.1, can be considered to be radiation in the sense that energy is transmitted into a medium (the water) and radiates out from the point where it is disturbed. Such waves in shallow water (here a few mm deep) travel slowly (here, about 15 cm/s).

In the wave picture of light, it spreads out from a source, but at a speed that is roughly 100 million times faster than shallow water waves, so that we cannot see the spreading. Earlier, we did see that GPS radio signals spread out as spherical wavefronts if in flat space, so that the intersection of spheres can be used to determine location. We have also noted that diffraction, a characteristic of waves, gives evidence that light has wave-like aspects. "Light" has many possible wavelengths and is only one type of electromagnetic radiation. When electric and magnetic fields interact with one another, such a propagating wave can arise, going out from a source, be that a quantum radiator like an atom, or one in which classical concepts are sufficient, like an antenna. Electromagnetic radiation has some characteristics of particles even in classical theory: it can carry energy, momentum, and angular momentum. Its quantum aspect is that it comes in quanta called photons, but these are massless and travel at the speed of light. The most energetic photons, gamma rays, show more particle-like aspects than the much longer wavelength radio waves.

We saw in Chapter 4 that the unexpected discovery of various sorts of emissions from matter led to the concept of radioactivity. Certain substances were "active" in the sense that they "radiated" what relatively quickly were found to be alpha, beta, or gamma "rays". It was later found that nuclear reactions resulted in both direct radiation from the reactions themselves, and possible formation of radioactive substances. For example, it was discussed in Chapter 5 that radioactive carbon-14 results from reactions with neutrons in the atmosphere (although without too many details, so far, about where the neutrons come from). A surprising aspect of radioactivity in space is that it can attain energies typical of nuclear processes, but usually does not arise from them. Early in the space age, American researchers famously realized that "space is radioactive", but in fact there are no "active" substances emitting high-energy particles, and they arise in new ways. Energetic particles in space can include alpha particles, but often protons dominate. The alpha particles are *not* newborn ones from decaying heavy elements, and the protons are *not* the result of nuclear reactions.

In Chapter 3, it was noted that the energy output of the Sun, most of which is in the form of visible light, requires a matter destruction rate of about $4.4 \times 10^9$ kg/s. The

DOI: 10.1201/9781003451433-8

**FIGURE 8.1** Innocuous radiation: waves in shallow water made by a finger in a shallow layer of water in a baking pan (author photo).

mass of the Sun is very close to $2 \times 10^{30}$ kg, so, if present conditions are a good average of what characterizes its whole lifetime, then that lifetime will be $2 \times 10^{30}/4.4 \times 10^9$, ~$4.5 \times 10^{20}$ s. With $3.2 \times 10^7$ seconds in a year, this is very roughly $10^{13}$ years. In fact, nuclear reactions take place in only a very small central core of the Sun, so it will never burn its whole mass. Changes in its structure will limit its life to about 0.1% of the value just calculated, about $10^{10}$, that is ten billion, years. The evidence of the nuclear reactions in the Sun is contained in some of the most invisible constituents of the Invisible Solar System. Although Chapter 4 ended on a rather satisfactory note, with the discovery of the neutron in 1932 allowing the understanding of nuclear structure, and with alpha decay explained, there was a big problem with beta decay.

## 8.1 THE INVISIBLE SOLUTION TO THE BETA DECAY PROBLEM

The simplest beta decay is that of the neutron (Figure 8.2). This leaves a proton, which is of lower mass than the neutron, plus an electron (beta), but we *now* know that there is an extra particle that can carry off energy and momentum. The mass of a neutron is 1,838.7 times that of an electron, while a proton is about 1,836.1 times as massive, the difference being 2.6 electron masses. Since only one electron is created in this decay, there is the energy equivalent of 1.6 electron masses left over in the

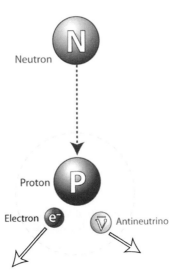

**FIGURE 8.2** Neutron beta decay. Free neutrons turn into protons, releasing a beta (electron) and an invisible antineutrino (Z. Deretsky, US National Science Foundation).

form of energy and the extra particle. The latter is called a *neutrino* generically and, more correctly in this case, an *antineutrino*. Neutrinos have very small rest mass and hardly interact with matter, so are very difficult to detect. Protons have insufficient rest mass energy to change into neutrons on their own. This does happen in certain nuclei where other sources of energy can be available and is called inverse beta decay. The free proton is regarded as stable, but proton inverse beta decay is a fundamental process, the intricate details of which, when it can take place, allow the universe as we know it to exist. More on that to follow! Although neutrons have the simplest beta decay, they were not discovered until 1932. Well before then, a deep problem with beta decay of known radioactive substances had been found.

Beta decay can occur in many nuclei and, in some ways, can be regarded as being due to a neutron going rogue and turning into a proton. This tends to happen in nuclei which have an excess of neutrons. An example of such a nucleus is that of radioactive lead-214 in the radium decay chain of Figure 4.19. With a Z of 82, like all lead, this isotope has A = 214, thus the number of neutrons, 132, is the difference as marked on the vertical axis. The end of the decay chain is also an isotope of lead, lead-206, which with A = 206 has only 124 neutrons. Lead-210 is also a beta emitter with 128 neutrons, so, in lead, more than 124 neutrons seem to be too much, at least based on what we see there. This is sort of correct, but one should not draw conclusions based on limited data. Lead-208 turns out also to be stable, having 126 neutrons, and is the most abundant isotope of lead, and the heaviest stable nucleus to exist. However, lead-208 does not arise in this decay scheme, which is why it is not shown in Figure 4.19. Returning to lead-214, which is in the radium decay chain shown, making a beta spectrum experimentally is not too difficult and can be related to Mme. Curie's sketch of Figure 4.17. One can spread the emissions out by bending them in a magnetic field, and what one sees is a bit complex, as seen in Figure 8.3.

The surprising and puzzling result of recording such a spectrum (say, on a photographic plate) is that there is a line spectrum of beta particles, but also a large continuous spectrum. The line spectrum arises from transitions in the inner electron shells, it was later realized. Gamma-ray lines, also shown, originate from energy level transitions in the nucleus, much as light originates in the electron clouds. However, the beta decay spectrum, unlike the alpha decay spectrum, which is only at specific energies, is dominated by continuous emission, spread out in energy (black line). There is a fixed maximum energy or endpoint, but it was very puzzling how the nucleus could have such continuous behavior. A nucleus turning into another nucleus while giving off an electron should give a definite energy to the electron for reasons of both conservation of energy and momentum. If an electron at fixed energy, and possibly a gamma ray line, was observed, all would work out. However, a continuous electron energy spectrum was the last thing expected in this simple scenario. Much like when the prodigious energy output of radium was discovered, energy did not seem to be conserved. To make matters worse, if one could measure the velocity of the emitted electron and of the original and product nuclei, neither was momentum.

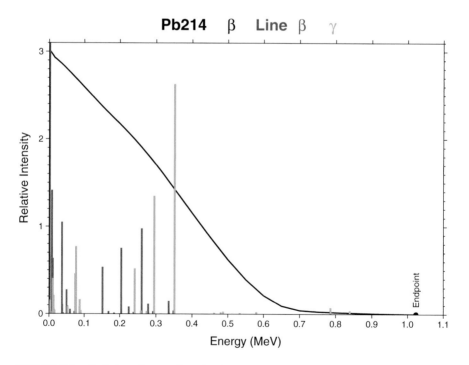

**FIGURE 8.3** Emissions from radioactive lead-214. There is a continuous spectrum of electrons (black), a line spectrum of electrons (red), and also emission of gamma ray lines (green) (author).

Again, some very fundamental physical laws seemed to get broken when dealing with the new science of the nucleus.

To resolve the problem, the German theoretical physicist Wolfgang Pauli had proposed, in a letter to the "radioactive ladies and gentlemen" written in 1930, the existence of a neutral particle he called a "neutron", which would have had about the mass of an electron. Pauli was so prominent in theory, and so distant from experimentation, that it was joked that he could make experiments fail by merely passing through the town in which they were being performed in a train. In a twist, it was a famous experimenter, Enrico Fermi, who first completely explained beta decay theoretically while living in Rome, in 1934. Only once he left Italy for the USA did Fermi become known as a leading experimentalist. Fermi's solution was to posit the existence of the new neutral particle. It could carry away energy and momentum, sharing with the other product particles, depending on the exact directions they left in. Since neutrons had been discovered and named by the time of his paper, he gave the new neutral particles the name "neutrinos", from the Italian meaning "*small neutral one*" in contrast to the more massive neutrons. This was somewhat along the lines of Pauli's idea, but the neutrinos could have very small mass, carrying basically only energy and momentum (as do, for example, massless photons). Assuming their presence allowed the accounting of energy and momentum to be corrected. However, neutrinos have so little interaction with matter that they elusively avoided direct experimental detection until 1956. Fermi's theory was pleasingly complete and resembled electromagnetic interactions so that despite the lack of direct detection for over 20 years, neutrinos were almost immediately accepted into the particle physics family.

## 8.2 SOLAR ENERGY IS IMPOSSIBLE

The Sun's prodigious energy production was discussed in the chapter on light since most of it arrives at Earth in that form. Once one has a mass of gas like the Sun, thermal energy (causing pressure) in the gas is needed to support it. Most of the energy flowing to the surface in the outer part of the Sun is carried by gas motion, called convection. The surface arrives at a temperature that allows getting rid of that energy into space by emitting electromagnetic radiation appropriate to that temperature, and, in fact, remarkably like that of a "blackbody". The pressure support requires the gas to get hotter and denser as one goes down into the star. A "solar model" obeying the laws of thermodynamics and gas physics is easy to construct and, with those considerations *alone*, no nuclear power source is needed to have the model closely resemble the Sun. The problem is that it cannot be a static model since energy leaves the system. Energy must be liberated inside the Sun to keep it nearly static, making up for the loss at the surface. For example, the Sun's luminosity could be accounted for with contraction, in which energy is liberated as matter moves down in its own bulk gravity field. As we saw, the problem with this model is that, over tens of thousands of years, it changes much more than that allowed by geology and biology, which suggest that Earth has existed in a form resembling its present one for hundreds of millions of years. In his 1926 book "The Internal Constitution of the Stars",

Sir Arthur Eddington, mentioned here in Chapter 2 with regard to General Relativity, managed to push the Sun's lifetime to 20 million years, using models which still hold up today, but without an energy source. Realizing that this was still grossly too short, he rationally discussed processes to provide energy, including nuclear processes. Even radioactive decay of materials then known (such as radium) would not provide enough energy, but he calculated that the conversion of hydrogen to helium could. He could also calculate why this was *impossible* at the temperature of the center of the Sun under then-known theories. Nevertheless, he was so confident that this must be the mechanism that he stated that his critics should "go and find a *hotter place*", which appears to be a euphemism suited to an English gentleman. The conversion of hydrogen to helium is known as *fusion*, and indeed was impossible in the Sun under the laws of physics known in 1926. Our daily experience of the Sun passing over-head, providing light and heat, depends on remarkable phenomena discovered only years after Eddington's book. The process of fusion also has a fingerprint in the most invisible particles in the Solar System, which are of course neutrinos.

The Sun's energy production, allowing it to have a long life, is in fact totally dominated by nuclear fusion. "Fusion" implies joining things together, and what gets fused are nuclei. The "other" energy-releasing form of energy release from interacting nuclei is nuclear fission, which is instead a breaking apart. Fission plays such a small role in this story that it is mentioned here only for completeness. Very heavy nuclei can often release energy by splitting into two roughly equal-sized parts plus some debris which often includes neutrons, as shown in Figure 8.4. This

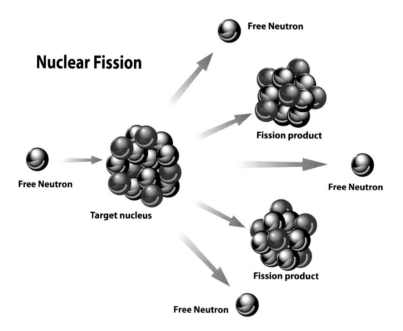

**FIGURE 8.4** Nuclear fission "splits" atoms and is an uncommon process in the Solar System (Shutterstock).

rarely happens spontaneously, but numerous elements can be stimulated to fission by incoming neutrons. If one gets to a state in which more neutrons become available to stimulate more fission, one can attain a "chain reaction". In fission bombs, the chain reaction develops very quickly, causing an explosion since every split atom gives energy, and there are literally exponentially more of them until the materials separate and neutrons no longer can interact with enough fissionable matter. In fission reactors, the only kind of nuclear reactor in non-experimental operation now, the reaction is controlled by carefully moderating neutron activity. Again, each fission gives energy. In both fission and fusion, we find the amount of energy released per atom involved to greatly exceed that available from chemical interactions. Early observations of the first known nuclear processes amazed physicists by the amount of energy released per gram of material, and this is typical of nuclear processes, due to the very strong binding of nuclei.

The charged nuclei left (usually two of them) from fission are close to each other but no longer bound by the strong nuclear force. These "fission product" nuclei repel each other very strongly, gaining a large amount of kinetic energy. When slowed down, they create heat, a desired outcome for power production, and, in a negative way, for "atomic" bombs. Heat is less important than neutrons in maintaining fission reactions, since neutrons can go on to produce other fission reactions, causing the chain reaction. Another important factor with fission is that if the nuclei produced as fission products are more or less random, chances are that they will have something "wrong" with them and eventually remedy that by radioactive decay. For this reason, fission reactions produce products that are usually radioactive, an effect that can be enhanced by neutrons being abundant. In this sense, fission weapons are "dirty" and sometimes made intentionally so. Fission reactors contain their radioactive products, but, eventually, they must be removed and stored when the fuel is replenished. Storage of such nuclear waste is an issue for fission reactors, and, although their activity goes down with time, some of it has long half-lives on human or even geological timescales.

In contrast to fission, which usually depends on neutrons, easily able to enter nuclei, fusion can happen only if suitable nuclei are brought very close to each other, since the range of nuclear forces is very short. For fusion to occur, high temperatures are needed so that charged nuclei move very fast to overcome the force of repulsion that their positive charges feel. In contrast, neutral neutrons involved in fission have no issue penetrating nuclei and usually must be slowed down (moderated) so that they remain near the nuclei long enough to cause a reaction. For fusion, nuclei must get close enough that nuclear forces act, which takes place on a scale not much larger than that of the nuclei themselves, that is, of the order of $10^{-15}$ m, which has the SI unit of fm or femtometer. This unit is also sometimes referred to as a "fermi" in honor of the Italo-American physicist Enrico Fermi, who, after his pioneering work on beta decay theory mentioned earlier, created the first nuclear reactor and played a major role in mid-twentieth-century nuclear physics. The early stellar modelers could easily determine that the temperature in the core of the Sun is about 15 million degrees. While this sounds hot, and certainly is by Earthly standards, the mid-1920s understanding of nuclear reactions showed that particles with the energy this implies (only about 1 keV) would be repelled from other nuclei at distances far too great to have nuclear interactions. In Chapter 4, alpha decay was explained by alpha particles escaping the barrier due to the attractive nuclear force through quantum mechanical tunneling, using their wave-like properties. Gamow's 1928 explanation of tunneling also works in reverse, allowing particles a finite probability of getting much closer than classical theory would permit. In the case of fusion, the barrier is due to repulsion between particles of the same (positive) sign of electric charge. Tunneling into each other's space can allow the attractive nuclear force to take over to enable a nuclear fusion reaction. In contrast to the case with fission, in which abundant neutrons are the main factor, high temperatures are needed to give fusing nuclei the speed needed to overcome the repulsion barriers, even when tunneling helps. To generate significant amounts of power through fusion, many interactions per second must take place, and this is favored by a high density in the material. High temperatures and densities are needed, preferably involving nuclei with low charge, to enable fusion. These are very difficult to attain on Earth, with artificial attempts to get more energy from a fusion reactor than was put in still a tantalizing goal in the early 2020s. In the core of the Sun, temperatures and pressures are the dream of fusion reactor experimenters and are maintained over very long periods of time. It would seem that there is lots of low-charge fuel around in the form of hydrogen, but, surprisingly, even when protons can get very close to each other, they are *very* unlikely to have a fusion reaction.

In the core of the Sun, the conditions nominally favorable for fusion are a temperature of about 15 million degrees and a density 150 times that of water (much denser than any solid on Earth, with uranium being about 20 times denser than water). It bears repeating that it is not nuclear reactions that keep the Sun hot, rather its hot gas supports the Sun and permits reactions to occur. The main role of reactions is to come into balance with the heat loss from the surface, allowing the Sun to be stable and long-lived. It has existed in much the same configuration for most of its 4.6 billion years and has about the same amount of time left before changes due to nuclear burning affect its structure and end its life. The Sun is by mass about 73% hydrogen, 25% helium, and 2% heavier elements (generically called "metals" by astronomers). There are lots of protons to react to form helium through fusion, although the helium in the Sun is largely *not* a result of fusion reactions, but is left over from the Big Bang. The heavier elements have highly charged nuclei and, in fact, even helium cannot react at the temperatures found in the Sun due to its four times higher mutual Coulomb barrier than hydrogen pairs have. Despite the proton abundance and tunneling, there is a serious impediment to the sorts of proton-to-helium reactions that Eddington envisaged. Something extraordinarily unusual, and in fact impossible under most circumstances, must happen to allow protons to have a fusion reaction at energies found in stars, even if they get very close to each other. One of them must change into a neutron!

## 8.3 THE BIG BANG WAS THE BIG EASY

The type of fusion taking place in the Sun today has little in common with that which formed the elements in the Sun in the first place. Its hydrogen and helium are the results of "incomplete combustion" in the Big Bang which took place 13.8 billion years ago. The universe supported vigorous nuclear reactions once it cooled down to the point that nuclei formed from more fundamental particles, but its cooling and expansion were so rapid that they did not take place for very long. The reactions were mostly fusion, making helium (two protons, usually two neutrons) from hydrogen (protons) and neutrons via routes involving deuterium (heavy hydrogen with one proton and one neutron) and tritium (heavy hydrogen with one proton and two neutrons). These reactions started when the universe was about two minutes old and were over within a minute. In terms of nuclear reactions, not too much happened in the universe after the first three minutes, until likely millions of years later with the formation of the first stars. The 2% of heavier elements, which astronomers call "metals", are the results of the later nuclear reactions in many generations of massive stars that took place between the time of the Big Bang and the formation of the Sun "only" 4.6 billion years ago. Heavier elements are "ashes" of nuclear "burning", but of a type that does not significantly take place in the Sun. They will not be discussed much here, but why the initial Sun had 73% hydrogen and 25% helium is both interesting and informative to our discussion about present nuclear fusion in the Sun and the invisible evidence for it. That invisible evidence is the neutrinos discussed above, proposed slightly after neutrons were found. Neutrons are the key to understanding the current composition of the Sun and thus its available nuclear fuel because the hydrogen/helium content of the Sun was basically determined by the Big Bang, in which fusion took place in a very neutron-rich environment.

As noted in Chapter 4, neutrons produced by alpha particles bombarding beryllium were discovered due to not deviating in a magnetic field and being able to transfer momentum to protons in substances containing hydrogen. Gamma rays would also not be deviated, so the "beryllium radiation" was initially thought to be that. However, gamma rays would not be able to eject energetic protons. This is possible only if particles of about the same mass impact the protons, which are originally at rest (Figure 8.5). Simple classical mechanics arguments allow this effect to precisely determine the mass of the original incoming particles if the mass of the target atoms is varied, which was done by James Chadwick in discovering neutrons in 1932. The initial guess that the neutrons had close to the same mass as protons was confirmed by him by also using other materials as targets. It may seem strange that a neutral particle could interact with a charged one like a proton. These particles should not be thought of as balls, despite the impression the figure gives. Really, the only interactions possible are electromagnetic and nuclear force. The latter does not happen in this case of "scattering" without reaction. It turns out that, although neutral in bulk, neutrons have a magnetic moment. This is consistent with neutrons not being fundamental particles but rather having internal constituents, as suggested by Figure 4.1. In any case, a magnetic interaction, which is the electromagnetic force, can occur and, despite its low efficiency, can eject protons as observed. The nuclear force cannot kick in very much despite neutrons possibly getting quite close to the protons, since they are moving quite fast. Slow neutrons, in contrast, react very strongly with protons or other nuclei. As mentioned above, this may induce fission. With lighter nuclei, "neutron activation" instead may occur, in which materials are made

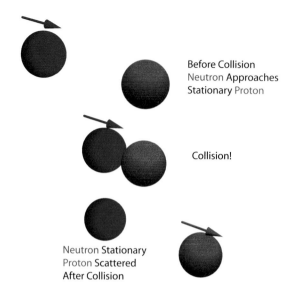

FIGURE 8.5   Collision of two equal mass particles in three time steps top to bottom. Initially the proton is at rest with an incoming neutron. After the collision the proton takes all the momentum (bottom). (author).

radioactive by exposure to neutron beams from reactors. In the Big Bang, there were initially only protons with which neutrons could interact, so we proceed to discuss that reaction. As it turns out that more complex forms of hydrogen than protons also took place in the Big Bang, and do so now in the Sun, we must discuss isotopes in more detail first.

Not too much attention has been paid to basic concepts about isotopes so far, although several of them were mentioned in discussing radium decay. Prout's hypothesis of the early nineteenth century was mentioned, that all elements are built up from protons. Certain problems came from this, for example, the common element chlorine has an average atomic mass (determinable for example from HCl electrolysis) which is 35.45, obviously not an integer multiple of the unit mass of hydrogen. With mass spectroscopy, elements such as this resolve into two or more clear signatures, each of which is very close to a multiple of the proton mass. Chlorine itself has the stable isotope $^{35}_{17}$Cl with about 76% abundance in nature, with $^{37}_{17}$Cl making up the remaining 24%. These have 18 and 20 neutrons, respectively, accompanying the 17 protons in the nucleus, and thus different masses, so that bulk chlorine found in nature has a non-integer mass number. Normal chemical processes do not respond much to mass, so for example in HCl electrolysis, one would get 35.45 grams of chlorine for every gram of hydrogen.

Once the existence of many isotopes was revealed by mass spectroscopy in the 1920s, and with improvement in lab techniques, it was perceived that there was a slight problem that even hydrogen did not seem to weigh one hydrogen mass, but was slightly heavier. Thus, it seemed logical to see if hydrogen had a heavier isotope. Indeed it does, as shown by American chemist Harold Urey, who got the Nobel Prize in 1934, but may have become more famous for later experiments showing that amino acids, the building blocks of life, could have been generated under the conditions found on the early Earth. This heavy isotope, $^2_1H$, is called deuterium, and its nucleus, with one proton and one neutron, is called a deuteron (sometimes denoted as "d"). After deuterium was identified in 1931, the neutron was found in 1932, so its configuration as a bound proton and neutron was fairly obvious. They are not, however, bound very tightly, so that deuterium is not very abundant in nature, constituting only about 0.01% of hydrogen atoms. Deuterium is stable, but a third isotope of hydrogen, tritium, with 2 neutrons, undergoes beta decay. Nevertheless, these neutron-rich forms of hydrogen are the most reactive for fusion, and most human attempts at energy-producing fusion are based on deuterium-tritium reactions. Early hydrogen bombs used this reaction, with compression and heating arising from encasing them in a standard fission "atomic" bomb. Controlled fusion reactors usually try to get these fuels very hot in a plasma, or very hot and dense by compression using intense laser light. Proton-proton reactions themselves are impossible to reproduce on Earth at low energies but do occur in the Sun with perhaps a surprising amount of difficulty, as will be discussed below. Let us first explore fusion to understand the Big Bang, at least in terms of where the bulk of the mass in the Solar System comes from.

For a reaction to take place, the products must have lower energy than the original reactants. If energy flows out of the system (an exothermic reaction, the most common kind) to make this the case, it would have to be put back in if one wanted to break up the final product. In that sense, the product nucleus is bound, although other conditions than energy may play a role in determining which reactions are possible. It is common to give the binding energy per nucleon, the total energy liberated by a reaction divided by the number of nucleons in the final nucleus, as a positive number, but in fact we should regard nuclei as "being in a hole" once formed, needing energy to be broken up again. We have noted this to be the case for bodies in orbits and for electrons in atoms, and it is also true of stable products of nuclear reactions.

Figure 8.6 shows the binding energy per nucleon for light nuclei. The atomic number is shown on the Z scale, while the mass number is shown on the A scale. Binding energy per nucleon is on the vertical scale and indicated also by a color scheme. The many energetically possible isotopes, not all of which are stable, are arrayed by atomic number, forming bands in the Z-A plane but shown in 3-D on lines above it. The single proton is the leftmost point and has zero binding energy (the element H is not labeled). The deuteron is the next point to it and is seen to be weakly bound (about 1 MeV). Having an extra neutron makes the tritium nucleus at Z = 1, A = 3 ($^3_1H$) more tightly bound still, yet it is unstable to beta decay.

Present nuclear reactions in the Sun rely on inverse beta decay, which is needed in order to allow protons to combine and eventually form helium. The generation of helium by fusion allows the Sun to be very long-lived. However, by the middle of the twentieth century, it was clear that the observed proportion of roughly 71% H and 27% He by mass in the Sun and local stars could not be attained by slow formation of helium from hydrogen. Many lines of proof converged by the late twentieth century

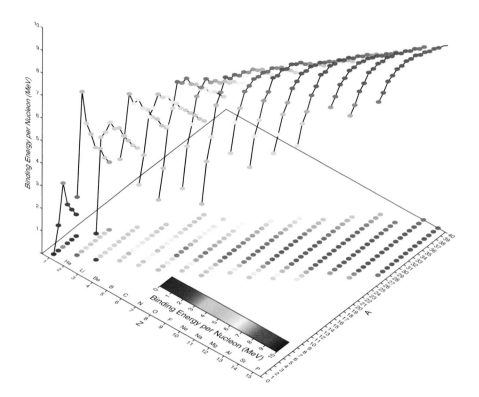

**FIGURE 8.6**  Binding energy per nucleon. More-tightly-bound nuclei are higher and indicated by red. Black lines join isotopes, and the projections of their positions in Z vs. A are in the bottom flat plot. Note the isolated peak corresponding to helium-4 but general upward trend with Z (author, based on data in Huang et al., 2021).

to indicate that the universe began with a hot and dense early stage. Originally a derisive name, the term "Big Bang" stuck to describe the beginning of the universe.

We will succinctly discuss the brief nuclear fusion phase of the Big Bang, which describes where the hydrogen and helium come from, and illustrates fusion reactions which differ from those currently going on in the center of the Sun. The earlier phases were extremely hot and dense but characterized by rapid expansion, cooling, and decrease in density. Much as we could consider blackbody walls to be in equilibrium with light in earlier discussions, we can regard high-energy photons to be interacting with particles in the first few seconds of the universe. These early times were dominated by photons, with quarks and antiquarks forming once the temperature got low enough. Free quarks are no longer observed as they are all bound strongly in aggregates (see Figure 4.1) which we know as particles, including neutrons and protons. Once neutrons and protons formed, they interacted to come into equilibrium with the radiation, a process that favored the existence of protons since neutrons are more massive. It is only after much cooling that the temperature came into the range of "only" billions of kelvin at 100 seconds after the start of the Big Bang. This time

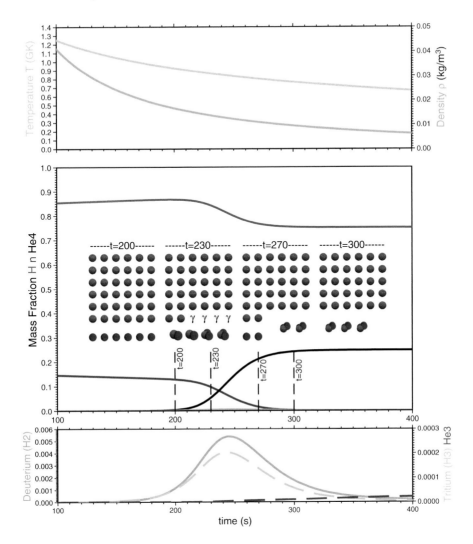

**FIGURE 8.7** Big Bang physical conditions in the first few minutes (author, based on calculations from F. X. Timmes's code at https://cococubed.com/code_pages/net_bigbang.shtml).

corresponds to the beginning of Figure 8.7. At that time, the density was in the kg/m³ range, about that of air and much less than in the center of the present Sun. However, the temperature of about 13 billion kelvin (13 GK) was about 1,000 times that in the Sun. Under these conditions, "normal" nuclear reactions involving the strong force could occur directly between neutrons and protons. Since the product of these reactions, deuterium, is a weakly bound nucleus, it tended to disrupt initially but could exist in small quantities at one point, with a maximum though small abundance about 250 seconds after the Big Bang.

The main reason for the small abundance of deuterium then and now is its rapid further reaction, mainly to form tritium, which in turn reacts to form helium. These reactions slowed down rapidly after by the 250-second mark, the "first three minutes", so that most of the original hydrogen remained, not having had the chance to participate in a reaction . The reaction product, helium, is very tightly bound and stable, while the intermediate products are largely consumed, coming to near their present very low levels by about the 360 second, or six-minute mark. As a result, a solar-like composition of 75% H, 23% He by number is common in the universe, with local exceptions where further nuclear burning has taken place, usually in stars. The Sun is only about one-third the age of the universe, and the remaining 2% of its mass has mostly come from nuclear burning in heavy stars which managed, through explosion or other means, to get their nuclear-burning products into the "enriched" gas which eventually formed the Sun(.

The Sun itself has negligible generation of new heavy elements in its core, since at its comparatively low temperature, mostly the "proton-proton" (p-p) reaction takes place and does not go beyond helium. Some variants of it do involve light elements, briefly discussed below. To contrast this reaction with the dominant reaction in the Big Bang, and to introduce the concept of using invisible neutrinos to verify solar interior processes, we will now examine the basic p-p cycle. It is worth noting that fusion (hydrogen) bomb fusion reactions have more in common with the Big Bang than with current reactions in the Sun, because either by setting them off with a neutron-rich fission explosion or by using substances that generate neutrons, they generally involve neutron-proton reactions, at least in part. They and attempts at fusion reactors often also use deuterium and tritium, which are most reactive among the hydrogen isotopes. Protons do not react easily with other protons for the two reasons which start off our next discussion.

## 8.4  THE IMPOSSIBLE MADE POSSIBLE

First, protons repel each other. It can easily be shown that at the speed protons move in the solar interior, despite its temperature of 15 million degrees being high by Earth standards, they have little energy compared with that needed to bring their nuclei close enough for nuclear forces to become effective. In classical physics, their repulsion is too strong for them to approach to the femtometer separation needed. It is only the quantum mechanical phenomenon of tunneling, discussed in the context of alpha decay in Chapter 4, that allows such overlap. The tunneling works in the opposite sense to that for alpha decay, since it allows interpenetration of the proton's wave functions as they come close to each other. Without what may appear to be an esoteric quantum effect, the Sun would not shine. The other remarkable, and very demanding, requirement is that one of the protons must undergo inverse beta decay *while near the other*. This happens in only one in $10^{23}$ proton-proton interactions. Basically, one proton must turn into a neutron while near the other one, which then reacts with the neutron to form deuterium. The protons must be so close that they are interacting via the strong nuclear force when this happens and are sometimes regarded as being a "diproton" or a helium-2 nucleus. However, in this configuration, the electrostatic repulsion is stronger than any strong nuclear force, meaning that the configuration is unstable and does not last long. As a result, the probability of a rare inverse beta decay occurring, although it is energetically possible, is vanishingly small. The lifetime of an individual proton at the center of the Sun is about $10^{10}$ years before it interacts with another proton in this manner. When it does occur, this first step of the p-p chain liberates a positron and an electron neutrino. While the positron almost immediately annihilates with an electron, of which many are available in the dense plasma near the solar core, the antineutrino has almost no interaction with matter, and leaves the Sun at almost the speed of light. The energy produced by annihilation is gamma radiation (in fact two photons of 0.511 MeV) but the solar interior is opaque for all electromagnetic radiation. It takes thousands of years for light to diffuse to the surface, being degraded, of course, from the very high gamma energies to the eV range of visible light. On the way, the initial two gamma rays give rise to many more of the lower-energy photons that eventually escape from the solar surface.

Two amazing statements are contained in the above paragraph. If either effect were not so tiny and improbable, the Sun would not exist as we know it. First, quantum mechanical tunneling is needed for ordinary protons to even get near enough to each other to do the reactions they have in the Sun. Second, the vastly improbable need to have one of them have an inverse beta decay while close to another is *very* constraining. If either process was a bit more probable, the Sun (and all stars) would have gone out in a blaze of glory shortly after forming. On the other hand, if these processes were of *zero* probability, there would now be no light emission from the Sun, or stars at all, and they all would have shrunk to cold dark bodies within a few thousand years of forming. If you are a fan of country and western music, you might feel that people are doomed to be unlucky. However, consideration of these two paragraphs should convince you that some lucky balanced numbers got us to where we are now.

## 8.5 NEUTRINO DETECTION

The details of beta decay are intricate enough that it is best to consider simply that, in a nucleus with excess neutrons (including the isolated neutron), they can decay into protons. Inverse beta decay, which is so important in our lives in allowing one proton near enough to another to form a deuteron, can also be viewed simply as protons turning into neutrons. Both processes involve neutrinos, the most invisible constituents of the Invisible Solar System. The almost impossible to detect is our strongest evidence that the almost impossible to happen actually does happen (Figure 8.8).

Despite the very long time that an individual proton can resist reacting, the proton-proton chain initiated by protons combining to form deuterium is the dominant reaction set in the Sun. This first step does not resemble the proton-neutron reactions which characterized the nucleosynthesis stage of the Big Bang, involving only the strong force. The inverse beta or "beta plus" decay of a proton releases a positron (to carry off the positive charge) and an electron neutrino. Also known as "positron emission", in more complex nuclei it characterizes those with "too many" protons. Unlike beta decay, in which an electron is emitted, accompanied by an electron antineutrino, in inverse decay their antiparticles are involved, the positron and electron neutrino. The reaction may be written as $^1_1H + ^1_1H \rightarrow ^2_1H + e^+ + \nu_e$. As for beta

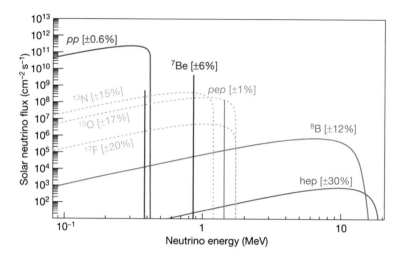

**FIGURE 8.8** Neutrino flux at Earth from the Sun from *pp* (red) and other burning cycles. For continuous curves, the flux labeling is, in addition, understood to be per MeV (D'Angelo et al., 2016: CC BY-4.0).

decay electrons, the positrons have a continuous spectrum. Unlike in laboratory beta decay, those in the Sun cannot be observed and annihilate immediately with readily available local electrons. Instead, we expect that the neutrino involved in carrying off energy will have a continuous spectrum, as shown by the red line in Figure 8.8. In total, 1.442 MeV of energy is released in this "p-p" reaction, including the annihilation energy. Both the positron and the electron it annihilates with have a mass of 0.511 Mev/c², so 1.022 MeV is accounted for there, leaving a maximum of 0.42 MeV as the maximum possible for *pp* neutrinos, as shown in the figure where the red curve intersects the bottom axis. Please note that, in the logarithmic energy scale shown, 0.1 or 10⁻¹, which is labeled, has 0.2, 0.3, 0.4 as the first tick marks right of it, which are not labeled. Noting that the flux peaks at about 2 × 10¹¹ neutrinos/(cm² s MeV) and that the sharply peaked spectrum is about 1 MeV wide, we find that *pp* neutrinos of all energies are passing through Earth at a rate of about 2 × 10¹¹ neutrinos/(cm² s). Converting from the figure to SI values, there are about 2 × 10¹⁷ neutrinos traversing each square meter of Earth's cross-section each second. Due to the very low rate of interaction with matter, it is accurate to say "traversing" since almost all of them pass through the whole Earth with no interaction at all. Since your body has about one square meter of cross-section, 2 × 10¹⁷ of these *pp* neutrinos pass through *you* each second, unnoticed. Very rough mental arithmetic suggests that one atom in your body might be transformed by this most benign form of radiation in your lifetime!

In the main *pp*-chain, the only neutrino energy loss is in the initial reaction. Neutrinos interact so little with matter that they leave the Sun at almost the speed of light. Light itself is very heavily absorbed and the initial gamma rays given off are degraded to visible light by the time their energy makes it to the surface. Neutrinos tell us the state of the Sun as it is, while gamma rays are treated as heat, which makes its way outward on a time scale of thousands of years. At one point, initial problems with interpreting neutrino measurements gave rise to a consideration that the Sun might be taking a pause in its nuclear burning. Since nuclear burning does *not* give rise to the Sun's heat, this was an admissible, although extreme, hypothesis. The other aspect of neutrinos' lack of interaction with matter is that they are very difficult to detect. Detectors based on the interaction of photons with material rely on the fact that this electromagnetic interaction is strong. In the days of chemical-based photography, chemicals were affected by light, with the main commercial performer in this field being silver halide, mentioned previously in the context of the discovery of UV light. Now, most photographic detection is done with the photoelectric effect, in which the electric field of light moves electrons around, for example, in the detector chip of a digital camera. To illustrate why electromagnetism is stronger than the weak force by which neutrinos interact, consider the modern view of electromagnetic interaction shown at the left in Figure 8.9.

The view of an electron repelling another electron is that they interact through exchange of photons. One photon interaction is shown, but a real event would consist of many such interactions, and with the important quantum consideration that these photons are *virtual*. This esoteric concept is related to the uncertainty principle mentioned previously with regard to the structure of atoms; position and momentum cannot simultaneously be known with great accuracy, so the best we can do is know them on the scale of an atom. Similarly, energy and time cannot be localized exactly; one can create a photon "out of nothing" as long as it does not exist for very long. Photons have no rest mass, making them easy to make, real photons have an infinite lifetime. Making photons as virtual particles is "easy", so exchanging them gives rise to a comparatively strong force which can act over large distances. Geometric considerations alone make this force fall off with the inverse square of distance. On the other hand, the weak force for neutrinos deviating electrons uses the virtual force carrier of the (neutral) weak force, which is a $Z^0$ boson. Although we are used to speaking of four forces of nature, the electromagnetic and weak interactions can

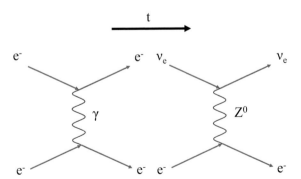

**FIGURE 8.9**   Feynman diagrams for electromagnetic interaction mediated by virtual photon (left) and neutrino-electron scattering by virtual Z particle (right). Virtual particles shown in blue, real in red. Time increases to right in each panel (author).

be combined into the *electroweak* theory, and its prediction in 1968 called for the existence of the $Z^0$ boson as well as $W^\pm$ bosons involved in beta decay. $Z^0$ was discovered in nuclear accelerator physics experiments in 1983, and the discovery was so important in bolstering the so-called "Standard Model" of particle physics that the 1984 Nobel Prize in Physics was awarded for it. Already within 30 years, the results of neutrino-electron interaction mediated by virtual $Z^0$ bosons were an "application" of this knowledge, confirming our basic knowledge of the Sun. On the right in Figure 8.9, a neutrino is shown exchanging a virtual $Z^0$ with an electron, causing both to deviate in their paths. The mechanism is very similar to that of electrons interacting with each other via virtual photons. However, a major consideration illustrates why weak interactions are in fact so weak. Unlike the massless photon, the $Z^0$ boson has a mass of about 91 GeV/$c^2$, which is over ninety times that of a proton! Having an intrinsic half-life of about $10^{-25}$ s, it is already an ephemeral particle, and the huge mass makes it difficult to create virtual $Z^0$s "out of nothing". The result is that this mechanism gives a very low possibility of interaction with matter. The relative strengths of electromagnetic and weak interactions are illustrated by the fact that the distance a gamma ray from fusion in the Sun travels before being absorbed is at most 1 cm, while rough estimates of the path length of a neutrino, if traveling all the time in gas like that at the center of the Sun, give about $10^{16}$ m, or one light year! On a more personal level, it was neutrino pioneer Raymond Davis Jr. who stated in his Nobel Lecture (2002) that "the atoms in the human body capture a neutrino about every seventy years, or once in a lifetime", as mentioned above. We now turn to how such elusive particles can be detected.

Going from the quantum electrodynamical approach to the classical, when a neutrino-electron interaction as discussed takes place, the electron is accelerated, and in Chapter 3 it was noted that acceleration makes electrons emit electromagnetic radiation, which, in some cases at least, would be photons of light. *Photomultiplier tubes* can amplify the tiny current of even one electron's emission from a target due to the photoelectric effect, through a cascading system of metal re-emitters of electrons, with high voltage between them, to make a detectable current. To have a reasonable rate of detection, large amounts of target material are needed, and in the case of the "Borexino" (Figure 8.10) experiment under the Apennine mountains of central Italy, this is highly purified water which lets photons get to photomultipliers on the walls of the chamber. Both the extreme purification of the water and the subterranean location are to reduce spurious emission sources so that the very low signal from the neutrinos can be detected. Borexino reported the detection of the *pp* neutrinos in 2014 and has subsequently detected neutrinos from other reaction chains in the Sun (for example, the $^{13}$N and $^{15}$O orange dotted traces of Figure 8.8, associated with "CNO" burning, which supplies a very small percentage of the Sun's energy). Neutrino "lines" of only one energy are associated with reactions in which an electron (of basically zero energy) is captured by the nucleus, so that a definite energy is associated with the process.

Historically, solar neutrino astronomy began with Raymond Davis Jr.'s detector, based on 100,000 gallons (nearly 400,000 liters) of cleaning fluid deep (1.5 km) underground in the Homestake Gold Mine in South Dakota. Davis stated that "after

**FIGURE 8.10** Borexino neutrino detector ca. 2022. The exterior thermal control blanket is seen, which surrounds a large water tank with photomultiplier detectors to pick up light from neutrino-electron interactions. Note man on ladder for scale (Orebi Gann, 2020).

the first run, it was clear the solar neutrino flux was lower than predicted", which precipitated the "solar neutrino problem", only solved at the turn of the twenty-first century. Although it is a property of the Solar System to be permeated by neutrinos from the Sun, and the structure of the Sun has been understood for about a century, the solution was that neutrinos come in three "flavors", associated not only with electrons, but also with unstable related particles called muons and tauons. The former has a mass about 200 times that of an electron, and the latter about 3,500 times. Only electron neutrinos can take part in the beta (i.e., electron) decay processes used to detect neutrinos, or evidently the electron deflection method discussed above. To various degrees, "neutrino oscillations" change what left the Sun as electron neutrinos into other types that are not detected. The solar neutrino problem is thus regarded as solved but at the expense of neutrinos being more complex than imagined before, and likely having non-zero, although very small, mass.

## 8.6 ENERGIZATION OF PARTICLES BY WAVES

Sir Arthur Eddington gave a series of lectures at Cornell University in New York in 1934, and a rough Atlantic crossing (in a ship) had him ask "Is the ocean composed of water or of waves?". Certainly, in a storm at sea, one's attention focuses on the waves. Space is full of waves of many different types, some of them having some analogy to ocean waves. However, let us first focus on the characteristics of waves. We encountered electromagnetic waves in Chapter 3 as a way to represent light.

Light has a very high frequency, so that many waves traverse space in a short period of time. We might therefore think of waves as always being in long trains of oscillations. However, strictly speaking, a function representing wave is simply something that satisfies a "wave equation". Surprisingly, any function can do this if it has as variables the sum or difference of position and the product of speed and time. This means that if one travels along with the wave as time advances, the wave retains its shape. Having mentioned electromagnetic waves, they do obey the wave equation, but we cannot imagine following along, at least not physically, to chase them at light speed. Slower waves, however, can be treated classically and it does make sense to think of following along with them. Figure 8.11 shows two cases of a wave moving toward the right. This single "bump" moving along retains its shape if one shifts the observing position along with it. The horizontal axis is distance in arbitrary units, but as one goes forward in time (upward in each panel) the wave moves to the right the same amount but it still looks the same.

Waves usually are in some medium: in this case we would like to make an analogy with surfing and thus will regard this single wave as being a swell of water, and it would not be unrealistic to consider the vertical axis as the height of the water in meters. A surfer or other object merely

sitting still in the water is on average simply raised and lowered as waves pass. This situation is illustrated in the bottom panel of Figure 8.11. However, a particle already moving in the same direction as the wave can take energy from it and come to have its speed. A "perfect" example of wave capture of a particle is shown in the top panel, in which the particle starts moving with the wave. In water waves, due to drag, this is not possible. Surfers do paddle hard to get some speed so that the wave picks them up, but they end up on the leading edge of the wave, not the top. They are essentially continuously going downhill. While the largest waves are found under special circumstances on the shores of large oceans, the longest surf ride was on a Canadian tidal bore, a solitary wave progressing up a river, and was for 29 kilometers!

Another characteristic of waves is the transport of energy. Ocean waves may travel thousands of kilometers, but eventually hit a coastline, with interesting characteristics like "breaking", which is a violent process that dissipates energy. Directly hitting a coastline, waves can deposit enough energy to break down rock. We saw in Chapter 3 that electromagnetic waves also transport energy, with the most notable number being the solar "constant" of approximately 1,340 W/m$^2$ on the cross-section of the Earth exposed to sunlight. Exchange of energy between waves and particles is largely responsible for making space "radioactive". Often, we can consider there to be two "populations" of particles in space, one of low energy, behaving like a fluid despite having very few collisions, and one of high energy that can be higher than the energies emitted by radioactive nuclei. As was mentioned above, high-energy particles in space are dominated by protons as well as electrons, while radioactivity commonly involves alpha particles, which are helium nuclei. The simple examples of Figure 8.11 illustrate that a particle can best take energy from a wave if it is traveling at about the same speed as the wave. The red particle, initially at rest, may get a small forward push from the wave, but about the same push in the other direction as

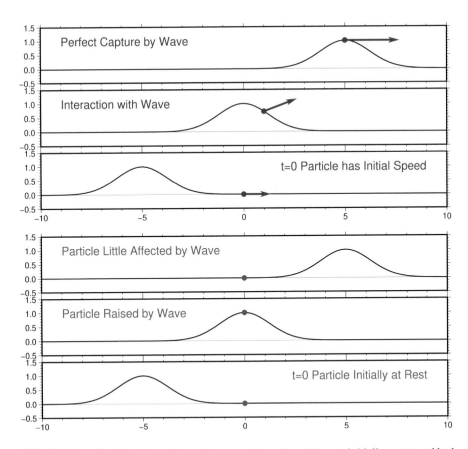

**FIGURE 8.11** Particles interacting with waves. The red particle was initially at rest and had minimal interaction with the moving wave. The blue particle, initially moving at nearly the speed of the wave, can "surf" on it (author).

the "downhill" part of the wave goes by. As a result, it does not interact much with the wave. On the other hand, the blue particle in the top panels was already going about the same speed as the wave. In this contrived example, the particle picks up speed from the wave to the point that it can climb to the top and then goes along exactly in phase with the wave. Surfers can wait for the perfect wave by remaining still in the water, much like the red particle in Figure 8.11, but when the perfect wave approaches, they paddle quickly to get up near its speed and "ride" it.

Although Figure 8.12 gives an idea of raw power in a wave, and that real waves must be ridden by constantly "going downhill", the simpler situation of Figure 8.11 gives more of an idea of the transfer of energy. The red particle at rest is still at rest, so it has not picked up much energy from the wave. The blue particle, however, is shown going about twice its initial speed once riding the wave, then having four times its initial energy since energy scales as the square of speed. The *gain* in energy is three times the initial energy. Furthermore, if there were a group of many waves going at different speeds, the blue particle would be well positioned to pick up more speed from yet faster ones. Interaction with many waves can enhance how much energy is gained. The loss of energy from the wave is not considered in this simple case. In the real case of surfing, the energy gained from going continually down the moving slope is dissipated as friction. Once a surfer or a particle is trapped in a wave, the wave itself may speed up, further enhancing the energy gain.

Since electromagnetic waves travel at the speed of light, they are not efficient in accelerating particles, since no matter how fast particles go, they will generally be affected by the averaged effect of many wave crests passing by them, and these averages come out to zero. There are, however, many types of plasma waves which can interact with particles to energize them.

Ocean waves are usually generated by a "wind-over-water" process called Kelvin-Helmholtz driving. There is a favored wavelength for the maximum growth of waves, which is why sea waves generally come in regularly to shore. Similarly, the solar wind interacts with the magnetopause to frequently cause Kelvin-Helmholtz waves on its flanks (Figure 8.13). Here the restoring force is magnetic: rippled magnetic field lines have higher energy than straight ones. In sea waves, it is simply gravity attracting the water in the waves downward that is the restoring force. Much as the speed of the wind determines the wavelength of water waves, the speed of the solar wind determines the characteristics of Kelvin-Helmholtz waves. In this case, more complex than wind on water, the direction of magnetic fields is also important in determining whether waves will grow. In turn, the surface waves can couple to field lines inside the magnetosphere. Wiggling field lines are equivalent to waves (in some cases, like standing waves found on musical instrument strings). In turn, they

FIGURE 8.12    A surfer travels with a wave, but not at the top! (Shutterstock).

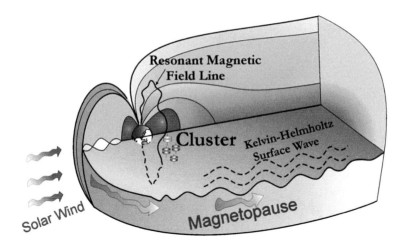

**FIGURE 8.13** Surface waves similar to water waves make magnetic field lines resonate and, in turn, cause particle acceleration. Cluster is a spacecraft constellation (Zong, 2022: CC BY-4.0).

can give energy to particles. In this way, energy from the solar wind can be coupled into the magnetosphere, energizing particles which are then trapped in the radiation belts. In Figure 8.13, the "Cluster" spacecraft constellation is shown. By having multiple spacecraft near one another, waves going by can have their wavelength determined, and associated energetic particles can be measured. The basic mechanisms of energy increase for particles interacting with waves usually have to do with "riding along" with the wave. Bouncing particles can sometimes do this multiple times. In some cases, so many particles get energized that they start to influence the wave, making the entire interaction quite complex.

## 8.7 SOLAR ENERGETIC PARTICLES

In discussing magnetospheres, it was seen that a basic feature of solar wind interaction with magnetized planets is the formation of a shock wave, known, by analogy to the waves made by ships moving through water, as a "bow shock". Although Figure 8.13 does not explicitly show the bow shock, it serves as a reminder that there is a transition from the solar wind to a region dominated by the planetary magnetic field. The bow shock responds, but generally slowly, to changes in the solar wind. The real bow shock of a moving ship would be moving with the ship, and some shock waves in space are not stationary but move out from sources of disturbance at supersonic speed. They are then capable of accelerating particles.

Explosions in the solar corona, sometimes but not necessarily associated with solar flares, cast large volumes of heated gas into the solar wind. These coronal mass ejections often move much faster than the already supersonic solar wind. Images of coronal mass ejections were seen in Chapter 6, which discussed the solar wind, and Figure 8.14 shows them schematically along with the two types of observed

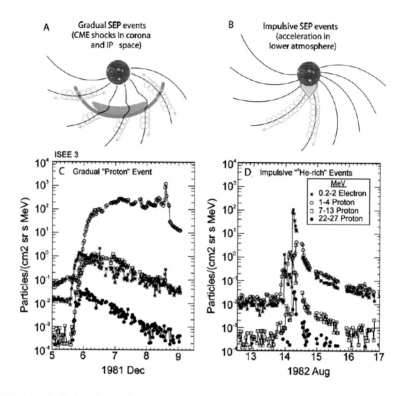

**FIGURE 8.14**   Solar Energetic Particle events formed by protons riding on shock waves from coronal mass ejections (McComas et al., 2016).

solar energetic particle (SEP) events. Recall that alpha particles from radioactive decay may have energies in the several MeV range yet are not very penetrating. They are not biologically dangerous since they deposit all of their energy very quickly in skin, the outer part of which is already made from dead cells impervious to damage. Similarly, protons of a few MeV are not dangerous. Although SEP events are characterized by such low-energy protons, they also contain much more energetic ones, which do have penetrating power and are dangerous to biological life in space. Space travelers receive a steady background dose due to cosmic rays, discussed below, but SEP events, which rise quickly and then may last several days, may be far more dangerous to those voyaging in space during them. In terms of origin, it is clear that simple surfing on a shock wave cannot accelerate particles to MeV energies. The solar wind speed of about 400 km/s corresponds to about 1 keV, so even a fast shock wave at several times that speed falls short by a factor of about a thousand for acceleration through simple surfing. Likely, complex dynamic magnetic fields are in the shock waves, possibly with other types of waves embedded in them. Although we can clearly associate SEP events with the coronal mass ejections (CMEs) mentioned in Chapter 6, the mechanism of particle acceleration remains unclear. A further mystery of SEP events is that the small ones often seem to be enriched, sometimes by quite large factors, in the rare helium isotope $^{3}$He.

## 8.8 RADIATION BELTS

The famous discovery that "space is radioactive" was made by James Van Allen's team in 1958. Once some puzzling aspects of the data were correctly interpreted, it was initially realized that radiation in space increased with height. Later, a more complex distribution of "radiation belts" was mapped out. In fact, the dipole-dominated shape of the Earth's magnetic field, with its toroidal symmetry, makes the zones of enhanced radiation more toroidal, or doughnut-shaped. A cutaway view, schematically showing the two main radiation belts, with red indicating maximum intensity, is shown in Figure 8.15. The Van Allen Probes (initially known as RBSP) spacecraft, which explored the belts from 2012 to 2019, found a third belt that could be distinguished in active times. We have noted that, although the low-energy plasmas of the Solar System can usually be considered collisionless unless near a major body (inside the Sun's critical point or near an ionosphere), they behave like fluids. The high-energy particles in radiation belts of the Earth or of other planets are not only collisionless, they are little affected by the low-energy plasma. As such, many of their properties can be considered using single-particle motion laws. Once that is done, some aspects may in turn reflect the bulk properties of the high-energy

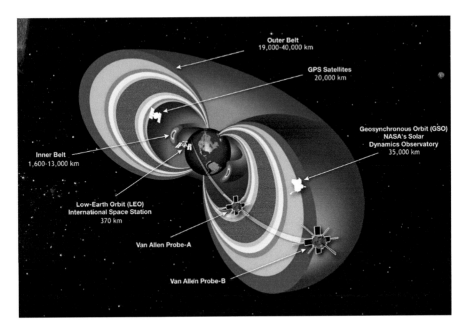

**FIGURE 8.15** Earth's radiation belts in cutaway view. Heights (not radii) of satellite orbits are shown. The Van Allen Probe pair were on a highly elliptical orbit from 600 to 30,400 km to map the radiation belts in detail (NASA, modified)

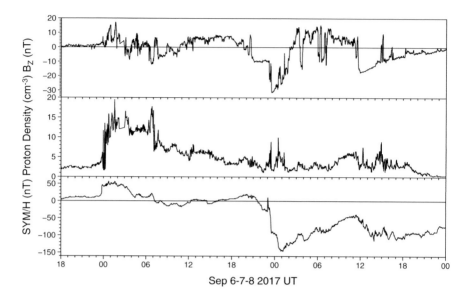

**FIGURE 8.16** Storm response from September 6, 2017, 18:00 UT to the end of September 8, under varying solar wind conditions. Bottom panel is the SYM/H index reflecting the magnetic effect of the ring current. Middle panel is proton density, and top panel is the Bz. The large increase in density at 0 UT on September 6 results in compression and a rise in magnetic field, while the ring current causes a large decrease almost 24 hours later (author, from NASA OMNI 1-minute data).

particles. One important and interesting aspect of this is the "ring current" which characterizes magnetic storms. Although the magnitude of this current is often larger than the current in substorms which were discussed in the previous chapter, it is more distant from the surface. Its magnetic effects are usually smaller in magnitude, but they are felt worldwide.

Figure 8.16 shows such magnetic storm effects as indicated by the ground magnetic field in the bottom panel, with the averaging of several magnetic observing stations creating an index called SYM/H. The middle panel indicates the proton density, which is normally about 5 per $cm^3$. The top panel shows the vertical component of the solar wind magnetic field, which, in the previous chapter, was shown to enhance reconnection and allow substorms. Although that also happened in this strong event, we emphasize only the storm as indicated by the SYM/H index. It had a positive excursion at 00 UT on September 7, and a much larger negative excursion almost exactly 24 hours later at nearly 00 UT on September 8. The positive excursion is associated with the simple pushing back of the dayside magnetopause by solar wind pressure that went up due to the density increase being rammed into the dayside by the ca. 500 km/s solar wind. Not much reconnection took place since the magnetic field was positive; the solar wind could not reconnect and one simply had fluid and magnetic pressure pushing back and increasing the magnetic field in the magnetosphere, including (by indirect means) at the surface of Earth. Already at about 20 UT on

September 7 (middle of plot), the magnetic field became negative (southward) as seen in the top panel. The almost immediate effect on the SYM/H was a decrease from slightly positive values. When the field dropped to a (rare) value of −30 nT slightly later, the SYM/H went quickly to very large negative values, −150 nT, indicating a major storm. Such behavior was already observed in the 1830s by the "magnetic society" organized by the same C. F. Gauss who helped recover Ceres, but only in 1911 did the Norwegian Carl Størmer (sometimes written in German fashion as Störmer) propose a convincing reason for it by effectively discovering the radiation belts.

Scandinavians have made scientific contributions to the study of auroras and thus space throughout the scientific age. The original association of magnetic fields with auroras was made in Scandinavia. This is likely due to the frequent presence of auroras at those northern European latitudes, coupled with modernization and prosperity that lead to having educated researchers. Possibly due to seeing auroras frequently, the "Scandinavian school" had not much problem envisaging them as being made by particles coming into the atmosphere from space.

On the other hand, the "British school" believed that space really was a vacuum apart from possible intermittent times when sunspots shot material outward (which was observed in the form of prominences). The British tended to attribute changes in ground magnetic fields as being due to ionospheric currents in closed loops (which is indeed true of certain current systems such as those caused by solar heating effects). Having arisen in the early twentieth century, this debate was not resolved until satellites in the late 1960s detected electric currents in space through their magnetic effects which showed that particles flowed basically along magnetic field lines, which intersect the atmosphere in the auroral zones. By then, the main proponents of the two sides were Englishman Sydney Chapman and Swede Hannes Alfvén, so the stand-off had become known as the "Alfvén-Chapman" debates.

Størmer proposed that the decrease in the horizontal part of the Earth's magnetic field at low latitudes was due to a "circular stream in the geomagnetic equatorial plane of the Earth", and that the intensification of that stream would make the auroras move away from their normal band roughly 23° from the magnetic pole. By using the Ampère Law right-hand rule, one can see in Figure 8.17 that an electric current

FIGURE 8.17    Ring current (blue) and resulting magnetic field (red) around and near Earth (author).

around Earth in the direction shown would give rise to a southward magnetic field inside the loop, including at the surface. The actual motion of the particles making up such a current is subtle, but Størmer was a mathematician and up to the task of writing equations to describe it. The calculations were incredibly laborious in the days before computers, but the particle motion described is still the basis of radiation belt theory.

To begin to understand Størmer's work, we will consider the motion of particles, both electrons and protons, in a magnetic field considered to come up out of the page. The motion of particles is dictated by the *Lorentz force*, in which the electric field moves them directly and has strong analogies to the effect of gravity, and the magnetic field deviates them at right angles to the direction they are moving in. If there are both electric and magnetic forces, the motion is more complex, because the electric force accelerates particles, which then are deviated by the magnetic field. Thus, we consider only magnetic fields initially. Since the force is at right angles to the particle motion, magnetic fields deviate particles without a change in energy. Again, this has analogies to gravity in that a satellite in a circular orbit only has its direction changed by gravity and neither gains nor loses energy. In a uniform magnetic field, all orbits are circular. The Lorentz force is also proportional to speed, and, like any force, the acceleration (which is toward the center of the orbit) is inversely proportional to mass. The inward force is also proportional to charge, which for protons and electrons is the same.

Figure 8.18 shows the resulting orbits for protons in its upper left and upper middle panels. For the same proton energy (about 1 keV), the orbit radius is larger in a weak field than in a strong field. The field used is about the same as in the radiation belts, so proton orbital radii are on a scale of kilometers. When the lower mass of electrons, 1/1836 that of protons, is taken into account, their orbits are $\sqrt{1836} \approx 43$ times smaller than those of protons. A comparison of electron and proton orbit sizes is given at the upper right (Figure 8.18), which is a blow-up (scale is in m, not km) of the very small box near $x = 1$ km, $y = 0$ km of the bottom panel. The most important lesson from this simulated motion is that in a field gradient, shown in the bottom panel by a color gradient from weak (blue) field at the bottom to strong (red) field at the top, is that the varying orbit radius in each strength of field causes the overall motion to move, to the left for protons and to the right for electrons.

Such motion in opposed directions for the particles of opposite charge gives rise to an electrical current. Since the direction of conventional current flow is that of positive particles, this current is toward the left (both positive protons going left give this, and negative electrons going right). It is quite clear that from the initial position at $x = 1$, the electrons moved 6.5 km: this is approximately how far the protons moved as well. To a first approximation, the speeds of electron and proton motion are the same. The motion overall is called a drift, in this case, *gradient drift* due to the variation (gradient) of the upward magnetic field's strength. The drift speed is larger for higher-energy particles and in larger field gradients, and smaller if the field strength is small or the particle charge is large. The latter effect is not too important since most ions are protons and most negative charges are electrons, with the same

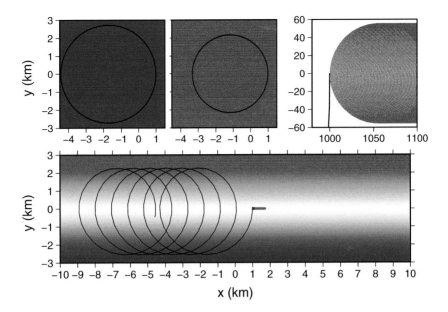

**FIGURE 8.18** Orbits in a weak (blue) magnetic field (upper left), a strong (red) magnetic field (upper middle) and in a changing magnetic field (bottom) for electrons (red) and protons (black). Upper right is a blow-up of the region at $x = 1$ km, labelled in meters. (author).

charge. It can be shown that if the field has curvature, as a dipole field does (so that overall motion is in a ring as shown in Figure 8.17) there is also a *curvature drift*. However, the net effect is that there is a westward ring current as shown in Figure 8.17, with magnetic effects inside it as are observed when there is a ring current intensification during a magnetic storm, exactly as shown by the data in Figure 8.16. If one reconciles the gradient direction of Figure 8.18 with the increase in strength toward the middle in Figure 8.17, one can see that the current arises from changes in radius of the individual particle orbits, and since those are of kilometer scale, there are many such small orbits making up the overall particle path around Earth. This discussion is based on motion in a plane, while Figure 8.17 gives the impression that the ring current extends a bit above the equatorial plane, and the view of particle density of Figure 8.19 shows particles far away from the plane, extending all the way to the poles. Størmer developed a theory that allowed the tracing of particle paths in the more complex three-dimensional fields that actually exist in space. In it, we find that particles, in addition to a drift motion, can also move along magnetic field lines, "bouncing" when they get near Earth. In this sense, the radiation belts are "magnetic bottles", storing particles, including those of high enough energy to be considered "radioactive", near Earth and other magnetic planets in space.

Figure 8.18 depicts a two-dimensional situation. One is looking into the page along the magnetic field lines, and they vary in only one direction, from bottom to top in the figure. Since the Lorentz force produces no effect due to motion parallel to the field lines, particles with some initial speed out of the plane of the page would simply continue along the field lines at their original speed. Of course, field lines for Earth begin in the southern magnetic Hemisphere and enter the northern Hemisphere, so they are not straight. The magnetic field strength also varies greatly along them since it depends on the third power of distance for a dipole field. This now allows a discussion of the three-dimensional path of particles associated with the radiation belts, as depicted in Figure 8.19.

For clarity of how these paths in the dipole field arise, field lines have been "flattened" as shown in Figure 8.20. This magnetic field configuration could arise from two coils, one at each end where the field lines are close together (strong field). It was illustrated in Figure 8.18 that particles will orbit in circles in a uniform magnetic field and that they will drift perpendicular to a gradient in the field. In Figure 8.19, the field is circularly symmetric but motion around only one field line is shown. The spiral motion of a particle goes alternately in front of and behind the field line illustrated. If a particle has some motion along the field line when at the equator, it will move toward the pole that motion takes it toward. However, it will slow down as it approaches the stronger field near the pole and turn around. This allows particles to bounce, in principle forever, between the hemispheres. They also drift, with the gradient/curvature drift directions for electrons and protons shown near the Earth in the figure. In addition to bouncing, a particle would move around the Earth following such a drift. The part of Earth's magnetosphere corresponding to the radiation belts is thus effectively a magnetic bottle. Particles cannot move too far out or they will cross the magnetopause and be lost; this does happen and can be an important depletion mechanism for the radiation belts.

Returning to Figure 8.20, we examine magnetic forces responsible for both cyclic motion and bounce. Particle orbits are shown where particles bounce at both ends of a magnetic mirror field configuration. At other positions between these points, the orbits are helical as shown in Figure 8.19. The magnetic field is shown only in the XZ

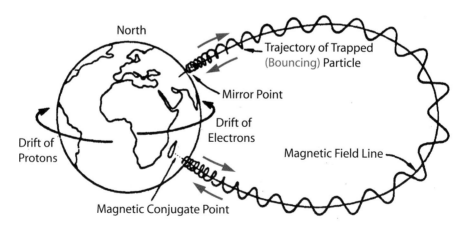

**FIGURE 8.19**  Trajectory of mirroring particle along Earth's magnetic field (NASA 1965, Hess and Mead).

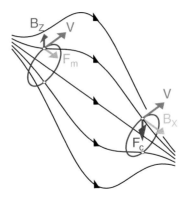

**FIGURE 8.20** Cut in XZ plane of magnetic field lines symmetric around the X axis (middle black arrow). Particle paths are shown in blue near mirror points, velocity arrows at top of loops are shown as dark cyan arrows; for other vectors, see text (author).

plane and the X axis is along the central straight field, but it is symmetric much as a real bottle is. At the right in the figure (note the top field line is "cut" for clarity), the particle is orbiting since the Lorentz force on it makes a force at right angles to both its velocity and the magnetic field. The orbit is dictated only by the parallel or X part of the magnetic field and for a positive particle the force is shown by a green arrow pointing inward. The "right-hand rule": if one swept the fingers of the right hand from the $\vec{V}$ to the $\vec{B}_X$ vector, one's thumb would be pointing along the direction of the Lorentz force. This is labeled as $\vec{F}_c$ since it is a "centripetal" force much as we encounter with gravity force in circular orbits. As in that case, the force applies at all points around the orbit and circular motion is accelerated motion. By considering the $\vec{V}$ vector and inferring the $\vec{B}_X$ vector at the left end, one can see that they are identical so that the orbital motion there is identical.

At the left end of the figure, the force resulting from the vertical (radial) part of the magnetic field is shown. Since the direction of $\vec{B}$ is upward, even though it is diminishing in strength, that part ($\vec{B}_Z$) of the field creates a Lorentz force parallel to the X axis, which is shown as $\vec{F}_m$ for the mirror force. One can imagine that, with the direction of $\vec{B}_Z$ reversed at the right end, while $\vec{B}_X$ is the same, the direction of the mirror force reverses. In between the bounces, there is also a mirror force and the result is a helical trajectory much like that shown in Figure 8.19.

Before the Nuclear Test Ban treaty of 1963, a nuclear detonation of 1.4 megatons was set off in space over the Pacific Ocean by the USA (the then-Soviet state also conducted space nuclear testing). In the inner part of the artificial radiation belt thus created, electrons were found to have a lifetime of months. Although Figure 8.16 does not show the entire duration of the magnetic storm discussed, clearly the ring current decayed in a few hours in response to the $B_Z$ becoming positive and got stronger relatively quickly when negative $B_Z$ prevailed. Clearly, there are both injection mechanisms (thankfully, not nuclear weapons anymore) and decay mechanisms for the radiation belts.

One decay mechanism mentioned above is for outer radiation belt particles to simply hit the magnetopause and leave into interplanetary space. Another is a bit more subtle. If, during the bounces, something increases the parallel speed of particles, their bounce point may be lowered into the atmosphere. The agent might be waves, and indeed sometimes particles come down showing the wave frequency imposed. Figure 8.21 shows the highly original Norwegian scientist Kristian Birkeland experimenting with a "terrella" in the laboratory. The electric currents along magnetic field lines in Earth's auroral zone are named in his honor, and the terrella shows that, in addition to observing auroras, Scandinavians were inspired by the Crookes tube techniques of the late nineteenth century to experiment with particles in a vacuum. Størmer was Birkeland's student and of course, also extended the studies with advanced mathematics. The terrella was a metal sphere with a coil electromagnet in it to simulate a "small Earth" (basically the meaning of the word). Birkeland could simulate the aurora zones as preferred places for particles to hit the sphere, although that is not clear in this photo. Although the particles making most bright auroras originate from the magnetotail, some can come from the radiation belts as well.

The question of how to get particles *into* the radiation belts remains unsettled. Since substorms are known to generate energized particles when reconnection

**FIGURE 8.21**   Kristian Birkeland with his terrella producing artificial radiation belts in the laboratory ca. 1900 (public domain).

changes the stored magnetic energy of the magnetotail into particle energy, they may be a source of ring current particles. Others maintain that, instead, storms energize substorms. So, there is a bit of a "chicken and egg" problem. It was alluded to above that waves can energize particles, then mentioned that they can cause particles to bounce lower and thus leave the radiation belts. Of course, the waves themselves can also be generated by instabilities in the particle populations.

The ring current problem is an important part of space weather. In particular, when electrons get energized into the MeV range, they can be very damaging to electronics or even to biological systems (like astronauts) in space. The ability to predict space weather reliably is an important goal of space research. Often, protective measures involve turning off systems that generate revenue, so there is a cost associated with them. For this reason, predictions have to be reliable. For both intellectual and practical reasons, research on near-Earth space being "radioactive" is an important field of study. This is especially important for spacecraft (see Figure 8.16) near the radiation belts, such as GPS and those in synchronous orbit. The International Space Station is in a rather low orbit and astronauts are not normally exposed directly to radiation belt particles.

As a final note on radiation belts, much as we saw that Earth's magnetosphere is both small and pervaded by a weak field compared with those of the giant planets, whose radiation belts vastly exceed ours. Spacecraft going near Jupiter are heading for a very radioactive environment. For example, the Juno spacecraft, which started orbiting Jupiter in July 2016 (and as of this writing has been there nearly eight years) must carry much of its electronic equipment in a titanium vault about the size of a refrigerator, with 1-cm-thick titanium walls. Jupiter's radiation belts are made worse by an abundant supply of sulfur coming from the volcanic moon Io. It was known to be problematic, leading the Galileo mission, first to orbit Jupiter, about 20 years before Juno, to use an arcane microprocessor called RCA COSMAC for many of its functions. However, the Voyager flyby ran into many problems with radiation, while Galileo's launch was delayed. Two other critical microcomputers were replaced by radiation-hardened versions at the last minute and at a cost of $5 million. Were it not for this, in the words of a NASA Technical Report, "a doomed spacecraft carrying an unknown time bomb would have been traveling toward an unfriendly Jupiter waiting to hurl ion thunderbolts at it".

## 8.9 COSMIC RAYS

In discussing the discovery of radioactivity, it was mentioned that it was initially puzzling that radiation appeared to be coming from the sky, where it might have been expected to come from the ground given that its known sources were radioactive ores. Furthermore, some particles, such as the positron, were first discovered in what had by then come to be known as the "cosmic radiation". Although some part of this originated within the Solar System, it was eventually realized that a steady component was mainly due to sources likely in our Galaxy. For this reason, this part is called "galactic cosmic rays". These have now been seriously studied for about a century from the surface of Earth and from space, and with a variety of techniques.

They form an important part of the radiation environment, even at the surface of Earth.

Two aspects stand out from our accumulated knowledge about the energies of cosmic ray particles (gamma rays are also present). One is that they may have truly enormous energies. Rarely, individual particles have been observed with over $10^{20}$ eV, which is 16 J. Recalling that a smallish apple falling a meter has 1 J of energy, this is equivalent to an apple falling 16 m, about the height of a three-story building. This brings out the amazing aspect of the energy distribution shown in Figure 8.22: there is a steep fall-off of power of $-2.6$ (note that in a log-log plot, like the figure, a straight line is a power law relation) over many orders of magnitude from about a GeV (rest mass energy of a proton) to over $10^{20}$ eV. This is why one need not worry too much about being hit by a particle bringing in the energy of a well-tossed apple: the chances of this are even less than those of the proverbial one neutrino interaction you may expect to experience in your lifetime.

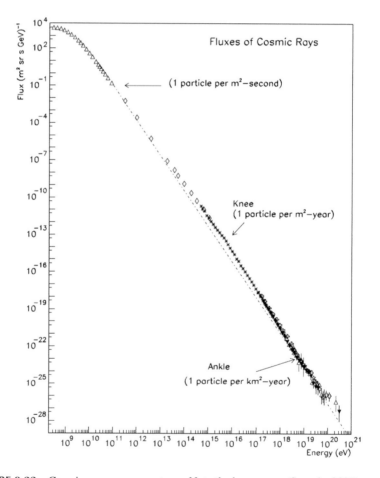

**FIGURE 8.22**   Cosmic ray energy spectrum. Note the huge range (Swordy, 2001).

We infer that these particle energies arise in the most energetic processes that occur in our galaxy: likely, things to do with supernova explosions and collapses to black holes. Why, therefore, can we not simply detect particles coming from the regions known to harbor such entities? The answer, as it has been to so much in the last two chapters, is magnetic fields. The magnetic field in interstellar space is small, at most of the order of nT like it is in the solar wind. One would think that its ability to make a super-energetic particle spiral around would be small. This is true, but the scale of the galaxy is huge. It does not take much bending to make the source of energetic particles blur out. As a result, we really have no solid idea, from an observational basis, about where galactic cosmic rays come from. Clearly, the exotic sources mentioned are about the only things that could give rise to such energies. However, how, why, and where remain hidden behind a magnetic veil.

Cosmic ray abundances (Figure 8.23) are roughly those found in the solar system with some notable exceptions. The most stunning of these are in the light elements lithium, beryllium, and boron (Li, Be, B). These are elements that were not made in the Big Bang and are fragile and not made to last inside stars either. As a result, they are rare in the Solar System. In fact, in the modern age, these rare elements are among the most economically valuable. Lithium, for example, is becoming a valuable part of our battery-based power storage systems for periodic power sources like solar and wind energy. However, there is no shortage of these light elements in cosmic rays! Similarly, but not by a factor of one million

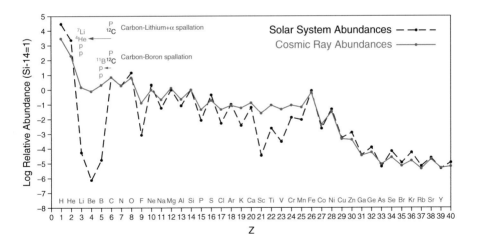

**FIGURE 8.23** Cosmic ray abundances given on a logarithmic scale with data values ranging over a ratio of about 100 billion ($10^{11}$). Solar system abundances are shown by a black dashed line with dots, while cosmic ray abundances at typical energies are shown by a solid red line with red dots. Common spallation reactions for carbon are shown producing the elements boron and lithium, which are rare in the Solar System. Abundances are relative to silicon-14 (author, inspired by work of Dr. Martin Israel at https://izw1.caltech.edu/ACE/ACENews/ACENews83.html but remade from data sources).

as for the light metals, certain higher mass elements are more abundant in cosmic rays than in the Solar System. We have every reason to expect that other regions of the galaxy have about the same composition as the Sun: stellar spectra, for example, show that. The come-from-afar, highly energetic elements apparently do not resemble the composition of any place they could have come from! The best explanation is that having come a long distance and being very energetic, they reacted with other elements on the way despite interstellar space being highly devoid of matter. In fact, if one factors in our knowledge about the density of space, the energy of the cosmic rays, and their possible reactions at that energy, one finds that they have traveled a much greater distance than the size of the galaxy. How can that be? Again, we appeal to magnetic fields, which in bending their trajectories may have made them go around the galaxy several times before getting to us.

In coming to Earth, cosmic rays are also highly reactive with our atmosphere. Figure 8.24 shows the result of a very high-energy "primary" cosmic ray hitting

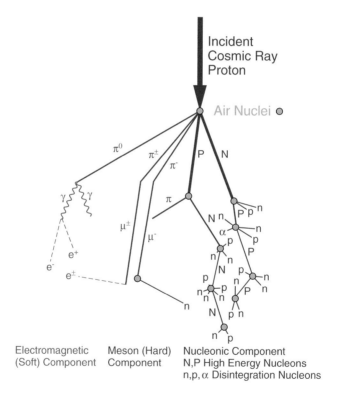

**FIGURE 8.24** Cosmic ray interaction with Earth's atmosphere upon entering from above. Reactions beget reactions in what ends up as a "shower" of secondary particles and gamma radiation. The exotic particles known as pions ($\pi$), which may be charged or neutral, and muons ($\mu$), which are leptons like electrons, are short-lived but form the "hard" component (author, inspired by a diagram made by J. A. Simpson in 1953).

a nucleus in the atmosphere (most likely nitrogen or oxygen). Secondary particles including positrons (discovered in cosmic rays), electrons, protons and neutrons of all energies, and exotic particles such as pions and muons (whose normally short lifetime is prolonged by special relativistic time dilation) are produced in a shower. Such showers allow indirect detection of cosmic rays. Much as neutrons were discovered due to their interaction with hydrogenous materials, cosmic ray secondary neutrons are detected using large blocks of paraffin, which contains a lot of hydrogen. Some of the energetic particles interact further with the atmosphere to emit light that can be detected from the ground. The highest-energy cosmic rays may generate millions of secondaries, mostly in the forward direction, much like a shotgun burst. In turn, some of this radiation does arrive at ground level. The radiation level in regions of the world at high elevations is measurably higher than at sea level since the secondary radiation is absorbed by air.

We now return to Figure 8.22 before returning to a recent and final amazing result of our technology with regard to galactic cosmic rays. Note that at the low-energy end, the power law (straight line) flattens out. There are not as many low-energy cosmic rays in the inner Solar System as one might expect by following the power law. Even at the extremely lower-energy range, protons of 100 MeV are not nice things to have around: able to penetrate electronics or biological systems and deposit significant amounts of damaging energy. If one continued the trend, there would be at least 100 times more of them, and we have reason to suspect that, outside the Solar System, the trend does continue: we are protected by being inside the heliosphere, likely mainly by its magnetic field. On a continuous basis, even this depleted input of galactic cosmic rays is a major source of radiation in space, especially outside Earth's protective magnetic field.

Voyager 2 left the heliosphere in 2018, traversing the boundaries discussed in Chapter 7. Figure 8.25 shows plasma data, which ended once in interstellar space where densities were too low for the detectors to operate. Before this, a "magnetic barrier" was found, although in fact interstellar magnetic fields of about 0.75 nT were higher than in the Solar System. Here galactic cosmic rays took a step upward and then ramped up further. We have finally detected the primary galactic cosmic rays *in situ* in the galaxy!

We have seen that magnetic fields make a magnetic bottle around the Earth, with highly radioactive regions in the radiation belts. However, magnetic fields also deviate cosmic rays from outside the Solar System. Those that make it through have the same Earth magnetic field system to navigate. On long space voyages, such as those envisaged to Mars, and even those to the proposed Lunar Gateway, well outside the radiation belts, galactic cosmic rays will be much more intense than they are on the protected surface of the Earth. Add in the unpredictable solar energetic particle events and gamma rays from flares, and this region of space seems quite threatening.

Space is, after all, radioactive.

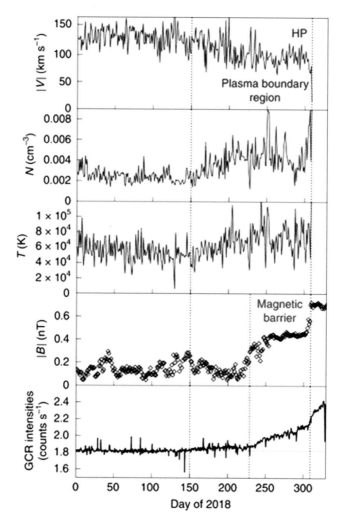

**FIGURE 8.25** Voyager 2 data as it left the solar system, with plasma data ending at the heliopause (HP). Inside this was a compression of solar wind magnetic field (bottom panel) and an abrupt jump in galactic cosmic rays (Richardson et al., 2019).

## 8.10 SUMMARY

Although space is "radioactive", the energetic particles are not of the same nature as those from the original radioactive materials investigated, consisting mainly of electrons and protons. Mechanisms to energize these particles involve changing magnetic fields, often in waves, and such fields in the right configuration can also store the particles. This storage is mainly in radiation belts which feature bouncing and drifting particles, which may have a large-scale magnetic effect. Cosmic ray origins are unclear, but we are protected from them by heliospheric and planetary magnetic fields.

# 9 Exploration of the Invisible Solar System

Almost all spacecraft launched into interplanetary space have carried detectors that did not return images, so by definition were sampling aspects of the invisible, even if they did also carry cameras. Images, of course, contain a large amount of data, but they generally do not reveal what we consider to be the "invisible" Solar System. In some cases, images were made at wavelengths that cannot be seen by human eyes or through our atmosphere. Some "invisible" radiations are detectable at the surface of the Earth with the proper instruments. In some cases, these instruments push advanced measurement techniques to the extreme.

Thus, we conclude our exploration of the Invisible Solar System by examining *how* we know things about it, by focusing on how we get to destinations using concepts of orbits already looked at, what otherwise empty destinations are desirable, and how we can sample multiple points in space. We do not focus much on instruments themselves, since that is a separate technical field unto itself. Non-imaging instruments are often referred to as "field and particle" instruments, but we may be more specific and say that usually aspects of plasmas, waves, fields, and particles are measured. Often the instruments to detect waves are different from those detecting more slowly varying fields, and high-energy particles are often detected in a different manner from the colder ones. Some aspects of the Invisible Solar System are best studied, or at least well studied, from the surface of the Earth. An example is the detection of magnetic effects from near-Earth space. However, magnetic detectors detect signals from all sources without discrimination, so that determining the source of magnetic signals can be a problem.In many cases, spacecraft instruments are miniaturized versions of instruments used in terrestrial laboratories, usually in addition run under microprocessor control and with challenges in the storage and transmission of data. Extreme reliability under extreme conditions is needed. For this reason, very carefully manufactured components must be used, which raises costs. Mass is a huge consideration since there is typically at least 100 times or more fuel mass needed for each unit of payload mass, and there is a limit to how large rockets can be. Since physical processes to be measured can take place rapidly, often it is necessary that a large amount of data be stored and then sent back to Earth. Storage is limited on a spacecraft. Data transfer is by radio, and the signal strength declines rapidly with distance. Now that we can put advanced computer systems on spacecraft, sometimes a large amount of data can be acquired, and the computer determines whether anything interesting is going on and only sends a fraction of the data, which could be of interest, to Earth. For spacecraft that go into deep space, the situation is more acute compared with operating in Earth orbit, which is relatively close. Mass being

DOI: 10.1201/9781003451433-9

a constraint, computers and memory are limited, but an even larger constraint is the decrease in the ability to transmit data. This, in turn, leads to the use of dish antennas, which have a small field of view, so that the spacecraft must be accurately oriented in space to point its antenna at Earth.

## 9.1 DEEP SPACE POWER AND DATA TRANSMISSION

Voyager 1, shown in Figure 9.1, is currently the farthest spacecraft from Earth, launched in September 1977. It famously flew by and photographed Jupiter and Saturn, in turn gaining energy from each so that it was placed on a hyperbolic orbit to leave the Solar System. While mission coverage was dominated by spectacular images sent back when near the giant planets, Voyager 1 has a full complement of plasma instruments. Its cameras were used briefly in the mission, but the plasma instruments continue to function, and have clearly shown that the outer boundary of the plasma Solar System (the heliopause) was passed in August 2012 at about 122 AU, or about 18 billion km from the sun, as discussed in the previous chapter. It continues its outward voyage at a speed of about 3.6 AU per year. Earlier, in December 2004, it had crossed the heliopause into subsonic solar wind flow at a distance of 94 AU. At these large distances from the Sun, available solar power is minimal, and the spacecraft uses a radioisotope thermal generator (RTG) based on the heat given off by decaying plutonium-238. Unlike uranium-238 of the same atomic mass, plutonium-238 (Pu-238) has a very short half-life of only 88 years and is produced in a nuclear reactor. This is a very neutron-rich environment, unlike what is usually found in nature. In that sense, plutonium is an entirely artificial element.

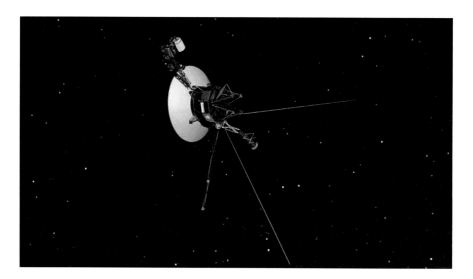

**FIGURE 9.1**  Voyager 1 showing its main dish pointing toward Earth, radioisotope generator on a boom above the spacecraft, and magnetometer boom to the lower right (NASA/JPL).

Plutonium-238 production from neptunium-237 is outlined in red in Figure 9.2. Although the paths to form neptunium-237 are not shown, they involve neutron addition to uranium in the neutron-rich environment of a fission reactor. Subsequent decays form neptunium-237. If a neutron is added to that long-lived isotope, neptunium-238 is formed, which is much less stable, and quickly beta decays

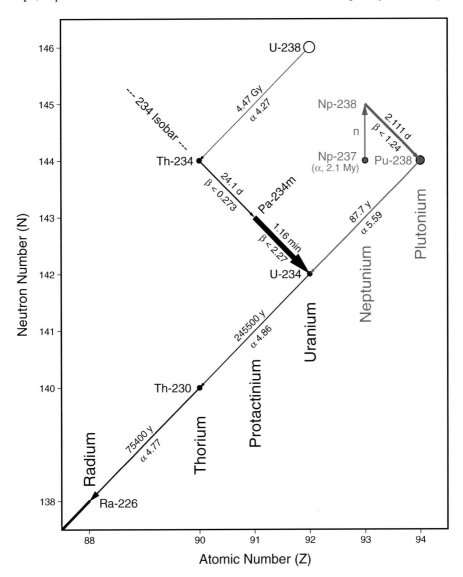

**FIGURE 9.2** Decay of uranium-238 (black) and synthesis of plutonium-238 by neutron absorption on neptunium-237 (red). Pu-238 has alpha decay on a timescale suitable for powering spacecraft. U-238 and Pu-238 decay paths converge at U-234 and make Ra-226, the starting point of Figure 4.19 (author).

to plutonium-238. Plutonium-238 is non-fissile, meaning that it cannot be used in bombs. Its usefulness is that it decays with emission of alpha particles of about 5.5 MeV energy to uranium-234, which, with a half-life of a quarter million years, may be basically regarded as stable in the context of an RTG. The alpha particles are easily stopped in the plutonium-238 mass itself or in its possible metal coating. Due to the high activity compared to radium, discussed before as being warm to the touch, plutonium-238 glows red hot due to alpha emission. The heat is applied to a thermocouple, which is a junction of two metals that produces a voltage if heat is applied. Each thermocouple gives a low voltage, and they are commonly used singly in thermometers. In a power generation application like an RTG, many of them are connected in series to give a usable voltage for equipment. Plutonium-238 is a good choice of alpha emitter for use in a long-duration spacecraft due to its long half-life compared with mission timeframe. Nevertheless, Voyager 1 and its twin Voyager 2 are approaching fifty years in space, and the RTG power will soon be too low to supply that needed for operation and communication. When new, the RTGs (three in total) produced about 500W, or about the output of a toaster. Even when relatively close to Earth, for example during their Jupiter encounters, the dish or "high-gain" antennas of the Voyagers, which are 3.7 m (12 feet) in diameter, had to be precisely pointed at Earth to relay data back with a data rate allowing images and other rich datasets to be transmitted.

On Earth itself, in turn, large dish antennas must be used to receive the data from distant spacecraft. A limited number of large dish antennas are available. NASA and Caltech's Jet Propulsion Laboratory (JPL) operate the Deep Space Network (DNS), which has one set of large dishes in the nearby Mojave Desert, with two other sets spaced one-third of the way around Earth from it, in Canberra, Australia (Figure 9.3), and Madrid, Spain. Each DSN site has a 70-meter diameter dish and some smaller ones. The spacing allows constant communication with deep space space-craft: as the spacecraft appears to set at one dish, it is rising at the one west of it. Most spacecraft are relatively near the equator in the sky, so the deployment of dishes at low and middle latitudes (two in the north, one in the south) allows them to be seen in sequence by all of the dishes. The dish on the spacecraft is fixed to it, with the slow rotational motion of the spacecraft keeping it constantly pointed at Earth, while the dishes on the ground are movable so that they can track spacecraft as Earth rotates, and switch from spacecraft to spacecraft as needed.

## 9.2 ORBITS AND GRAVITY ASSISTS

Earth orbits the Sun in a prograde manner (counterclockwise as seen from above the Sun's north pole) and is moving at 30 km/s. The ecliptic plane is defined by Earth's orbit, so most planetary spacecraft take advantage of Earth's motion, which gives them an initial velocity in the plane and moving prograde at about 30 km/s, and remain relatively close to the ecliptic plane. Of course, most planets also move very close to the ecliptic plane, so that spacecraft whose destination is one or more planets very naturally would stay in that plane anyway. Changes in direction and/or speed (taken together these are velocity) are often discussed in terms of the velocity

**FIGURE 9.3** Multiple large dishes of the Deep Space Network in Canberra, Australia. The largest dish is 70 m in diameter. An indication of scale is given by the buildings (NASA/JPL).

change or *delta-v*, and delta-v is "expensive" in the sense that it consumes fuel. This, of course, is something that most spacecraft have a limited amount of, so careful trajectory planning determines optimal paths in terms of fuel consumption, often sacrificing time in flight to minimize it. A rocket leaving Earth is largely a big fuel and oxidizer tank, with a tiny payload that includes extra fuel for maneuvering in space. Any increase in maneuvering fuel is amplified many times over in the size and fuel capacity of the rocket used for the initial launch.

Travel from one planet to another is done on a "transfer" orbit, and one of the most efficient is called the *Hohmann transfer*. In this, a spacecraft moves on an elliptical orbit with only small course correction "burns" in such a way that it is tangential to the origin planet's orbit at the start, and to that of the destination planet at the end. A larger rocket firing at the beginning injects the spacecraft onto the correct trajectory to arrive at a tangent to the target planet's orbit when that planet is at about the same location, and another large rocket firing is done to place it into orbit if that is desired. The elliptical orbit, under the influence of solar gravity, is called the *cruise phase*. Hohmann transfer orbits usually sacrifice time for minimum consumption of fuel. The orbital mechanics required for such orbits were worked out in Germany in the 1920s, at a time when rocketry was undergoing rapid development there.

The initial deep space flights were *fly-bys* or, in some cases (like the US "Ranger" program to the Moon that provided Figure 5.11), impacts. The term fly-by, or flyby, is still used despite of course the fact that spacecraft do not "fly". It rolls off the tongue better than "geodesic trajectory freefall near a planet". A benefit of flybys is that the "gravitational slingshot effect" or "gravity assist" may be used. In some cases, no investigations of the target planet of the flyby are taken, and it is used merely for its gravitational effect. The first such boost by a flyby was done unintentionally, although the concept was already familiar to the forward-looking rocket pioneers. The Soviet spacecraft Luna 1 missed its target of the Moon in 1959, and flew by it into interplanetary space. The first intentional flyby of a planet was of Mars by US Mariner 4 in 1964, and the use of a flyby for a gravity boost was pioneered, so to speak, by Pioneers 10 and 11 in the mid-1970s. Figure 9.4 shows the amazing trajectories that are possible by using the huge mass of Jupiter for gravity assist. Voyager 1 was launched into an elliptical orbit but without enough energy to leave the Solar System. By passing Jupiter in the proper way, it was able to boost, using gravity alone, onto a hyperbolic orbit that would in itself have taken it out of the grasp of the Sun. However, this was cleverly set up to allow it to pass close to Saturn. The difficult choice was made to fly by the giant moon Titan and forego a boost opportunity for Pluto. This trajectory also resulted in an increase of the inclination to about 35°. The twin Voyager 2 spacecraft visited the remaining giant planets using Jupiter, Saturn, and their gravity boosts and left the Solar System at an incredible 79° inclination. Voyager 1 not having been directed there, Pluto was visited by the New Horizons spacecraft, launched on January 11, 2006, with a very large rocket relative to the mass of the spacecraft, arriving at Jupiter only about a year later for a gravity boost to encounter Pluto in early 2015. By this time, Pluto had been "demoted" to dwarf planet status. The concept of a gravity boost is hard to fathom: where does one get energy for nothing? In orbit, a body goes in and out around its main body, trading off speed for gravitational potential energy. How does one go near a large planet and come out again with an energy boost? Part of the explanation, of course, is that the planet slowed down an almost infinitesimal amount, but it still seems magical. Let us examine in more detail what goes on in this situation.

Part of the ability to extract energy from the planet is due to *not* going into orbit around it. Coming in on a trajectory from deep space, it is not possible to go into orbit without braking. Only such a speed loss, possibly by "braking" rockets (although there are also "aerobraking" techniques a bit too complicated to discuss here) allows for being captured by the planet. If one does not brake, however, one still can enter a domain, if close enough to the planet, where its gravity dominates, and that of the Sun can be ignored. From the planet's point of view, the spacecraft approached on a hyperbolic (unbound) orbit, fell to its lowest point in the gravity well, and then traded kinetic energy for potential energy to climb back out again.

Considered only from the planet's point of view, something came in on a hyperbolic orbit, got deviated, and left at a different angle but at the same speed. The associated exit velocity is added to the planet's velocity around the Sun in considering the orbit within the Solar System. The result can be a large speed or direction change as viewed in that frame, which is the usual one we use to describe planetary motion.

It was pointed out in Chapter 4 that small bodies cannot change semimajor axis very easily. This is true if they are far from the large body. For example, asteroids

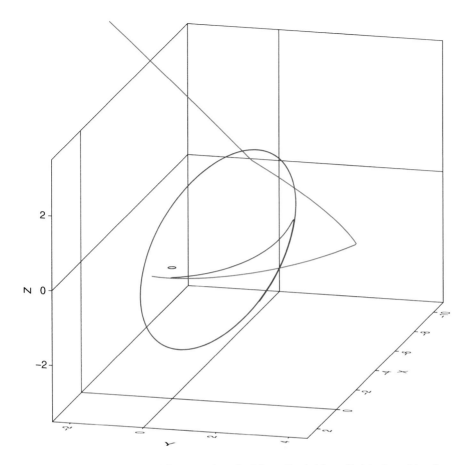

**FIGURE 9.4**   Voyager 1 (red line) was launched from Earth (about 8 o'clock position from the yellow Sun, shown much enlarged) on September 5, 1977, and reached Jupiter on March 5, 1979, where its trajectory was sharply bent to encounter Saturn on November 12, 1980. The slight bend visible there made it leave the Solar System at a steep angle. The NASA/ESA joint mission Ulysses (blue line) was launched on October 6, 1990, roughly one Jupiter orbit later than Voyager 1 and with Earth in roughly the same position. On February 8, 1992, its orbit was deviated extremely to make it fly over the Sun's south pole in 1994 and north pole in 1995, in the high inclination elliptical orbit that shows as an ellipse. Planets are not shown (author).

will temporarily change semimajor axis due to the influence of Jupiter, but then be affected in the opposite sense later. The net effect, when at a large distance, is no change in semimajor axis. When a close flyby is done, as we saw, the semimajor axis can undergo a large change. The same is true for natural bodies as long as they come close enough to a large planet (Figure 9.5).

The ability to navigatethe Solar System in this way is the culmination of hundreds of years of studies of celestial mechanics. In the cases shown, however, the voyages began with powerful launches followed by a quiescent cruise phase, leading

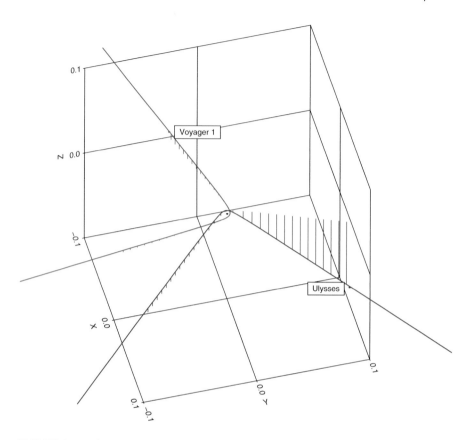

**FIGURE 9.5**   The paths of Voyager 1 (red) and Ulysses (blue) in the 0.1 AU cube centered on Jupiter (black dot, to scale, at center). The encounters were slightly over one Jupiter year apart, so the approach directions, from left, were roughly the same. Small bars show the distance to the ecliptic plane. Both orbits are hyperbolas representing fall in toward Jupiter and then escape. Voyager 1 remained near the plane of the ecliptic, encountering Jupiter close to its equatorial plane, while Ulysses passed to the north of it and thus got pulled down, exiting at an angle to the south that was even more steep when considered with respect to the Sun, as shown in Figure 8.3. This orbit was, however, bound to the Sun, whereas Voyager 1 was on an escape orbit (author).

to short periods of strong interaction with planets. A newer approach is to apply small amounts of power during the cruise phase, as exemplified for example by the Dawn mission (Figure 9.6) to large asteroid Vesta and then to dwarf planet Ceres. In most cases, the power for the cruise phase comes from solar panels. Compared with a chemical rocket, which is almost a controlled explosion, an ion engine has very low thrust (power output in a direction). It causes only small acceleration, but that acceleration can be sustained for long periods of time, building up to cause a significant change in orbital velocity. For applications such as entering an orbit, such as losing enough speed to be captured by a massive member of the Solar System,

**FIGURE 9.6**   Dawn spacecraft braking with its ion engine to enter orbit around dwarf planet Ceres on March 6, 2015 (NASA).

an ion engine has limits. It was shown to work with low-mass bodies such as Vesta and Ceres, which have slow speeds involved in orbiting, but, with planets, usually chemical rockets are used.

## 9.3  SPECIAL ORBITS

As mentioned in Chapter 4, it is possible for a spacecraft to occupy the $L_1$ or $L_2$ Lagrange point of a planet, with minor amounts of fuel needed. These points are respectively found between the planet and the Sun and slightly outside the planet but on the Sun-planet line. Strictly speaking, the planet's gravity plus that of the Sun make orbits with the same period as the planet possible near these points, but they are not stable against small variations. For this reason, as mentioned earlier, small amounts of fuel are needed to nudge the spacecraft back into place if it drifts. The $L_1$ point of Earth is very good for spacecraft that are imaging the Sun or the corona, since a spacecraft there has an unobstructed view in the inward direction. Since the solar wind has a radially outward flow, a spacecraft there can also detect solar wind disturbances before they get to the Earth. These Lagrange points are roughly 1% of the Earth-Sun distance from Earth, and, in the case of the $L_1$ point, that corresponds to about an hour's warning of a potential space weather event. The $L_2$ point is frequently used for spacecraft looking outward into space. For example, the James Webb Space Telescope (JWST) was placed at that point in 2022 for its astronomy mission. The $L_2$ point is not as relevant to understanding the plasma of the Solar System as is the $L_1$ point.

The $L_3$ Lagrange point is on the opposite side of the Sun from Earth and, apart from speculative science fiction, has little importance for space studies. The $L_4$ and

L$_5$ triangular Lagrange points are stable, as is evidenced, for example, by the very large number of Trojan asteroids trapped there along the Jupiter orbit (recall Figure 5.33). The corresponding points for Earth have been pointed out to be of interest in that events on the Sun could be observed from three points simultaneously if we had spacecraft there. One issue with coronal mass ejections directed toward Earth is that they are hard to see when coming directly our way. From the side at these Lagrange points, they could be more easily tracked. However, there are major communication problems for these points since they are 1 AU away from Earth so that signals would be faint. The closest we have come to three-dimensional images of plasma coming out from the Sun has been attained with the two spacecraft of the Stereo mission. Using very close flybys of the Moon shortly after their launch, these twin spacecraft were made to move along Earth's orbit slowly, and relative to Earth in opposite directions. They are frequently in positions to complement our Earth view.

A very important class of orbits attempts to deal with the problem that space is huge, with events involving blobs of plasma moving from place to place. It is hard to determine causality with one spacecraft. If one does detect a plasma concentration, it can be hard to link it to a source. This problem can be overcome to some extent using a spacecraft constellation, meaning that several spacecraft are in orbits that line up. Then, if something moves from spacecraft to spacecraft along the line, the speed and direction of motion can be clearly determined. Figure 8.13 illustrates the four European Cluster spacecraft, launched in 2000, studying waves from multiple vantage points. This concept is also behind the NASA Time History of Events and Macroscale Interactions during Substorms (THEMIS) constellation of five satellites launched in 2007, with sample data from a single spacecraft shown in Figure 7.15. The early mission orbits had periods of 1, 2, and 4 days, and at appropriate seasons lined up at various points of interest in and near Earth's magnetosphere, allowing coordinated studies. After several years, the two outermost probes were moved to orbit the Moon, which of course brings them monthly into Earth's magnetotail. This also allows studies of the interaction of the solar wind with the Moon, such as its plasma "shadow", as graphed in Figure 7.3. These two crafts were renamed ARTEMIS, which is an acronym but also the name of the Greek goddess of the Moon.

The remaining THEMIS probes still do line up in useful ways. For example, Figure 9.7 shows them in an ideal position to detect reconnection that took place during a substorm that was in turn during a major magnetic storm. The reconnection region is small, and the chances of being in exactly the right place to detect the magnetic field and particle events due to reconnection during a substorm are low. However, this was accomplished several times and has verified the basic picture that tail reconnection allows the conversion of magnetic field energy into particle heating and high-energy particles, powering space weather events. Similar studies on the dayside have shown that the reconnection that injects energy into the magnetosphere also occurs, mainly when the solar wind magnetic field is southward, as expected.

Electric fields in space are measured using long wires on spinning spacecraft. Most spin for stability, and this also allows certain instruments to measure in all directions in space. It is, however, difficult to measure electric currents in space.

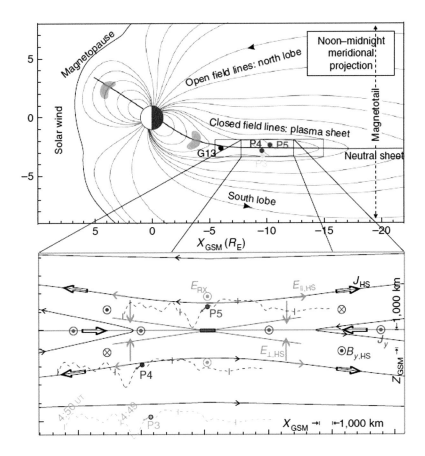

**FIGURE 9.7** Positions of three THEMIS (P3, P4, P5) probes and the GOES-13 (G13) geo-synchronous satellite during a magnetic storm on December 20, 2015, showing the position with respect to the overall magnetic field and radiation belts (pink/grey shading), and a blow-up near the reconnection region (red bar), in which electric fields (green arrows) are also shown. The "Neutral sheet" is equivalent to the plasma sheet (Angelopoulos et al., 2020).

Often, these are carried by low-energy particles that are affected by local electric fields. Sampling large-scale electric currents can be done by using the laws of elec-tromagnetics if one has magnetic field measurements at four points in a tetrahedron simultaneously. For this reason, the European Space Agency "Cluster" mission men-tioned above has four spacecraft which are usually not as spread out in space as those of THEMIS. After an initial launch disaster in 1996, replacement Cluster space-craft were launched in 2000. Figure 8.13 shows the approximate relative spacing of Cluster spacecraft, although at times they were made to come to only hundreds of kilometers from each other. Cluster also carried comprehensive plasma instruments comparable to those of THEMIS, and, when separated widely, could similarly be used to follow moving events.

## 9.4 SUMMARY

This final chapter has attempted to give some insight into how space missions are powered, communicated with, and placed into desired locations. Mission planning is an art and science of its own. Spacecraft are expensive, so it has not so far been possible to densely instrument space where data are needed. Ingenuity and patience using ensembles of spacecraft have, however, given us deep insights into physical processes in the Invisible Solar System.

# Bibliography

This book has made use of many sources, ranging from research papers to comic books. It has generally been found that the well-known web resource Wikipedia has reliable articles on astronomy, with varying quality of references. A professional web resource is the NASA Astronomical Data System, with the caveat that a search generally gives very many "hits", most of them to rather specialized material. A good source of online older material is the archive.org website. The main aim of this bibliography is to supply follow-up reading as "General Reading", and it includes some sources which were jumping-off points to other material inspiring this synthesis. For each chapter, specific references given by name and year (e.g. Asimov (1954)) mostly give figure credits, possibly with brief comments. Entries are given only the first time cited, although the citation may be repeated in other chapters.

## CHAPTER 1

### GENERAL READING

Daniel J. Boorstin, *The Discoverers*, Random House, New York, 1983. A sweeping study of intellectual history from ancient to recent times.

Michael J. Crowe, *Theories of the World from Antiquity to the Copernican Revolution*, Dover, New York, 1990. Short but readable summary, with some numerical techniques, up to the time of Galileo. Includes a section on Archeoastronomy, a topic lightly considered here.

Will and Ariel Durant, *The Story of Civilization*, Simon and Schuster, New York, 1975. With 14777 pages in 11 volumes, this "integral history" of the world cites numerous original sources and paints with a very broad brush, with an emphasis on intellectual threads of civilization placed in context. The first volume, *Our Oriental Heritage*, published in 1935, no longer reflects the state of the art in orientalist scholarship.

Andrew Fraknoi, David Morrison, Sydney C. Wolff, *Astronomy*, OpenStax/Rice University, 2016, updated 2022. A web and PDF resource with print edition available, generally nonmathematical and well-illustrated with Creative Commons 4 attribution licence. Available at https://openstax.org/details/books/astronomy.

Robert M. Geller, Roger A. Freedman, William J. Kaufmann III, *Universe*, Freeman, New York, 2019. A very comprehensive introductory textbook at a "science stream" mathematical level. Modern textbooks are regularly updated and lavishly illustrated, and usually are aimed at science students, or with less mathematics, to support optional astronomy courses.

John Gribbin, *Science: A History 1543-2001*, BCA (Penguin), London, 2002. Starting with Copernicus, a general history of the modern sciences.

Sun Kwok, Our Place in the Universe: Understanding Fundamental Astronomy from Ancient Discoveries, 2nd Ed., Springer, Cham, 2017. Basic topics are explored largely from an observational and early historical perspective, although an appendix with equations is included.

## REFERENCES AND FIGURE SOURCES

Daderot (2016), Modified from public domain image by user Daderot at https://commons.wikimedia.org/wiki/File:King_Menkaure_and_two_goddesses,_plaster_cast_of_original_in_Museum_of_Fine_Arts,_Boston,_Egypt,_Giza,_Valley_Temple_of_Menkaure,_Dynasty_4,_c._2490-2472_BC_-_Harvard_Semitic_Museum_-_Cambridge,_MA_-_DSC06126.jpg

Fæ (2010), https://commons.wikimedia.org/wiki/File:Venus_Tablet_of_Ammisaduqa.jpg, user Fæ, cited August 2, 2023, Creative Commons Attribution-ShareAlike 3.0 https://creativecommons.org/licenses/by-sa/3.0/

Galileo Project (2023), http://galileo.rice.edu/sci/observations/sunspot_drawings.html, cited August 2, 2023. See copyright information on page, but two dimensional renderings of drawing of this age are considered to be in the public domain).

Newton, I. (1731) *A Treatise of the System of the World*, 2nd ed., F. Fayram, London. https://archive.org/details/1728-newton-a-treatise-of-the-system-of-the-world/page/5/mode/1up. Cited August 2, 2023. Now public domain.

Stellarium (2023) https://stellarium.org/, cited July 23, 2023. Stellarium is excellent sky display software, a free open source program available for many types of computer.

Wikimedia-Galileo (2023), public domain, from https://commons.wikimedia.org/wiki/File:Galileo%27s_sketches_of_the_moon.png, cited July 23, 2023.

## CHAPTER 2

### GENERAL READING

A. P. French, *Special Relativity*, Norton, New York, 1968. A clear exposition with an emphasis on mechanical effects. Only algebra is needed to understand special relativity quantitatively.

Douglas C. Giancoli, *Physics: Principles with Applications, 6th Ed.*, Pearson, Upper Saddle River, 2005. For those going beyond this book mathematically, but not familiar with calculus, this textbook is comprehensive, covering most of our topics with more mathematics, less flow, and more general context. Ideally, learning physics is "better" with calculus, and there are also many calculus-based textbooks.

Sun Kwok, Our Place in the Universe II: The Scientific Approach to Discovery, Springer, Cham, 2021. The development of astronomy after Newton is discussed, along with that of other sciences.

James Lequeux, *Le Verrier: Magnificent and Detestable Astronomer*, Springer, New York, 2013. An up-to-date summary of Le Verrier's life and work, including the discovery of Neptune.

Samuel Eliot Morison, *The Great Explorers*, Oxford, New York, 1978. Written by a great admirer of Columbus, this and Morison's other books on Columbus bear a close read to see how misinterpreting Ptolemy changed the world.

Edwin F. Taylor, *Exploring Black Holes*. Although this does exist as a print book, most of the subject matter is about much stronger gravitational fields than are found in the Solar System. Most relevant here is the section on GPS found (with indirect access to the whole book) at

https://www.eftaylor.com/exploringblackholes/Ch04GlobalPositioningSystem 160401v1.pdf

John Archibald Wheeler, *A Journey into Gravity and Spacetime*, Scientific American Library, New York, 1990. A beautifully illustrated guide to general relativity by a master, also including his poetry, with a good section on tides as a result of spacetime curvature.

### REFERENCES AND FIGURE SOURCES

Bremiker, C. (1845) Star chart (annotated, possibly by Galle in 1846, portion, contrast enhanced), https://commons.wikimedia.org/wiki/File:Bremiker_Hora_XXI .jpg, cited August 2, 2023.

Campbell, W. W., and Trumpler, R. J. (1928) Observations made with a pair of five-foot cameras on the light-deflections in the Sun's gravitational field at the total eclipse of September 21, 1922, Lick Observatory bulletin no. 397, 130-160. University of California Press, Berkeley.

von Laue, M. (1921) *Die Relativitätstheorie*, Band II, 1921. Vieweg, Braunschweig. Public domain.

Wikimedia-Brahe (2023) https://commons.wikimedia.org/wiki/File:Tycho-Brahe -Mural-Quadrant.jpg, cited July 23, 2023 (public domain).

NASA figures and images are from the U.S. government and are in the public domain.

Tide data such as that used to make Fig. 2.5 may be found at tides.gc.ca and other government sources.

## CHAPTER 3

### GENERAL READING

Jay M. Pasachoff, *The Complete Idiot's Guide to the Sun*, Alpha, New York, 2003. A very comprehensive basic guide to the Sun, written by a renowned eclipse chaser.

### REFERENCES AND FIGURE SOURCES

Arimatsu, K., K. Tsumura, F. Usui, Y. Shinnake, K. Ichikawa, T. Ootsubo, T. Kotani, T. Wada, K. Nagase, J. Watanabe (2019) A kilometre-sized Kuiper belt object discovered by stellar occultation using amateur telescopes, Nature Astronomy 3, 301-306, https://www.nature.com/articles/s41550-018-0685-8.

Baird, C. S. (2013). https://www.wtamu.edu/~cbaird/sq/2013/07/03/what-is-the-color-of-the-sun/ (Public Domain based on data from American Society for Testing and Materials Terrestrial Reference). Dr. Baird's website features many science topics and the ability to submit questions.

Buie, M. W., B. A. Keeney, R. H. Strauss, T. E. Blank, J. G. Moore, S. B. Porter, L. H. Wasserman, R. J. Weryk, H. F. Levison, C. B. Olkin, R. Leiva, J. E. Bardecker, M. E. Brown, L. B. Brown, M. P. Collins, H. M. Davidson, D. W. Dunham, J. B. Dunham, J. A. Eaccarino, T. J. Finley, L. Fuller, M. L. Garcia, T. George, K. Getrost, M. T. Gialluca, R. M. Givot, D. Gupton, W. H. Hanna, C. W. Hergenrother, Y. Hernandez, B. Hill, P. C. Hinton, T. R. Holt, R. R. Howell, J. L. Jewell, R. L. Kamin, J. A. Kammer, T. Kareta, G. J. Kayl, J. M. Keller, D. A. Kenyon, S. R. Kester, II, J. N. Kidd, Jr., T. R. Lauer, C. W. S. Leung, Z. R. Lorusso, C. B. Lundgren, L. O. Magana, P. D. Maley, F. Marchis, R. L. Marcialis, A. E. McCandless, D. Joy McCrystal, A. Marie McGraw, K. E. Miller, B. E. A. Mueller, J. W. Noonan, A. M. Olsen, A. R. Patton, D. O'Conner Peluso, M. J. Person, J. G. Rigby, A. D. Rolfsmeier, J. J. Salmon, J. Samaniego, R. P. Sawyer, D. M. Schulz, M. F. Skrutskie, R. J. C. Smith, J. R. Spencer, A. Springmann, D. R. Stanbridge, T. J. Stoffel, P. Tamblyn, B. Tobias, A. J. Verbiscer, M. P. von Schalscha, H. Werts, Q. Zhang (2021) Size and Shape of (11351) Leucus from Five Occultations, *The Planetary Science Journal*, 2:202 (38pp) https://doi.org/10.3847/PSJ/ac1f9b. (CC BY4.0).

Habbal, S. R., M. Druckmüller, N. Alzate, A. Ding, J. Johnson, P. Starha, J. Hoderova, B. Boe, S. Constantinou, M. Arndt (2021) Identifying the Coronal Source Regions of Solar Wind Streams from Total Solar Eclipse Observations and in situ Measurements Extending over a Solar Cycle, *Astrophysical Journal Letters* 911, L4, https://doi.org/10.3847/2041-8213/abe775 , CC BY 4.0. (Note: material from this source was also used in preparing cover art).

Hertz, H. (1893) *Electric Waves*, translated by D. E. Jones, Macmillan, London. (Public domain).

Lockyer, J. N. (1869) The Recent Total Eclipse of the Sun, *Nature*, 1, 14-15. (Public domain).

Marchant, W. H. (1914) *Wireless Telegraphy: A handbook for the use of operators and students*, Whittaker and Co., New York. (Public domain).

McNish, R. Larry (2013) RGB Spectrum Generator V1.3, used by author, https://grumpyoldastronomer.com/rgbspectrum/index.htm, by permission of R. Larry McNish

Wikimedia-Roemer (2007) https://commons.wikimedia.org/wiki/File:Illustration_from_1676_article_on_Ole_R%C3%B8mer%27s_measurement_of_the_speed_of_light.jpg, cited August 2, 2023. (Public domain).

# CHAPTER 4

## GENERAL READING

M. Korsunsky, *The Atomic Nucleus*, Dover, New York, 1963. A very readable older book on radioactivity and nuclear physics with extensive historical discussion, with a Russian spin on mid-20th century nuclear technology, including the type of reactor leading to the Chernobyl nuclear disaster.

E. Rutherford, *A Newer Alchemy*, Cambridge U. P., Cambridge, 1937. Reprinted in 2014, this classic was published in the year of Rutherford's death and explains his discoveries for the general public.

Eyvind H. Wichmann, *Quantum Physics (Berkeley Physics Course, Vol. 4)*, McGraw-Hill, New York, 1971. While this is a quantum mechanics textbook, it has a wealth of pictorial and historical material. An older book, it does not use SI units,

## REFERENCES AND FIGURE SOURCES

Becquerel, H. (1896), https://commons.wikimedia.org/wiki/File:Becquerel_plate .jpg. (Public domain).

Curie, M. Skłodowska (1903), *Recherches sur les Substances Radioactives,* Gauthier-Villars, Paris. Figure retouched by Author and D. Connors from public domain (French) version at https://commons.wikimedia.org/wiki/File:Doctoral_the-sis_by_Marie_Curie_(1903).pdf

Gamow, G. (1928) *Zur Quantentheorie des Atomkernes*, Zeitschrift für Physik 51, 204-212, Figs. 1 and 3. (Public domain, montage by author).

NOAA SEC (2003) National Oceanic and Atmospheric Administration Space Environment Center, US government. (Public domain).

NOAA SWPC (2022) National Oceanic and Atmospheric Administration Space Weather Prediction Center, US government. (Public domain).

Röntgen, W. (1896) https://commons.wikimedia.org/wiki/File:X-ray_by_Wilhelm _R%C3%B6ntgen_of_Albert_von_K%C3%B6lliker%27s_hand_-_18960123-02 .jpg. (Public domain).

# CHAPTER 5

## GENERAL READING

Patrick Michel, Francesca E. Demeo, William F. Bottke (eds.), *Asteroids IV*, University of Arizona Press, Tucson, 2015. At over 900 total pages, this is not a "light" read but is in general not mathematical, and as other planetary books from this press, authoritative.

Donald K. Yeomans, *Comets: A Chronological History of Observation, Science, Myth, and Folklore*, Wiley, New York, 1991. Comes through as promised in the subtitle, although obviously the chronology ends in 1991 (with a naked eye comet list to 1700).

REFERENCES AND FIGURE SOURCES

Classical Numismatic Group (2006) https://commons.wikimedia.org/wiki/File
:S0484.4.jpg, cited August 3, 2023. https://creativecommons.org/licenses/by-sa/3.0
/deed.en.

ESO/NAOJ/NRAO (2014) https://www.eso.org/public/ireland/images/eso1436a/
?lang, cited August 3, 2023. ESO is the European Southern Observatory, NAOJ is
the National Astronomical Observatory of Japan, and NRAO is the (US) National
Radio Astronomy Observatory.

Kirkwood, D. (1867) *Meteoritic Astronomy: A Treatise on Shooting-Stars, Fire-
Balls, and Aerolites*, Lippincott, Philadelphia.

Newton, I. (1687) Principia mathematica, comet. https://hdl.handle.net/1911
/78833 (permalink).

Plekhanov, N. (2013) https://en.wikipedia.org/wiki/Chelyabinsk_meteor#/media/
File:Chelyabinsk_meteor_trace_15-02-2013.jpg, cited August 3, 2023.

Popular Science (1910) January 1910, p. 10, drawing of original, https://commons
.wikimedia.org/wiki/File:PSM_V76_D014_Halley_comet_on_the_bayeux_tapestry
.png, cited August 3, 2023. (Public domain).

Sandby, P. (1783) https://commons.wikimedia.org/wiki/File:Paul_Sandby_-_The
_Meteor_of_August_18,_1783,_as_seen_from_the_East_Angle_of_the_North
_Terrace,_Windsor_Castle_-_Google_Art_Project.jpg. (Public domain; cropped).

Timeline (2005) https://commons.wikimedia.org/wiki/File:Crookes_radiometer
.jpg, cited August 3, 2023.

Vollmy, A. (1888) https://commons.wikimedia.org/wiki/File:Leonids-1833.jpg,
cited August 4, 2023.

Wikipedia-Inner (2006) https://commons.wikimedia.org/wiki/File:InnerSolar
System-en.png, cited August 3, 2023.

# CHAPTER 6

GENERAL READING

Carrington, R.C. (1859) Description of a Singular Appearance seen in the Sun on
September 1, 1859, *Monthly Notices of the Royal Astronomical Society*, 20, 13–16.
(Used by permission of Oxford University Press).

Stuart Clark, *The Sun Kings*, Princeton University Press, Princeton, 2007. The
story of the long-fought establishment of the Sun-Earth connection, with emphasis
on developments by and after Richard Carrington in 1859.

Michael J. Carlowicz, Ramon E. Lopez, *Storms from the Sun: The Emerging
Science of Space Weather*, Joseph Henry Press, Washington, 2002. A mostly historical and nontechnical description of practical aspects of the Sun-Earth interaction.

Wilmot N. Hess, Gilbert D. Mead, *Introduction to Space Science*, Gordon and
Breach, New York, 1968. An advantage of a 50+ year old book is that the *basic* science remains valid, and the clarity of exposition is good, including what are now
historic aspects.

REFERENCES AND FIGURE SOURCES

Baliukin, I., J.-L. Bertaux, E. Quémerais, V. Izmodenov, W. Schmidt (2019). SWAN/SOHO
Lyman- mapping: The hydrogen geocorona extends well beyond the Moon. *Journal of Geophysical Research* 124, 861–885. https://doi.org/10.1029/2018JA026136

Galli, A., I. I. Baliukin, M. Bzowski, V. V. Izmodenov, M. Kornbleuth, H. Kucharek, E. Möbius, M. Opher, D. Reisenfeld, N. A. Schwadron, P. Swaczyna (2022) The Heliosphere and Local InterstellarMedium from Neutral Atom Observations at Energies Below 10 keV, *Space Science Reviews* 218:31, https://doi.org/10.1007/s11214-022-00901-7.

Habbal *et al.* (2021) See full reference under Chapter 3.

Hale, G. E. (1927) The Fields of Force in the Atmosphere of the Sun, *Nature*, 119, 708-714.

Hale, G. E., F. Ellerman, S. B. Nicholson, A. H. Joy (1919) The Magnetic Polarity of Sunspots, *Astrophysical Journal* 49, 153-178 and Plates (of which Fig. 6.10 is a part of Plate V).

McComas, D. J., B. L. Barraclough, H. O. Funsten, J. T. Gosling, E. Santiago-Muñoz, R. M. Skoug, B. E. Goldstein, M. Neugebauer, P. Riley, A. Balogh (2000) Solar wind observations over Ulysses' first full polar orbit, *Journal of Geophysical Research* 105, 10,419-10,433.

Orcinha, M., N. Tomassetti, F. Barão, B. Bertucci (2019) Observation of a time lag in solar modulation of cosmic rays in the heliosphere, *Journal of Physics Conference Series* 1181, 012013, doi:10.1088/1742-6596/1181/1/012013, CC BY 3.0.

Papatashvili, N., J. H. King (2020) OMNI Combined Heliospheric Observations (COHO), Merged Magnetic Field, Plasma and Ephemeris, Definitive Hourly Data. NASA Space Physics Data Facility, https://doi.org/10.48322/6ffx-3441, accessed through CDAWeb, January 3, 2023.

Richardson, J. D., L. F. Burlaga, H. Elliott, W. S. Kurth, Y. D. Liu, R. von Steiger (2022) Observations of the Outer Heliosphere, Heliosheath, and InterstellarMedium, Space Science Reviews 218:35, https://doi.org/10.1007/s11214-022-00899-y.

## CHAPTER 7

GENERAL READING

Committee on Solar and Space Physics, *Plasma Physics of the Local Cosmos*, National Academies Press, Washington, 2004. Short, nontechnical monograph, like many NAP publications, available for free download. Recommended.

C. T. Russell, J. G. Luhman, R. J. Strangeway, *Space Physics: An Introduction*, Cambridge University Press, Cambridge, 2016. The standard textbook on space physics, at an upper undergraduate level.

REFERENCES AND FIGURE SOURCES

ChamouJacoN, (2009) https://commons.wikimedia.org/wiki/File:Reconnection.gif, cited August 5, 2023. Animated GIF, of which one frame used here.

   Fujimoto, M., W. Baumjohann, K. Kabin, R. Nakamura, J.A. Slavin, N. Terada, L. Zelenyi, Hermean Magnetosphere-Solar Wind Interaction (2007) *Space Science Review* 132: 529–550,
   DOI 10.1007/s11214-007-9245-8

## CHAPTER 8

### GENERAL READING

Steven Weinberg, *The First Three Minutes: A Modern View of the Origins of the Universe*, Basic Books, New York, 1977. Topics covered in the present book occurred in the last of these minutes, with the basic physics well understood by 1977. Weinberg's very readable book also presents other material that was updated in 1993 but likely must now be supplemented from other sources in the fast-changing domain of cosmology.

   Michael W. Friedlander, *A Thin Cosmic Rain*, Harvard University Press, Cambridge MA, 2000. A readable introduction although now nearly 25 years old in a rapidly changing field.

REFERENCES AND FIGURE SOURCES

D'Angelo, D., M. Agostini, K. Altenmüller, S. Appel, G. Bellini, J. Benziger, D. Bick, G. Bonfini, D. Bravo, B. Caccianiga, F. Calaprice, A. Caminata, P. Cavalcante, A. Chepurnov, S. Davini, A. Derbin, L. Di Noto, I. Drachnev, A. Etenko, K. Fomenko, D. Franco, F. Gabriele, C. Galbiati, C. Ghiano, M. Giammarchi, M. Goeger-Neff, A. Goretti, M. Gromov, C. Hagner, E. Hungerford, A. Ianni, A. Ianni, K. Jedrzejczak, M. Kaiser, V. Kobychev, D. Korablev, G. Korga, D. Kryn, M. Laubenstein, B. Lehnert, E. Litvinovich, F. Lombardi, P. Lombardi, L. Ludhova, G. Lukyanchenko, I. Machulin, S. Manecki, W. Maneschg, S. Marcocci, E. Meroni, M. Meyer, L. Miramonti, M. Misiaszek, M. Montuschi, P. Mosteiro, V. Muratova, B. Neumair, L. Oberauer, M. Obolensky, F. Ortica, M. Pallavicini, L. Papp, L. Perasso, A. Pocar, G. Ranucci, A. Razeto, A. Re, A. Romani, R. Roncin, N. Rossi, S. Schönert, D. Semenov, H. Simgen, M. Skorokhvatov, O. Smirnov, A. Sotnikov, S. Sukhotin, Y. Suvorov, R. Tartaglia, G. Testera, J. Thurn, M. Toropova, E. Unzhakov, A. Vishneva, R. B. Vogelaar, F. von Feilitzsch, H. Wang, S. Weinz, J. Winter, M. Wojcik, M. Wurm, Z. Yokley, O. Zaimidoroga, S. Zavatarelli, K. Zuber, G. Zuzel (2016) Recent Borexino results and prospects for the near future, *EPJ Web of Conferences* 126, 02008, DOI: 10.1051/epjconf/201612602008

   McComas, D. J., N. Alexander, N. Angold, S. Bale, C. Beebe, B. Birdwell, M. Boyle, J. M. Burgum, J. A. Burnham, E. R. Christian, W. R. Cook, S. A. Cooper, A. C. Cummings, A. J. Davis, M. I. Desai, J. Dickinson, G. Dirks, D. H. Do, N. Fox, J. Giacalone, R. E. Gold, R. S. Gurnee, J. R. Hayes, M. E. Hill, J. C. Kasper, B. Kecman,

J. Klemic, S. M. Krimigis, A. W. Labrador, R. S. Layman, R. A. Leske, S. Livi, W. H. Matthaeus, R. L. McNutt Jr, R. A. Mewaldt, D. G. Mitchell, K. S. Nelson, C. Parker, J. S. Rankin, E. C. Roelof, N. A. Schwadron, H. Seifert, S. Shuman, M. R. Stokes, E. C. Stone, J. D. Vandegriff, M. Velli, T.T. von Rosenvinge, S. E. Weidner, M. E. Wiedenbeck, P. Wilson IV (2016) Integrated Science Investigation of the Sun (ISIS): Design of the Energetic Particle Investigation, Space Science Reviews 204:187–256, DOI 10.1007/s11214-014-0059-1

Orebi Gann, G. D. (2020) Neutrino detection gets to the core of the Sun, *Nature* 587, 550-551.

Richardson, J. D., J. W. Belcher, P. Garcia-Galindo, L. Burlaga (2019) Voyager 2 plasma observations of the heliopause and interstellar medium. Nature Astronomy 3, 1019–1023, https://doi.org/10.1038/s41550-019-0929-2.

Zong, Q. (2022) Magnetospheric response to solar wind forcing: ultra-low-frequency wave–particle interaction perspective Ann. Geophys., 40, 121–150, (Fig. 3) https://doi.org/10.5194/angeo-40-121-2022.

Swordy, S. W. (2001) The Energy Spectra and Anisotropies of Cosmic Rays, *Space Science Reviews* 99**:** 85–94.

## CHAPTER 9

### GENERAL READING

Archie E. Roy, *Orbital Motion, 4<sup>th</sup> Ed.*, IOP Publishing, Bristol, 2005. This textbook covers all of orbital mechanics at a fairly advanced mathematical level, including several chapters on interplanetary navigation.

### REFERENCES AND FIGURE SOURCES

Angelopoulos, V., A. Artemyev, T. D. Phan, Y. Miyashita (2020) Near-Earth magnetotail reconnection powers space storms. *Nature Physics* 16, 317–321. https://doi.org/10.1038/s41567-019-0749-4. Quite readable, as is the case with most "Nature" articles, but with a detailed Supplemental Material section. Also gives credits to investigators on THEMIS/ARTEMIS, data sources for some figures in this book.

# Index